Environmentally Benign Catalysts

Anjali Patel
Editor

Environmentally Benign Catalysts

For Clean Organic Reactions

Springer

Editor
Anjali Patel
Department of Chemistry
Maharaja Sayajirao University of Baroda
Gujarat, India

ISBN 978-94-007-6709-6 ISBN 978-94-007-6710-2 (eBook)
DOI 10.1007/978-94-007-6710-2
Springer Dordrecht Heidelberg New York London

Co-published by Springer Science+Business Media B.V., Van Godewijckstraat 30, 3311 GX Dordrecht, The Netherlands with Panch Tattva Publishers, No. 777, Vikas Nagar, Dehu Road, Pune-412113, India.

Sold and distributed by Springer Science+Business Media in all countries except Malaysia and SAARC countries – Islamic Republic of Afghanistan, People's Republic of Bangladesh, Kingdom of Bhutan, Republic of India, Republic of Maldives, Federal Democratic Republic of Nepal, Islamic Republic of Pakistan, and Democratic Socialist Republic of Sri Lanka.

Sold and distributed by Panch Tattva Publishers in Malaysia and SAARC countries – Islamic Republic of Afghanistan, People's Republic of Bangladesh, Kingdom of Bhutan, Republic of India, Republic of Maldives, Federal Democratic Republic of Nepal, Islamic Republic of Pakistan, and Democratic Socialist Republic of Sri Lanka.

Library of Congress Control Number: 2013945854
A catalogue record for this book is available from the British Library

Printed on acid-free paper
Springer is part of Springer Science+Business Media (www.springer.com)

Preface

Catalysis by heteropoly acids (HPAs) has received wide attention in the past two decades. During this time new and promising developments have been reported at both the academic and industrial levels. HPAs chemistry is exciting and undergoes continuous revision and actualization. Heterogeneous catalysis is an attractive tool, particularly because it often satisfies some of the principles of green chemistry. Work on heterogeneous catalysis by supported HPAs has greatly expanded during the past few years and has recently emerged as an innovative method for green chemistry practices. There are a number of publications on HPAs/polyoxometalates by renowned scientists. Even though catalysis by HPAs, especially supported heteropoly acids, is an emerging field, no book is available on this aspect.

This book consists of general introduction (Chap. 1) followed by 11 chapters which is an overview given by several leading national and international scientists. Chapter 2, by Sai Prasad and co-workers, gives an excellent overview of synthesis of various types of ammonium salt of molybdophosphoric acid, vanadium incorporated molybdophosphoric acid and supported vanadium in ammonium salt of molybdophosphoric acid, their characterization and use for ammoxidation of 2-methyl pyrazine. K. Parida (Chap. 3) has reviewed nicely the synthesis and characterization of Cs salt of phosphotungstic acid, supported Cs-PTA, Fe and Pd modified PTA and their use as catalysts for carrying out acylation, Heck vinylation, bromination of phenol, oxidation of *trans*-stilbene and hydrogenation of *ortho-nitrophenol*, respectively.

Lingaiah (Chap. 4) demonstrates synthesis, characterization of supported vanadium-substituted tungstophosphoric acid and their use as catalyst for the selective oxidative cleavage of olefins to carbonyl compounds at room temperature. Halligudi and co-workers (Chap. 5) have discussed in detail the supported HPAs (silicotungstic acid and phosphotungstic acid) and their applications in acid-catalyzed reactions as well as immobilized vanadium-substituted phosphomolybdic acid and applications in oxidation reactions. They have covered nicely a wide range of industrially important reactions such as alkylation, acylation, allylation and oxidation. In addition, they have also reported the use of new catalyst, molybdovanado phosphoric acid supported on ionic liquid-modified SBA-15 for the oxidation of broad range of alcohols.

Chapter 6, by Nadine Essayem, focuses on the use of tungstophosphoric acid supported onto different supports for glycerol etherification followed by alkoxylation of terpenes over supported tungstophosphoric acid by Jose Castanheiro (Chap. 7).

Jose Dias, in Chap. 8, demonstrates effect of acidity, structure and stability of supported tungstophosphoric acid on different catalytic reactions such as *trans-alkylation* of benzene with aromatics, esterification of acetic acid as well as oleic acid and ethanol and cyclization of (+)-citronellal. Chapter 9, by Anjali Patel, describes biodiesel production over 12-tungstophosphoric acid anchored to different mesoporous silica supports. Further, one of the most recent emerging category of catalyst, supported lacunary polyoxometalate-based catalysts, was also covered by the same author (Chap. 10). It describes solvent-free oxidation of benzyl alcohol over supported mono lacunary phosphomolybdate. It is worth to notice that all the reported catalysts in the book are reusable and are promising environmentally benign catalysts.

Last two chapters cover two important areas of HPAs: recoverable homogeneous catalysts and electrocatalysts. Chapter 11, by Marico Jose daSilva, describes a versatile bifunctional reusable homogeneous phosphomolybdic acid for esterification of fatty acids and oxidation of camphene. Prof. B. Vishwanathan (Chap. 12) covered very emerging and highly attempted area of research for HPAs, electrocatalysis by HPAs.

Overall, best efforts have been made to cover a number of industrially important organic transformations using all types of catalysts based on HPAs (parent, salts, modified, lacunary). However, homogeneous catalysts are not included as they are beyond the scope of this book. The editor hopes that the present book is intended to open a new direction for both academic and industrial research by giving a major impetus for the development of third-generation catalysts. It will be of immense value to postgraduates, researchers and chemists, especially working in the field of heterogeneous catalysts.

Last, but not the least, the editor is thankful to all the authors for their hard work, timely submission and cooperation in spite of their busy schedule. The editor also thanks her research group, especially Dr. Pragati S. Joshi and Mr. Soyeb Pathan, for helping in various stages.

Vadodara, Gujarat, India Anjali Patel

Contents

1 **Introduction** 1
Anjali Patel

2 **Heteropoly Compounds as Ammoxidation Catalysts** 11
K. Narasimha Rao, Ch. Srilakshmi, K. Mohan Reddy, B. Hari Babu, N. Lingaiah, and P.S. Sai Prasad

3 **Transition Metal-Substituted Salt of Tungsten-Based Polyoxometalate-Supported Mesoporous Silica as a Catalyst for Organic Transformation Reactions** 57
Surjyakanta Rana and Kulamani Parida

4 **Vanadium-Substituted Tungstophosphoric Acid Supported on Titania: A Heterogeneous Catalyst for Selective Oxidative Cleavage of Olefins to Carbonyl Compounds at Room Temperature** 91
N. Lingaiah, K.T. Venkateswara Rao, and P.S. Sai Prasad

5 **Supported Heteropoly Acids and Multicomponent Polyoxometalates as Eco-Friendly Solid Catalysts for Bulk and Fine Chemicals Synthesis** 105
G.V. Shanbhag, Ankur Bordoloi, Suman Sahoo, B.M. Devassy, and S.B. Halligudi

6 **Glycerol Etherification with Light Alcohols Promoted by Supported $H_3PW_{12}O_{40}$** 141
Rodrigo Lopes de Souza, Wilma Araujo Gonzales, and Nadine Essayem

7 **Alkoxylation of Terpenes over Tungstophosphoric Acid Immobilised on Silica Support** 153
M. Caiado, D.S. Pito, and J.E. Castanheiro

8 **Effect of Acidity, Structure, and Stability of Supported 12-Tungstophosphoric Acid on Catalytic Reactions** 165
José Alves Dias, Sílvia Cláudia Loureiro Dias, and Julio Lemos de Macedo

9 **Biodiesel Production over 12-Tungstosilicic Acid Anchored to Different Mesoporous Silica Supports** 189
Varsha Brahmkhatri and Anjali Patel

10 **Solvent-Free Selective Oxidation of Benzyl Alcohol to Benzaldehyde over Monolacunary Phosphomolybdate Supported onto Hydrous Zirconia** 209
Soyeb Pathan and Anjali Patel

11 **$H_3PMo_{12}O_{40}$ Heteropolyacid: A Versatile and Efficient Bifunctional Catalyst for the Oxidation and Esterification Reactions** 225
Márcio José da Silva, Raquel da Silva Xavier, and Lidiane Faria dos Santos

12 **Electrochemical/Electro-catalytic Applications of Polyoxometalates** 245
B. Viswanathan

Index 257

Contributors

Ankur Bordoloi Catalysis Division, National Chemical Laboratory, Pune, India

Varsha Brahmkhatri Department of Chemistry, Faculty of Science, M. S. University of Baroda, Vadodara, India

M. Caiado Departamento de Química, Centro de Química de Évora, Universidade de Évora, Évora, Portugal

J.E. Castanheiro Departamento de Química, Centro de Química de Évora, Universidade de Évora, Évora, Portugal

B.M. Devassy Catalysis Division, National Chemical Laboratory, Pune, India

José Alves Dias Campus Darcy Ribeiro – Asa Norte, Instituto de Química, Laboratório de Catálise (A1-62/21), Universidade de Brasília, Brasília-DF, Brazil

Sílvia Cláudia Loureiro Dias Campus Darcy Ribeiro – Asa Norte, Instituto de Química, Laboratório de Catálise (A1-62/21), Universidade de Brasília, Brasília-DF, Brazil

Nadine Essayem Institut de Recherche sur la Catalyse et l'Environnement de Lyon (IRCELYON)-CNRS-Lyon1, Lyon, France

Wilma Araujo Gonzales Instituto Militar de Engenharia (IME), Rio de Janeiro, Brazil

S.B. Halligudi Centre of Excellence GIDC, Vapi, Gujarat, India

B. Hari Babu I & PC Division, Indian Institute of Chemical Technology, Hyderabad, India

Pragati S. Joshi Chemistry Department, Faculty of Science, M. S. University of Baroda, Vadodara, India

N. Lingaiah I & PC Division, Indian Institute of Chemical Technology, Hyderabad, India

Julio Lemos de Macedo Campus Darcy Ribeiro – Asa Norte, Instituto de Química, Laboratório de Catálise (A1-62/21), Universidade de Brasília, Brasília-DF, Brazil

K. Mohan Reddy Senior Research Associate, Center for Applied Catalysis, Seton Hall University, South Orange, New Jersey, USA

K. Narasimha Rao Chemistry Department, Faculty of Science, King Abdulaziz University, Jeddah, Saudi Arabia

Kulamani Parida Colloids and Materials Chemistry Department, CSIR-Institute of Minerals and Materials Technology, Bhubaneswar, Orissa, India

Anjali Patel Department of Chemistry, Faculty of Science, M. S. University of Baroda, Vadodara, India

Soyeb Pathan Department of Chemistry, Faculty of Science, M. S. University of Baroda, Vadodara, India

D.S. Pito Departamento de Química, Centro de Química de Évora, Universidade de Évora, Évora, Portugal

Surjyakanta Rana Colloids and Materials Chemistry Department, CSIR-Institute of Minerals and Materials Technology, Bhubaneswar, Orissa, India

Suman Sahoo Catalysis Division, National Chemical Laboratory, Pune, India

P.S. Sai Prasad I & PC Division, Indian Institute of Chemical Technology, Hyderabad, India

Lidiane Faria dos Santos Department of Chemistry, Exacts Science Center, Federal University of Viçosa, Viçosa, Minas Gerais, Brazil

G.V. Shanbhag Catalysis Division, National Chemical Laboratory, Pune, India

Poornaprajna Institute of Scientific Research, Bangalore, India

Márcio José da Silva Department of Chemistry, Exacts Science Center, Federal University of Viçosa, Viçosa, Minas Gerais, Brazil

Raquel da Silva Xavier Department of Chemistry, Exacts Science Center, Federal University of Viçosa, Viçosa, Minas Gerais, Brazil

Rodrigo Lopes de Souza Institut de Recherche sur la Catalyse et l'Environnement de Lyon (IRCELYON)-CNRS-Lyon1, Lyon, France

Instituto Militar de Engenharia (IME), Rio de Janeiro, Brazil

Ch. Srilakshmi Solid State and Structural Chemistry Unit, Indian Institute of Science, Bangalore, India

K.T. Venkateswara Rao Catalysis Laboratory, I&PC Division, Indian Institute of Chemical Technology, Hyderabad, India

B. Viswanathan National Centre for Catalysis Research, Indian Institute of Technology Madras, Chennai, India

Chapter 1
Introduction

Anjali Patel

Contents

1 Introduction 1
2 Advantages of Supported HPAs 7
References 8

Abstract In the twenty-first century, "Green and Sustainable Chemistry" has emerged as a new concept and scientific area in both academia and industry with the aim of providing environmentally friendly option for hazardous synthetic pathways to achieve clean sustainable environment, maintaining the health and safety of people. Almost all chemical processes, though beneficial, lead to unwanted wastes. The inefficient recovery of organic solvents and their disposal are problems which damage environment.

1 Introduction

Catalysis is a key technology to provide realistic solutions to such environmental issues [1]. The central tasks of catalysts are to lower the raw material and energy requirement of chemical reactions, the diminution of side products/waste products, as well as the control of hazardous effects of chemical substances for the sound establishment of chemical science and technology. To cope up with the problems, the development and implementation of eco-friendly as well as environmentally benign innovative catalysts as well as catalytic processes with high atom efficiency, simple operations, and simple work-up procedures is the prime objective for chemists and scientists.

A. Patel (✉)
Department of Chemistry, Faculty of Science, M. S. University of Baroda,
Vadodara 390 002, India
e-mail: aupatel_chem@yahoo.com

A. Patel (ed.), *Environmentally Benign Catalysts: For Clean Organic Reactions*,
DOI 10.1007/978-94-007-6710-2_1, © Panch Tattva Publishers, Pune, India 2013

Thus, the modern research on catalysts focuses on five main areas [2]:

1. Predicting and controlling catalyst structure
2. Improving the integration of catalytic processes
3. Energy consumption and raw material cost
4. Minimization of waste as well as by-products
5. Reducing the cost of the process

In this context, the use of heteropoly acids (HPAs) and HPA-based compounds as catalysts has become a very important field of research, industrially as well as academically [3–12].

In last two decades, HPA-based catalysts have played an important role in the field of acid as well as oxidation catalysis due to their high Bronsted acidity as well as their tendency to exhibit fast reversible multi-electron redox transformations under rather mild conditions and their inherent stability toward strong oxidants. HPAs are promising as environmentally benign catalysts as they possess a unique set of physicochemical properties, intrinsic multifunctionality, and the possibility of tailoring the composition to achieve defined acidic/redox properties. There are several advantages of using them as catalyst like high activity, selectivity, and stability. At the same time, they also become increasingly important for applied catalysis. They provide good basis for the molecular design of mixed oxide catalyst, and they have high capability in practical uses.

The structural reorganization of HPAs occurs in the solution state depending on the conditions such as temperature, concentrations, and pH of the solution. These compounds are always negatively charged although the negative density is widely variable depending on the elemental composition and the molecular structure.

The heteropolyanions have been known since the work of Berzelius [13] on the ammonium 12-molybdophosphate in 1826. After the discovery of this first heteropolyanion, the field of polyoxometalates chemistry progressed significantly [14].

1. About 20 years later, Svanberg and Struve showed that the insoluble ammonium salt of this complex could be used for the gravimetric analysis of phosphate [13].
2. However, the study of heteropolyanion chemistry did not accelerate until the discovery of the tungstosilicic acids and their salts in 1862 by Marignac [15]. He prepared and analyzed two isomers of 12-tungstosilicic acid, namely, tungstosilicic acid and silicotungstic acid now known as α- and β-isomers.
3. Thereafter, the field developed rapidly, so that over 60 different types of heteropoly acids (giving rise to several hundred salts) had been described by the end of first decade of this century.
4. In 1908, A. Miolati suggested a structural hypothesis for heteropoly compounds based on coordination theory. According to his hypothesis, the heteroatom was considered to have octahedral coordination with MO_4^{2-} or $M_2O_7^{2-}$ ligands
5. In the mid 1930s, A. Rosenheim had given a laboratory perspective for the synthetic and descriptive research of Miolati.
6. The first steps toward understanding the structure of polyoxometalate anions was taken by L. C. Pauling in 1929. Pauling [16] proposed a structure for 12:1 complexes based on an arrangement of 12 MO_6 octahedra surrounding a central

Table 1.1 Different types of heteropolyanions families [12]

Structure	General formula[a]	Charge	X^{n+}
Keggin	$XM_{12}O_{40}$	8-*n*	P^{5+}, As^{5+}, Si^{4+}, Ge^{4+}
Silverton	$XM_{12}O_{42}$	8-	Ce^{4+}, Th^{4+}
Dawson	$X_2M_{18}O_{62}$	6-	P^{5+}, As^{5+}
Waugh	XM_9O_{32}	6-	Mn^{4+}, Ni^{4+}
Anderson (type A)	XM_6O_{24}	12-*n*	Te^{6+}, I^{7+}

[a]Where *M* Mo^{VI}, W^{VI}, $V^{V,VI}$, etc.

XO_4 tetrahedron. He proposed the structure of 12-tungstoanions based on the central PO_4 or SiO_4 tetrahedrons surrounded by WO_6 octahedrons. In order to minimize electrostatic repulsions, he proposed that all the polyhedral linkages involved sharing of vertices rather than edges. As a result, the resulting formula required 58 oxygen atoms, i.e., $[(PO_4)W_{12}O_{18}(OH)_{36}]^{3-}$.

7. After Pauling's proposal, in 1933, Keggin [17, 18] solved the structure of $[H_3PW_{12}O_{40}]\cdot 5H_2O$ by powder X-ray diffraction and showed that the anion was indeed based on WO_6 octahedral units. As suggested by Pauling, these octahedra being linked by shared edges as well as corners. The application of X-ray crystallography to the determination of polyoxometalate structures accelerated the development of polyoxometalate chemistry.
8. A year later in 1934, Signer and Gross demonstrated that $H_4SiW_{12}O_{40}$, $H_5BW_{12}O_{40}$, and $H_6[H_2W_{12}O_{40}]$ were structurally isomorphous with Keggin's structure [19].
9. Bradley and Illingworth confirmed Keggin's work in 1936, by studying the crystal structure of $H_3PW_{12}O_{40}\cdot 29H_2O$.
10. These results of Bradley's and Illingworth's were largely supported by the single crystal experiments of Brown and co-workers, which were reported in 1977.

General information and polyhedral representation of various types of heteropolyanions are listed in Table 1.1 and Fig. 1.1.

The free acids or acidic forms of heteropolyanions are known as heteropoly acids (HPAs).

From the viewpoint of stability and acidity, Keggin-type heteropolyanion (POMs) is most important and hence most of the work was carried out with that. However, other heteropolyanions are also getting importance to some extent in the recent years. Generally, Keggin heteropolyanions have the general formula $[X_xM_mO_y]^{q-}$, in which X is the heteroatom, usually a main group element (e.g., P, Si, Ge, As), and M is the addenda atom, being a d-block element in high oxidation state, usually $V^{IV,V}$, Mo^{VI}, or W^{VI}.

The formation of Keggin-type heteropolyanions is represented in the following equation:

$$12MO_4^{\,m-} + XO_4^{\,p-} + 23H^+ \rightarrow [XM_{12}O_{40}]^{n-} + H_2O$$

The reasons why HPAs have attracted more attention are as follows [5, 10, 20–22]:

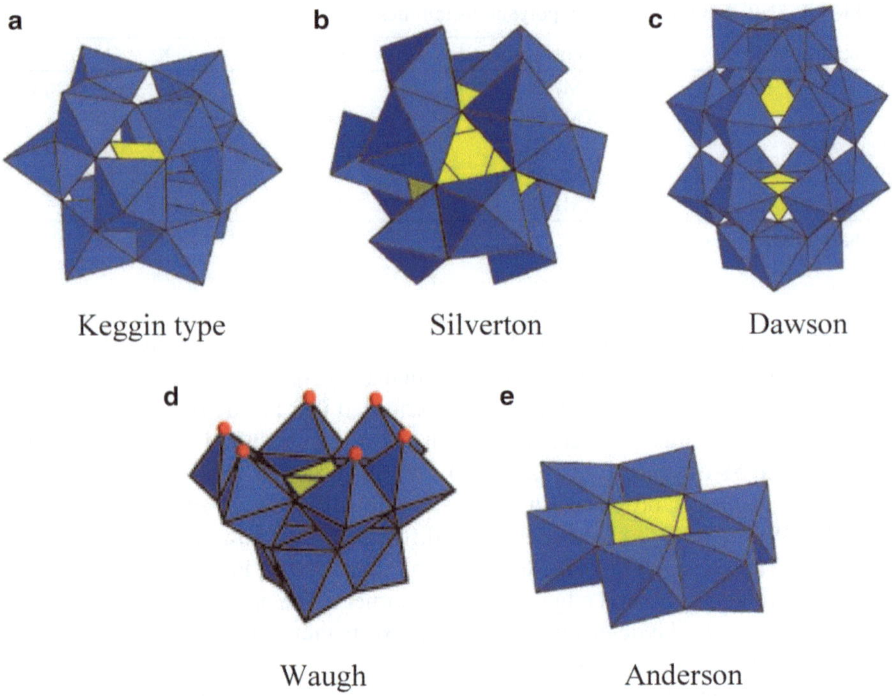

Fig. 1.1 Polyhedral representation of different types of polyanions

- Chemical properties of POMs such as redox potentials, acidities, and solubilities in various media (aqueous as well as organic) can be finely tuned by choosing constituent elements and counter cations.
- POMs are thermally and oxidatively stable in comparison with common organometallic complexes and enzymes.
- Metal-substituted POMs with "controlled active sites" can easily be synthesized.
- High activity and selectivity.
- Nontoxic.

HPA-based catalysts have played an important role in the field of acid as well as oxidation catalysis due to their high Bronsted acidity as well as their tendency to exhibit fast reversible multi-electron redox transformations under rather mild conditions and their inherent stability toward strong oxidants. In the reoxidation of heteropolyanions, reduced forms of heteropolymolybdates and heteropolyvanadates are stable and hardly reoxidized by molecular oxygen, whereas heteropolytungstates undergo facile reoxidation.

In spite of having these advantages, HPA-based catalysts suffer from the following disadvantages [20]:

- High solubility in polar solvents
- Low surface area
- Low thermal stability
- Difficulty of separation at the end of reaction

Fig. 1.2 Cs-salt-type heteropolyanions (Reproduced from Ref. [23])

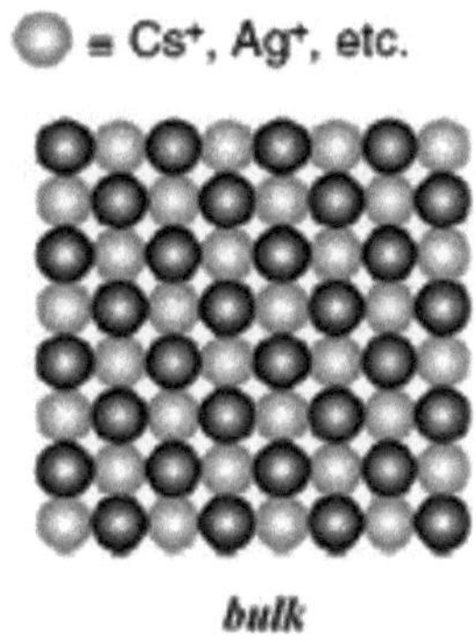

The above-mentioned disadvantages can be overcome by converting them into heterogeneous catalysts. Heterogeneous HPA-based catalysts are gaining importance as they can overcome the traditional problems of homogeneous catalysts. For the design of HPA-based heterogeneous catalysts, mainly two strategies can be implemented.

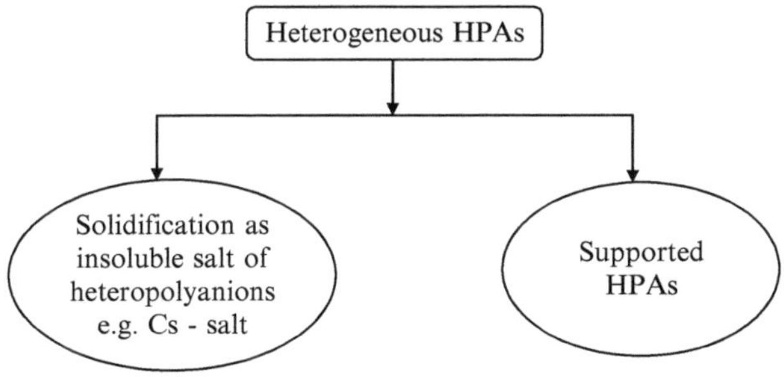

The structural as well as chemical properties of HPAs are strongly dependent on the counter cations. The solubility of the HPAs can be tuned by changing the counter cations. Generally, the Cs-salts are insoluble and can be used as catalysts (Fig. 1.2).

Supporting of HPAs can be done by using different techniques: "adsorption," "impregnation," "covalent interaction," "ion exchange," "intercalation," and "encapsulation." The preparation procedures as well as properties of HPA-based heterogeneous catalysts are much different from those of classical supported metal catalysts.

It is very interesting and important to note that the support does not always play merely a mechanical role but it can also modify the catalytic properties of HPAs. The support provides an opportunity to HPAs to be dispersed over a large surface area, which increases catalytic activity. A support can affect catalyst activity, selectivity, recyclability, material handling, and reproducibility. It is well known that basic support cannot be used for supporting HPAs, as it gets decomposed on the same. Hence, generally, acidic and neutral supports are used.

Depending on the nature of the support as well as method of supporting, the different interactions between HPAs and supports are expected as follows:

1. When the support is hydrous metal oxide/metal oxide, hydrogen bonding or adsorption type of interaction is possible (Fig. 1.3).

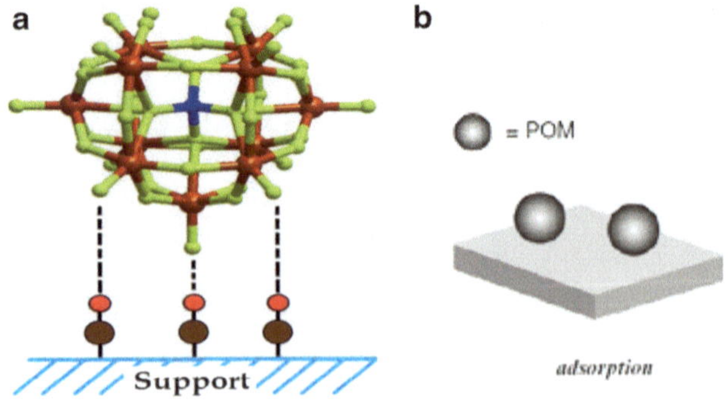

Fig. 1.3 (**a**) Hydrogen bonding [24a]. (**b**) Adsorption type (Reproduced from Ref. [23])

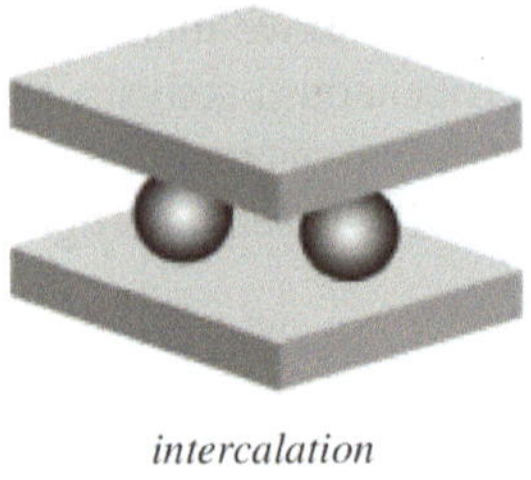

Fig. 1.4 Intercalation of HPA in clay type of supports (Reproduced from Ref. [23])

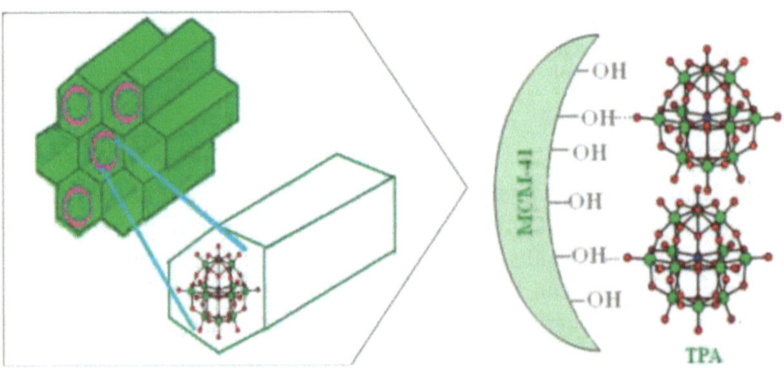

Fig. 1.5 Probable interaction of heteropolyanions with the mesoporous support [24b]

2. When the support is clay, intercalation is expected (Fig. 1.4).
3. In case of mesoporous materials, encapsulation takes place. A strong hydrogen bonding type of interaction exists between the terminal oxygens of heteropolyanions and the surface silanol hydroxyl groups of mesoporous materials (Fig. 1.5).
4. In addition, ion exchange and covalent bonding type of interactions are also possible as shown in Fig. 1.6.

Fig. 1.6 Ion exchange type of interaction for supporting HPA catalysts (Reproduced from Ref. [23])

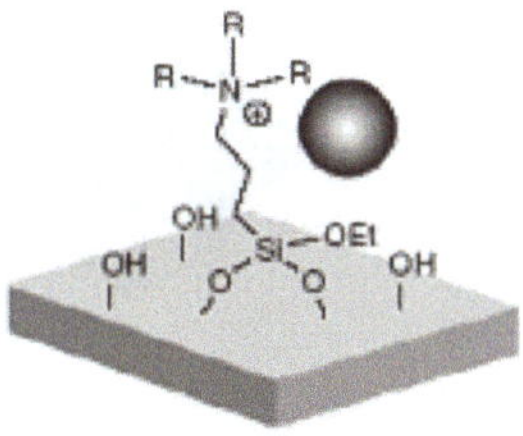

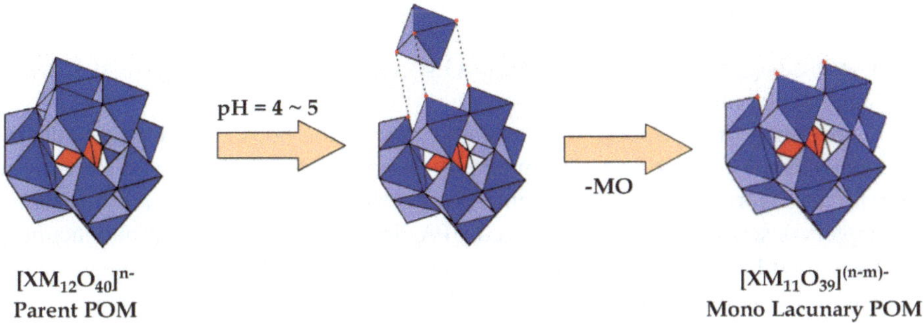

Fig. 1.7 Formation of mono lacunary polyoxometalate

2 Advantages of Supported HPAs

1. Increase in thermal stability as well as surface area.
2. They have high catalytic activity and selectivity.
3. Separation from a reaction mixture is easy.
4. Repeated use is possible.

Further as mentioned earlier, the acidic as well as redox properties of heteropolyanions can be tuned at molecular level which can lead to development of a new class of materials with unique structural as well as electronic properties. One of the most significant properties of modified precursors is their ability to accept and release specific numbers of electrons reversibly, under marginal structural rearrangement [25–27]. As a result, they are expected to play an important role in catalysis. Thus, the modification of parent heteropolyanion is likely to help in development of new generation catalysts with enhanced properties of acidity, redox potential, and stability.

The modification of properties can be basically done by tuning the structural properties at the atomic or molecular level in two ways:

1. By creating defect (lacuna) in parent heteropolyanion (POM) structures (i.e., lacunary polyoxometalates, LPOMs) (Fig. 1.7)
2. Incorporation of transition metal ions into the defect structures (i.e., transition metal-substituted polyoxometalates, TMS POMs) (Fig. 1.8)

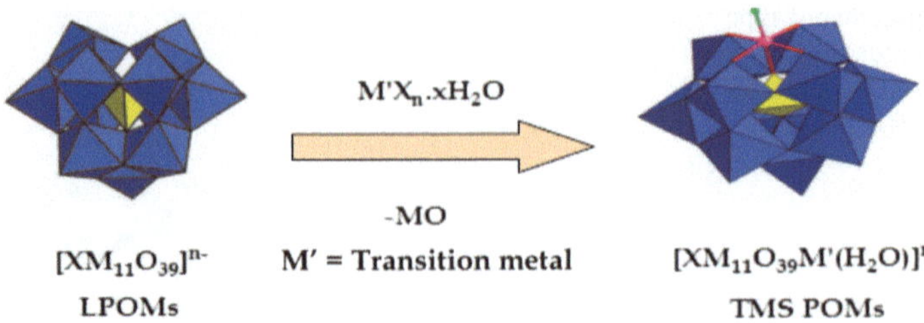

Fig. 1.8 Formation of TMS POMs from LPOMs

These designed LPOMs and TMS POMs can be used as catalysts as such or by supporting onto suitable supports.

Considering these aspects, the present version of the book is dedicated to development of third-generation environmentally benign HPA-based catalysts for various organic transformations. The present book consists of 11 chapters. Among them, 9 chapters describe the use of supported HPAs including salts of HPAs and lacunary and modified HPAs as catalysts for various organic transformations, 1 chapter deals with reusable homogeneous catalysts, and 1 chapter discusses electrocatalysis.

Acknowledgement The author is thankful to Dr. Pragati Shringarpure Joshi for assistance in the preparation of the chapter.

References

1. Centri G, Ciambelli P, Perathoner S, Russo P (2002) Catal Today 75:3
2. Tundo P, Mammino L (2002) Green Chem Ser 5:44
3. Okhuhara T, Mizuno N, Misono M (1996) Adv Catal 41:113
4. Mizuno N, Misono M (1998) Chem Rev 98:199
5. Neumann R (1998) Prog Inorg Chem 47:317
6. Hill CL, Chrisina C, Prosser-McCartha M (1995) Coord Chem Rev 143:407
7. Hill CL (2004) In: McCleverty JA, Meyer TJ (eds) Comprehensive coordination chemistry II, vol 4. Elsevier Pergamon, Amsterdam, p 679
8. Kozhevnikov IV (1998) Chem Rev 98:171
9. Kozhevnikov IV (2002) Catalysis by polyoxometalates. Wiley, Chichester
10. Mizuno N, Yamaguchi K, Kamata K (2005) Coord Chem Rev 249:1944
11. Mizuno N, Hikichi S, Yamaguchi K, Uchida S, Nakagawa Y, Uehara K, Kamata K (2006) Cat Today 117:32
12. Pope MT (1983) Heteropoly and isopoly oxometalates, vol 8, Inorganic chemistry concepts. Springer, Berlin
13. Pope MT, Muller A (2003) Polyoxometalate molecular science. Kluwer Academic Publishers, Boston/Dordrecht
14. Berzelius JJ (1826) Pogg Ann Phys Chem 6:369
15. Marignac C (1862) Ann Chim 25:362
16. Pauling LC (1929) J Am Chem Soc 51:2868
17. Keggin JF (1933) Nature 131:908
18. Keggin JF (1934) Proc R Soc A 144:75

19. Signer R, Gross H (1934) Helv Chim Acta 17:1076
20. Kozhevnikov IV (2002) Catalysts for fine chemical synthesis, catalysis by polyoxometalates, vol 2. Wiley, Chichester
21. Mizuno N, Kamata K, Yamaguchi K (2006) In: Richards R (ed) Surface and nanomolecular catalysis. Taylor & Francis Group, New York, p 463
22. Mizuno N, Kamata K, Uchida S, Yamaguchi K (2009) In: Mizuno N (ed) Modern heterogeneous oxidation catalysis–design, reactions and characterization. Wiley–VCH, Weinheim, p 185
23. Mizuno N, Kamata K, Yamaguchi K (2011) Top Organomet Chem 37:127
24. (a) Pathan S, Patel A (2011) Dalton Trans 40:348; (b) Brahmakhatri V, Patel A (2011) Ind Eng Chem Res 50:6096
25. Thouvenot R, Proust A, Gouzerh P (1837) Chem Commun 2008
26. Weinstock IA (1998) Chem Rev 98:113
27. Sadakane M, Steckhan E (1998) Chem Rev 98:219

Chapter 2
Heteropoly Compounds as Ammoxidation Catalysts

K. Narasimha Rao, Ch. Srilakshmi, K. Mohan Reddy, B. Hari Babu, N. Lingaiah, and P.S. Sai Prasad

Contents

1 Introduction 12
 1.1 Heteropoly Acids (HPAs) and Their Classification 13
 1.2 Heteropoly Acids as Useful Catalysts 14
 1.3 Modification of Heteropoly Acids 15
 1.4 Salts of Heteropoly Acids and Their Catalytic Functionalities 16
 1.5 Supported Salts of HPAs 16
 1.6 In Situ Synthesis of AMPA 17
2 Experimental 19
 2.1 Chemicals and Supports 19
 2.2 Preparation of Various Types of HPA Catalysts 19
 2.3 Characterization of HPAs 21
 2.4 Ammoxidation of MP over HPA Salts 22
3 Results and Discussion 23
 3.1 Studies on Bulk MPA and Vanadium Modified MPA (VMPA, VOMPA) Catalysts 23
 3.2 Studies on TiO_2-Supported MPA, VMPA, and VOMPA 24
 3.3 Studies on Bulk AMPA 26
 3.4 Studies on AMPA Supported on Nb_2O_5, SiO_2, TiO_2, ZrO_2, and Al_2O_3 35

K. Narasimha Rao
Chemistry Department, Faculty of Science, King Abdulaziz University, P.O. Box 80203, Jeddah 21533, Saudi Arabia

Ch. Srilakshmi
Solid State and Structural Chemistry Unit, Indian Institute of Science, Bangalore 560012, India

K. Mohan Reddy
Senior Research Associate, Center for Applied Catalysis, Seton Hall University, South Orange, New Jersey 07079, USA

B. Hari Babu • N. Lingaiah • P.S. Sai Prasad (✉)
I & PC Division, Indian Institute of Chemical Technology, Hyderabad 500607, India
e-mail: pssaiprasad@gmail.com

A. Patel (ed.), *Environmentally Benign Catalysts: For Clean Organic Reactions*, DOI 10.1007/978-94-007-6710-2_2, © Panch Tattva Publishers, Pune, India 2013

3.5 *In Situ* Synthesized AMPA-Based Systems 40
3.6 V, Sb, and Bi Modified AMPA-Based Systems 45
3.7 Supported Vanadium Incorporated AMPA Systems 47
4 Conclusions 52
References 53

Abstract This chapter gives an overview of the work carried out in our group on the application of various forms of heteropoly acids (HPAs) for ammoxidation of methylpyrazine (MP). It starts with a general description of HPAs, ammonium salts of HPA, and the metal substituted HPAs (combinedly called as heteropoly compounds, HPCs). The various methods adopted for the synthesis of these compounds have been described. The HPCs have also been dispersed on different supports, and the methodology of their preparation is presented. Different characterization techniques have been used to determine their physicochemical properties. The reactivity of the catalysts is evaluated for the ammoxidation of 2-methylpyrazine (MP) to 2-cyanopyrazine (CP). The enhanced thermal stability and activity of the supported systems compared with those of the bulk ones are highlighted. Two new methods – one for the preparation of catalysts by in situ synthesis of the ammonium salt of heteropoly acid on supports and the other for the determination of dispersion of HPC on the support by FTIR technique – developed in our laboratory are introduced. Finally, correlations drawn between the activity and selectivity with the physicochemical properties of HPCs are presented.

1 Introduction

The fine chemical industry has experienced tremendous growth over the past few years due to the high demand for products like pharmaceuticals, pesticides, fragrances, flavoring, and food additives. The production of these products, which require stringent specifications, is definitely nontrivial. The stoichiometric organic synthesis, largely followed so far, leaves huge quantities of inorganic salts as byproducts whose disposal is a serious problem due to the keen environmental awareness and tightened regulations. Furthermore, the increased competition in industry has pushed the research and development activity on fine chemicals toward finding more cost-effective catalytic routes. One specific area that has a large impact on the environment protection is the utilization of solid HPCs as alternate catalysts for conventionally used reagents such as HF, H_2SO_4, etc. In the area of environmental pollution abatement [1], HPCs have recently been identified as very active catalysts. The acidic and redox functionalities of HPCs are used in solution as well as in the solid state [2]. Heteropoly acids are well-known green catalysts for oxidation [3] and acid-catalyzed reactions [4]. The salts of heteropoly acids are more thermally stable than their parent acids [5] and are extensively used as catalysts to obviate solubility problems during the reactions. Many developments like incorporation of heteropoly acids into the channels of mesoporous materials have also been adopted

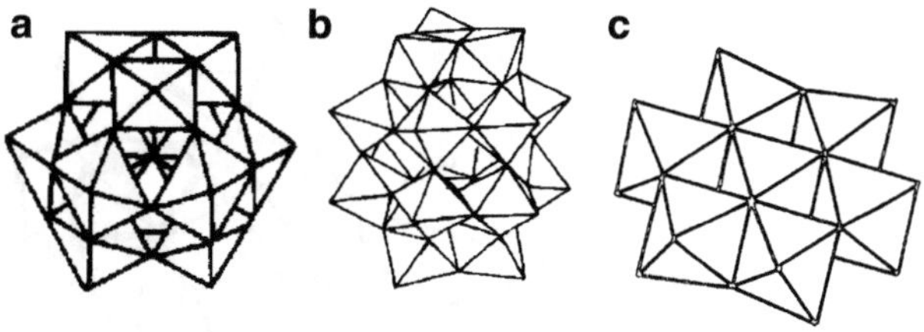

Fig. 2.1 Structures of heteropoly ion (**a**) Keggin, (**b**) Wells-Dawson, and (**c**) Anderson [8]

for improving product selectivity. There are still many areas for which new methods have to be explored. In this direction, concerted efforts have been directed toward developing more efficient HPC-based catalysts to be used in ammoxidation of aromatic and heteroaromatic compounds to their corresponding nitriles.

Single-step vapor-phase ammoxidation of different alkyl aromatics and heteroaromatics to their corresponding nitriles has become the topic of extensive research in recent times due to its usefulness as an essential commercial method. Selective synthesis of CP from MP by means of gas-phase ammoxidation is of particular industrial importance, because the resulting nitrile is a valuable intermediate for the production of an effective antitubercular drug, pyrazinamide. Pyrazinamide is a prodrug that stops the growth of *Mycobacterium tuberculosis*. Pyrazinamide is generally used in combination with other drugs such as isoniazid and rifampicin in the treatment of *Mycobacterium tuberculosis*. Initially, vanadia-based solids, either alone or in combination with other metals like Bi, Sb, and Nb particularly supported on TiO_2 [6], were used in the ammoxidation. However, all these catalysts were found to be active at high reaction temperatures (>430 °C) where the unwanted and more exothermic ammonia oxidation is favorable. Effective control of the exothermic reaction temperature is very critical, since there is every chance for reactor run away. Besides, in an exothermic reaction like that of ammoxidation, high reaction temperature always leads to drastic decrease in product selectivity. Efforts have been consistently made to develop low-temperature active catalysts. HPAs (e.g., 12-molybdophosphoric acid) and their ammonium salts are being viewed as promising low-temperature ammoxidation catalysts [7]. However, the low thermal stability of the bulk HPCs has become the major constraint for their applicability for vapor-phase reactions. Thus, efforts to develop novel catalyst systems with improved stability have become imminent.

1.1 Heteropoly Acids (HPAs) and Their Classification

The HPAs or polyoxometalates are made of a central or heteroatom linked with groups of peripheral atoms, again interlinked with oxygen atoms. These polyanions, having well-defined nanometer size structures, are classified based on the ratio between the polyatom and the heteroatom present in the total structure and the nature of arrangement of these atoms. These arrangements are classified mainly as Keggin, Well-Dawson, and Anderson structures [8]. The corresponding structures of heteropoly ions are shown in Fig. 2.1.

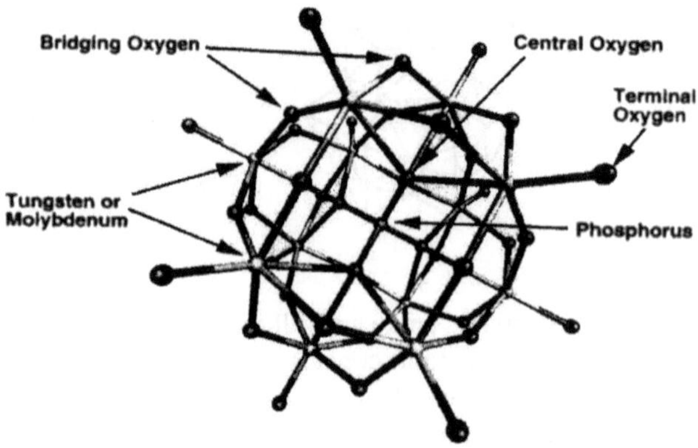

Fig. 2.2 The Keggin unit is the primary structure of the heteropoly acids and contains 12 transition metal atoms normally of tungsten or molybdenum, a central atom (usually phosphorus or silicon), and four types of oxygen atoms: central oxygens, terminal oxygens, and two types of bridging oxygens [3]

Among these compounds and their salts, those exhibiting Keggin-type polyanion structure are most extensively studied. The Keggin structures described by the general formula $H_nXM_{12}O_{40}$ ($M/X=12$) are also the most effective catalysts due to the unique combination of their stability, acidity, and structural accessibility. The central atom, X, is usually either P^{5+} or Si^{4+}, and the heteroatom, M, is usually W^{6+} or Mo^{6+}. The molecular Keggin structure consists of a central tetrahedron (XO_4^{n-}) surrounded by 12 linked octahedra containing the addenda atoms ($M_{12}O_{36}$). More than 65 elements from all groups of the periodic table can be found incorporated as heteroatoms [9].

Hereafter, the discussion in this chapter will be restricted to Keggin structures. As depicted in Fig. 2.2, there are four types of oxygen atoms in the Keggin unit: the central oxygen atoms (O_a), corner-sharing bridging oxygen atoms (O_b), edge-sharing bridging oxygen atoms (O_c), and terminal oxygen atoms (O_d). The central oxygen connects the heteroatom to a transition metal atom. The next two types of oxygen atoms bridge two transition metal atoms in adjoining octahedra. Finally, the terminal oxygen atom is bonded to only one transition metal atom [9]. The overall charge of the central tetrahedron is delocalized over the entire structure. Protons associate themselves with the exterior oxygen atoms (O_b, O_c, and O_d) and form acidic hydroxyl groups.

1.2 *Heteropoly Acids as Useful Catalysts*

The molybdo- and tungstophosphoric acids are found to be highly effective in catalyzing many reactions like olefin hydration and conversion of methanol to hydrocarbons [9]. Of late, the heteropoly acids are finding applications in the synthesis of fine chemicals too. They have also been widely involved in heterogeneous catalytic oxidation processes such as commercial vapor-phase oxidation of methacrolein, which produces more than 80,000 tons of methacrylic acid per year [10]. The bifunctional HPA catalysts can be used for sequential hydroformylation and oxidation of olefins [11]. HPAs also display catalytic activity in water splitting and oxygen transfer to alkanes [12]. The oxidation of alkenes and the coupling of aromatics utilize molybdovanadates as the reoxidants in the Wacker process [13]. Catalysts containing polyoxomolybdates are widely used for hydrodesulfurization, hydrodenitrification of fossil fuels [14]. The commercially successful direct oxidation of ethylene to acetic acid utilizes palladium and a HPC as catalysts producing 100,000 tons/year of the product [15]. There is another rapidly growing area, which is the heteropoly compound photochemistry and photo catalysis. Recently, their usefulness in bleaching wood pulp via oxidative delignification has been reported [16]. Molybdenum-based heteropoly compounds are better catalysts for oxidation reactions than their tungsten counterparts. Other applications of HPAs as ion-exchange materials, ion-selective membranes, and inorganic resistant materials have also been reported [17]. Thus, there has been a vigorous research activity taking place on expanding the applications of HPCs.

1.3 Modification of Heteropoly Acids

As stated earlier, the Keggin-type heteropoly acids are known to possess the unique features of strong intrinsic acidity and oxidizing property. Both the features can be suitably modified, by the variation of the cationic composition as well as by substitution of a few atoms of either molybdenum or tungsten in peripheral positions in the anion with other transition metals such as V, Sb, Fe, and Ti [3, 18]. Roch et al. [19] studied the addition of transition metals to HPAs as an important approach to control their redox properties and to impart higher thermal stability. Conclusions are drawn based on the studies of catalyzed oxidative dehydrogenation of isobutyric acid by $H_4[PMo_{11}VO_{40}]$ and $VOH[PMo_{12}O_{40}]$ at 320 °C. Thermal treatment of $H_4[PMo_{11}VO_{40}]$ at 320 °C leads to a complex mixture of different species with or without vanadium inside the Keggin structure. It has been reported that the protonic form, $H_3PMo_{12}O_{40}$, catalyzes oxidation of lower alkanes and that substitution of V^{5+} for Mo^{6+} improves the catalytic activity and selectivity [12]. Iron and copper have been the most widely used additives for enhancing the catalytic activity of HPAs, as reported in the literature [20]. Among the transition metals, the effect of metals like vanadium, iron, zinc, chromium, and nickel, in both cationic and anionic positions, on the catalytic performance of heteropoly compounds in gas-phase oxidations has been studied [21]. Catalysts containing V in combination with these elements are known to be effective for the ammoxidation of hydrocarbon reactions [21]. The addition of vanadium to a heteropoly compound improves its catalytic functionality with respect to partial oxidations as well.

Vanadium incorporated molybdophosphoric acid catalyst shows a unique bifunctional property, which arises due to the redox nature of vanadium and the acidic character of the molybdophosphoric acid [22]. The molybdovanadophosphoric acid-based catalysts are used commercially for the synthesis of methacrolein and in the conversion of isobutyric acid to methacrylic acid. Ressler et al. [23] studied the bulk structural evolution of a vanadium-containing heteropolyoxomolybdate [$H_4(PVMo_{11}O_{40})\cdot 13H_2O$], with vanadium substituting for Mo in the Keggin ion. Several authors have investigated molybdophosphoric acid (MPA) by replacing 1–3 Mo atoms with the corresponding number of V atoms [24, 25]. The main observation in the high vanadium incorporated systems is the expulsion of vanadium from the secondary structure and formation of a vanadyl (VO^{2+}) salt during catalysis [26]. Kozhevnikov [27] studied the applications of 40 Keggin-type heteropolyanions $PMo_{12-n}V_nO^{(3+n)-}$ as catalysts for aerobic liquid-phase oxidation.

1.4 Salts of Heteropoly Acids and Their Catalytic Functionalities

Berzelius [28] first synthesized ammonium phosphomolybdate in 1826, which is now known as ammonium 12-molybdophosphate. Because the salts of heteropoly acids possess higher thermal stability, higher microporosity, and insolubility in several solvents relative to the pure acid, ammonium salts of HPAs are used as catalysts for several high-temperature gas-phase reactions. Ammonium salt of 12-molybdophosphoric acid (AMPA), exhibiting the Keggin structure, has been studied for the oxidative dehydrogenation of isobutyric acid to methacrylic acid by McGarvey and Moffat [29]. The arrangement of NH_4^+ ion in heteropoly acid salt is shown in Fig. 2.3.

1.5 Supported Salts of HPAs

Initial catalytic studies were restricted to unsupported acids and their salts. However, as the bulk compounds had low surface area and inadequate thermal stability, it was thought that dispersing them on suitable supports would improve their properties. The supports range from γ-Al_2O_3, which contains considerable amount of basic centers, and SiO_2, SiO_2-Al_2O_3, and Nb_2O_5, which contain more acid centers. Misono [3, 4], in his review on catalysis by heteropoly acids, quoted that supports containing basic centers bring about the decomposition of the polyanion, and, hence, acidic supports are preferable.

There are conflicting reports on the thermal stability of the supported HPAs, particularly with reference of molybdophosphoric acid supported on Al_2O_3 and SiO_2-Al_2O_3 [31]. It is proposed that during the precipitation of the HPAs on supports containing Al_2O_3, the support interacts with the HPA, forming a compound between

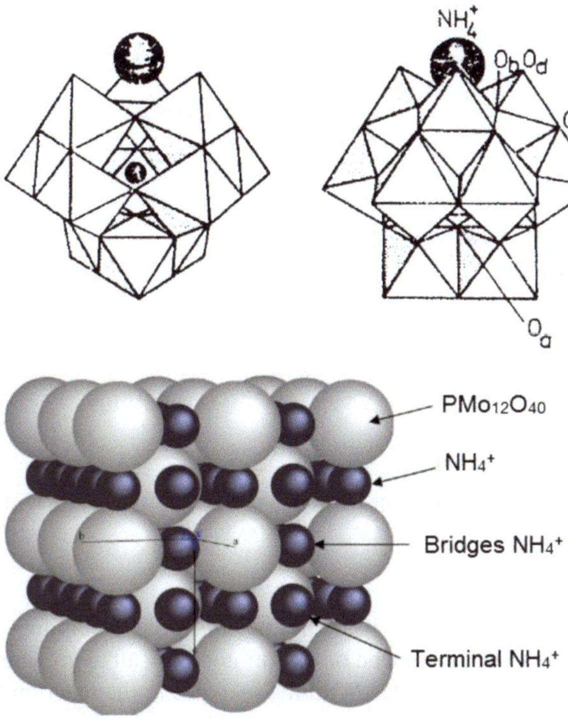

Fig. 2.3 The Keggin unit of AMPA with the position of NH_4^+ ion [30]

the two. In the case of acidic supports, the hydroxyls get protonated and interact with the negative heteropoly ion leading to electrostatic attraction which brings about better dispersion [32]. Better HPA-support interaction improves the thermal stability of the HPA when compared with the bulk acid.

The ammonium salt also contains strong acid sites. The super acidity is reported to be the manifestation of the presence of residual protons in the ammonium salt [33]. Hence, the ammonium salt is also expected to display variation in the nature of HPA-support interaction depending on the acid-base character of the support. Further, the bulk-phase ammonium salt is reported to be thermally more stable than its parent acid.

1.6 In Situ Synthesis of AMPA

The general procedure adopted by Lapham and Moffat [34] and Ito et al. [35] to synthesize ammonium salts of HPAs is a titrimetric method using $(NH_4)_2CO_3$, NH_4Cl, NH_4NO_3, NH_4SO_4, or $(NH_4)_2CO$ solutions and pure heteropoly acid. Ito et al. also described a homogeneous precipitation method to prepare the ammonium salt of HPA. Later, the same groups have studied the effect of temperature and time

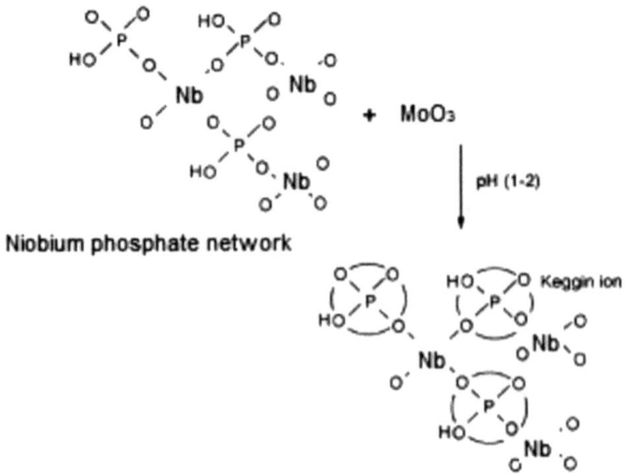

Scheme 2.1 Possible route to Keggin ion formation via grafting of MoO_x to the surface of niobium phosphate

of the reaction on the morphology of the acid. As the precipitation temperature increases from 0 to 95 °C, the shape of the ammonium salt aggregate changes from spherical to symmetric decahedral. With the increase in the reaction time from 3 to 24 h, the aggregates are found to be dodecahedral, microporous without mesopore, and highly crystalline. Other methods also exist in the literature for transformation of the hydrogen ions of the parent acid into ammonium ions. Following these methods, one could synthesize the salt with residual protons. In our group, we have synthesized the ammonium salt by using a novel in situ method of preparation (details *vide infra*). The phosphate ions on the support are made to react in situ with the ammonium heptamolybdate in the solution, and, thus, the heteropoly ion is grown at specific sites on the support.

Literature also reveals that ionic interactions between the support surface and HPA clusters help to generate the active species during the impregnation step. However, because of the weak nature of this interaction, it is possible for HPA clusters to leach from the support during reactions in polar solvents. In situ growth of a HPA cluster about a tethered Nb–PO_4 group, such as those found in the framework of a metal phosphate support (Scheme 2.1), could be used to form a bound Keggin-like cluster which may be stable under polar solvents.

Other advantages of in situ process are listed below:

- It avoids usage of corrosion medium like phosphoric acid and the cumbersome ion-exchange procedure.
- The formation of clusters of the salt on the support could be drastically reduced.
- Insolubility of AMPA in common solvents can be obviated in the preparation of supported systems.
- The method can be adapted to various organic and inorganic phosphate precursors.

2 Experimental

2.1 Chemicals and Supports

All solvents and acids used were of A.R. grade (99.9 %). $(NH_4)_6Mo_7O_{24}{\cdot}4H_2O$, $SbCl_3$, and $Bi(NO_3)_2{\cdot}5H_2O$ were supplied by S.D. Fine Chemicals, India. G.R. grade $(NH_4)_2HPO_4$, $(NH_4)H_2PO_4$, H_3PO_4 and NH_4VO_3 were purchased from Loba Chemie, India. All the supports used were commercial samples TiO_2-P-25, Degussa, Germany, Nb_2O_5-AD/1447, CBMM, Brazil, γ-Al_2O_3-3916R Harshaw, USA, ZrO_2, Zircon, India, and fluid SiO_2-53418, AKZO, Japan. These supports were used without further purification. Nb phosphate was supplied by (CBMM, Brazil). 2-methylpyrazine (99 %) was purchased from Aldrich and used without purification.

2.2 Preparation of Various Types of HPA Catalysts

2.2.1 Preparation of AMPA Using Different Phosphorous Precursors

The AMPA catalysts were prepared by dissolving ammonium heptamolybdate $[(NH_4)_6Mo_7O_{24}{\cdot}4H_2O]$ and diammonium hydrogen phosphate (DAHP) $(NH_4)_2H(PO_4)$ in stoichiometric ratio in minimum amount of water. The aqueous solution was first refluxed at 100 °C for 6 h and then concentrated on a water bath to reduce the initial solution to one-third of its volume. The mixture was then slowly heated to get the solid mass, which was dried, first at 120 °C (6 h) and then at 180 °C (6 h). The solid was divided into four equal parts, and each part was subsequently activated at four different temperatures, namely, 400, 450, 500, and 550 °C to get catalysts with different temperatures of activation designated as DAHP catalysts. The same procedure was repeated taking ammonium dihydrogen orthophosphate (ADHP), and the catalysts are designated as ADHP catalysts.

2.2.2 Preparation of Supported AMPA Catalysts

AMPA was first prepared using DAHP, adopting the method described above. The supported catalysts were prepared by impregnating commercial supports, namely, SiO_2, TiO_2, ZrO_2, Al_2O_3, and Nb_2O_5 with known amounts of aqueous solution of AMPA, followed by drying at 120 °C. All the samples were calcined at 350 °C for 4 h in presence of air. The extent of AMPA loading was thus varied between 5 and 25 wt%.

2.2.3 Preparation of Vanadium Incorporated MPA

Vanadium incorporation into the primary structure of molybdophosphoric acid, i.e., $H_4PMo_{11}V_1O_{40}$ ($MPAV_1$) was then taken up. Typically, calculated amount of disodium hydrogen phosphate was dissolved in water and mixed with required amount of sodium metavanadate that was already dissolved in boiling water. The mixture was cooled and acidified with concentrated sulfuric acid. An aqueous solution of sodium molybdate dihydrate was then added to this mixture. The addition of concentrated sulfuric acid with vigorous stirring resulted in a color change from dark red to light red. The VMPA formed was extracted with diethyl ether as the heteropoly acid was present in middle layer as heteropolyetherate, and the ether was removed by passing through air. The orange solid obtained was dissolved in water and concentrated until the crystals appeared. The prepared catalyst was subsequently calcined in air at 350 °C for 4 h.

In the case of vanadium in the secondary structure of MPA, i.e., $[(VO)_{1.5}PMo_{12}O_{40}]$ (VOMPA), $(VO)^{2+}$ ions were first exchanged with the protons of MPA. Required quantity of V_2O_5 was dissolved in oxalic acid at 100 °C and cooled to room temperature before adding this solution to the aqueous solution of MPA with constant stirring. The excess water was removed on a water bath and the solid sample dried at 120 °C for 12 h. The obtained catalyst was calcined in air at 350 °C for 4 h.

2.2.4 Preparation of Supported Vanadium Incorporated AMPA

The procedure adopted for the preparation of V metal substituted AMPA catalysts was as follows. Ammonium heptamolybdate $[(NH_4)_6Mo_7O_{24}\cdot 6H_2O]$ and diammonium hydrogen orthophosphate $[(NH_4)_2HPO_4]$ were added to water and completely dissolved at a temperature of 80 °C. To this solution, a calculated (depending on the number of V atoms to be incorporated) amount of ammonium metavanadate $[(NH_4)VO_3]$ was added. The contents were refluxed at 100 °C for 6 h, and then dilute HNO_3 was slowly added to the solution to maintain the pH of the solution between 1 and 2. The addition of acid yielded a reddish yellow precipitate. Excess water was then evaporated and the precipitate dried at 120 °C over night and calcined at 350 °C for 4 h in air. The catalyst was designated as vanadium incorporated ammonium salt of 12-molybdophosphoric acid (AMPV).

The supported catalysts were prepared by impregnating commercial supports, namely, SiO_2, TiO_2, ZrO_2, AlF_3, and CeO_2, and ceria containing mixed oxides, with known amounts of AMPV salt solution as prepared above followed by drying at 120 °C. All the samples were calcined at 350 °C for 4 h in presence of air. The extent of AMPV loading was also varied between 5 and 25 wt%.

2.2.5 In Situ Synthesis of AMPA

The catalyst was prepared by grafting the polyoxomolybdate species with the phosphate ion of the support. The methodology of preparation was as follows: Catalysts with varying MoOx loading were prepared by impregnating the metal phosphate supports with known quantities of ammonium heptamolybdate by the wet impregnation method. The amount of Mo in the impregnating solution was selected such that the nominal MoO_3 content varied between 5 and 20 wt%.

The impregnated samples were dried at 120 °C after evaporating the solution in a water bath. The dried masses were then carefully calcined in air at 400 °C over a period of 6 h and kept at that temperature for 4 h.

2.3 Characterization of HPAs

BET surface areas of the catalysts were determined by nitrogen adsorption, at liquid-nitrogen temperature, on a conventional all-glass high-vacuum apparatus. XRD patterns of the catalysts were obtained on a Siemens D-5000 diffractometer using Cu Kα radiation. FTIR spectra were recorded on a Biorad-175 C (USA) spectrophotometer following the KBr disc method. Raman spectra were recorded using a Raman microscope (InViaReflex, Renishaw) equipped with deep-depleted thermoelectrically cooled CCD array detector and a high-grade Leica microscope (long working distance objective 20×). Raman measurements were made on the sample spot irradiated by a visible 514.5 nm argon ion laser at a fixed laser power of circa 1 mW, exposure time of 10 s, and the spectral resolution of 1–1.3 cm^{-1} was used. ^{31}P MAS NMR spectra of solids were recorded in a 300 MHz Bruker ASX-300 spectrometer. A 4.5 ms pulse (90°) was used with repetition time of 5 s between pulses in order to avoid saturation effects. The spinning rate was 5 kHz. All measurements were carried out at room temperature using 85 % H_3PO_4 as standard reference.

Temperature-programmed reduction (TPR) of the catalysts was carried out in a flow of 10 % H_2/Ar mixture gas at a flow rate of 30 ml/min with a temperature ramp of 10 °C/min. Before the H_2-TPR run, the catalysts were pretreated with Argon gas at 250 °C for 2 h. The hydrogen consumption was monitored using a thermal conductivity detector. TPD of ammonia was carried out on a laboratory built apparatus equipped with a Q-mass detector. In a typical experiment, about 500 mg of oven dried at 110 °C overnight sample was taken in a U-shaped quartz sample tube. Prior to TPDA studies, the catalyst sample was pretreated at 200 °C for 1 h by passing pure helium (99.9 %, 50 ml/min). After pretreatment of the sample, it was saturated with anhydrous ammonia (in a flow of 10 % NH_3-90 % He mixture) at 80 °C with a flow rate of 75 ml/min and was subsequently flushed at 105 °C for 2 h to remove any physisorbed ammonia. Then the TPDA analysis was carried out from ambient temperature to 800 °C at a heating rate of 10 °C/min. The amount of NH_3 evolved (as the sum of ammonia and molecular nitrogen detected) was calculated using GRAMS/32 software. The acidic strength of the solid samples was measured by the potentiometric titration method employing a saturated calomel electrode (SCE). A known mass of solid suspended in acetonitrile was stirred for 3 h, and then the suspension was titrated with a solution of n-butylamine in acetonitrile (0.05N) using a flow rate of 0.05 ml/min. The initial electrode potential (Ei) was assumed to correspond to the maximum acid strength of the surface sites. The acid strength of surface sites was assigned as: very strong sites, $Ei > 100$ mV; strong sites, $0 < Ei < 100$ mV; weak sites, $-100 < Ei < 0$ mV; and very weak sites, $Ei < -100$ mV.

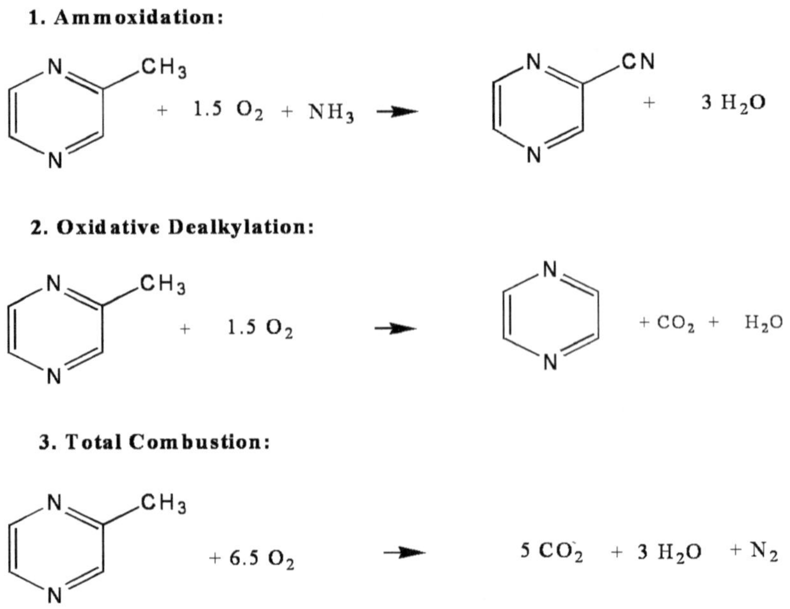

Scheme 2.2 The three possible reactions during ammoxidation of methylpyrazine

2.4 *Ammoxidation of MP over HPA Salts*

2.4.1 Ammoxidation of 2-Methylpyrazine (MP)

The ammoxidation is basically an oxidation reaction carried out in the presence of ammonia. The reaction is exothermic and is generally carried out in fixed bed reactors at atmospheric pressure. Cyanopyrazine (CP) is the main product of reaction with maximum selectivity approaching 85–100 %. Two other side reactions can also take place; the demethylation of MP gives rise to pyrazine formation, whereas total oxidation leads to formation of carbon dioxide and water vapor. The general scheme for the reaction is described in Scheme 2.2.

2.4.2 Evaluation of Catalysts

The ammoxidation reaction was carried out in a vertical fixed bed, continuous down flow quartz micro-reactor under atmospheric pressure. In a typical experiment, about 3 g of the catalyst (sieved to 18/25 BSS mesh, to avoid mass transfer limitations) diluted with an equal amount of quartz grains was packed between two layers of quartz wool in the reactor. The upper portion of the reactor was filled with quartz beads that served both as a preheater and a mixer for the reactants. Prior to introducing the reactant, MP, the catalyst was treated in ammonia flow at a rate of 20 ml/min for

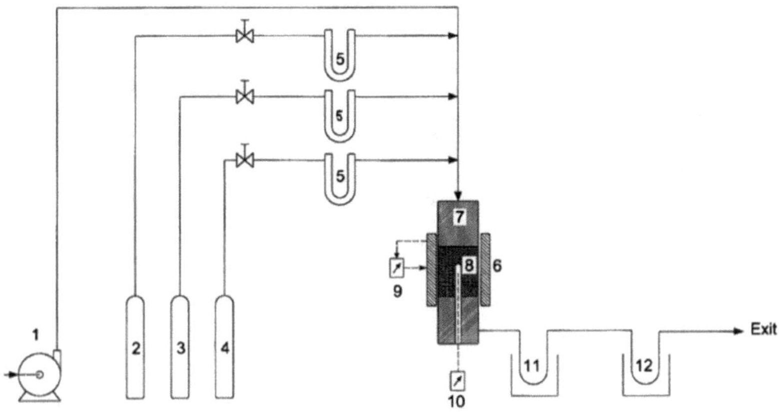

Fig. 2.4 Experimental setup for catalytic activity measurements

1 h. An aqueous mixture of MP (MP: water = 1:2.5, v/v) was fed into the reactor by means of a microprocessor-controlled metering pump (B. Braun, Germany). The molar ratio of the feed was kept at MP: water: ammonia: air = 1:13:7:38, maintaining a W/F_{liquid} ratio = 2.0 g cm^{-3} h. The reaction was studied in the temperature range of 360–420° C and monitored by a thermocouple with its tip located in the catalyst bed and connected by a PID-type temperature indicator-controller. After allowing the catalyst to attain steady state, at each reaction temperature for 30 min, the product was collected for 15 min and analyzed by gas chromatography, separating it on an OV-101 column (2 m long, 3 mm diameter) using an FID detector. From the analysis of non-condensable exit gas mixture, it was ensured that the quantity of any organic species was negligible. A schematic diagram of the experimental setup is shown in Fig. 2.4.

3 Results and Discussion

3.1 Studies on Bulk MPA and Vanadium Modified MPA (VMPA, VOMPA) Catalysts

Figure 2.5 shows the activity of various MPA catalysts (expressed as percent conversion of MP) during the ammoxidation at different reaction temperatures (360–420 °C). Introduction of vanadium into the Keggin structure (VMPA) is seen influencing the conversion of MP. The activity of the catalyst with vanadium incorporation into the primary structure of MPA (VMPA catalyst) is higher than that of the vanadyl salt (VOMPA) catalyst and pure MPA catalyst. Such a significantly different influence of vanadium atom due to the difference in its location in

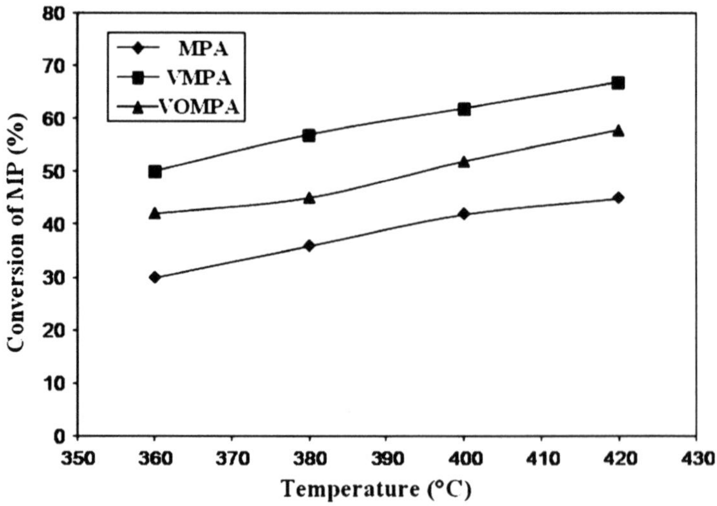

Fig. 2.5 Activity patterns of bulk MPA, VMPA, and VOMPA catalysts at different reaction temperatures (Ref. [36])

heteropoly compound may result in very different susceptibility for reoxidation process. Further evidence to this observation can be obtained by studying the supported catalysts, as stated below.

3.2 *Studies on TiO_2-Supported MPA, VMPA, and VOMPA*

The ammoxidation studies conducted on MPA/TiO_2, VMPA/TiO_2, and VOMPA/TiO_2 (with 20 wt% active component in the catalysts) expressed in terms of CP yield, at a typical reaction temperature of 380 °C, is shown in Fig. 2.6. The physicochemical properties of the catalysts can be found elsewhere [36]. The catalyst with vanadium incorporation in the primary structure shows higher yield of CP than the vanadium-free MPA catalyst and the catalyst with vanadium in the secondary structure (VOMPA). The difference in the behavior of the catalysts can be attributed to the difference in the location of vanadium atom in the heteropoly acid.

It has been reported that substitution of V^{5+} for Mo^{6+} in MPA results in the generation of more reactive lattice oxygen associated with the Mo–O–V species [37]. The oxidation reaction catalyzed by vanadium occurs by a redox reaction, and it normally follows the Mars-Van Krevelen mechanism. The lattice oxygen helps in the oxidation of the organic substrate, and the gas-phase oxygen supplements the loss of lattice oxygen. The lattice oxygen also plays an important role in the oxidation and reduction cycle of V during the ammoxidation reaction by enhancing its reactivity. The difference in the functionality of vanadium atoms in heteropoly compounds arises due to the very different susceptibility of

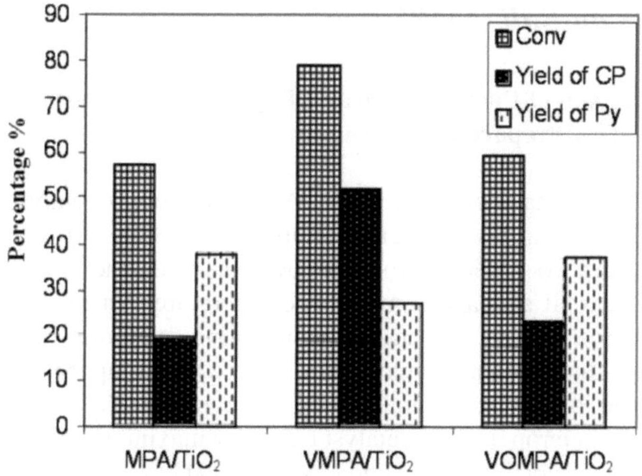

Fig. 2.6 Activity patterns of the TiO_2-supported MPA, VMPA, and VOMPA catalysts at 380 °C (Ref. [36])

Fig. 2.7 Activity patterns of the TiO_2-supported catalysts at various reaction temperatures

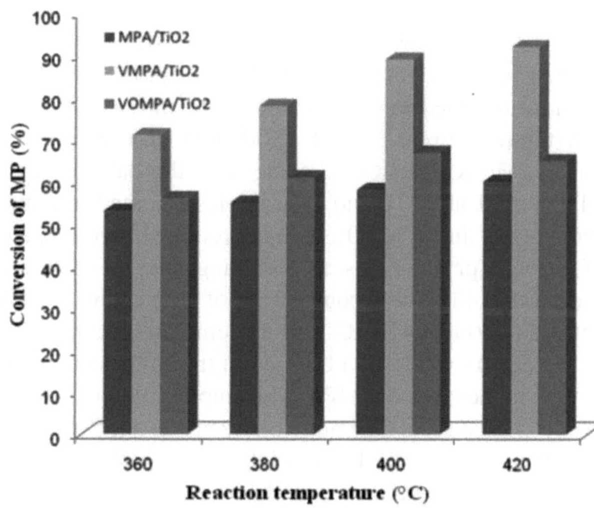

vanadyl cations and vanadium atoms in the Keggin structure for reoxidation process [38]. Garte et al. [39] reported that the vanadium atoms present in the Keggin structure, when reduced to V^{4+}, are easily reoxidized, while the reoxidation of $(VO)^{2+}$ vanadyl ions occurs at a much slower rate even at higher temperature. Thus, it is for this reason that the VMPA/TiO_2 exhibits higher activity than the VOMPA/TiO_2 in the ammoxidation reaction. The effect of reaction temperature is studied over these catalysts and the results are presented in Fig. 2.7. The activity of the catalyst increases with increase in reaction temperature. The activity is found to vary in the following order at any given temperature: VMPA/TiO_2 > VOMPA/TiO_2 > MPA/TiO_2.

3.3 Studies on Bulk AMPA

3.3.1 Comparison of the Structure and Reactivity of the MPA and AMPA Prepared by Precipitation Method

Bondareva et al. [7] have employed a MPA based HPC as catalyst and achieved about 75 % CP yield at lower reaction temperatures (380–390 °C). This catalyst, however, produces 10–25 % byproducts (pyrazine and others) depending on the process conditions. It appears from the open literature that in the ammoxidation reactions attention has been focused on obtaining maximum activity from the catalysts rather than obtaining maximum selectivity, an approach not favorable for development of environmentally acceptable processes. If the molybdenum phosphate based heteropoly acid catalyst can offer maximum selectivity to CP, the process could be acceptable [40]. Bondareva et al. [7] have noticed the formation of AMPA in the used catalysts even though the fresh catalysts were in the acid form. Therefore, it is interesting to observe the functionality of catalysts starting from AMPA itself, since AMPA is also reported to have better thermal stability than MPA [41].

The XRD patterns of the AMPA and MPA, pretreated at different temperatures, in the range 300–500 °C, are shown in Fig. 2.8. The XRD patterns are similar to those reported in our previous publication [42] giving the conclusions: (1) the low-temperature-calcined (300–400 °C) catalysts show the formation of $(NH_4)_3PMo_{12}O_{40}{\cdot}4H_2O$, agreeing with the data presented by Roch et al. [41] and Albonetti et al. [43], and (2) samples calcined at temperatures 450 °C and above show peaks due to MoO_3, with the release of ammonia either as such or in the form of nitrogen or its oxides, as also suggested by Damyanova et al. [44] and Hodnett and Moffat [45]. The decomposition of the Keggin ion of AMPA starts at a temperature between 400 and 450 °C, and its complete decomposition occurs at about 500 °C. Peaks due to MoO_3 are observed in the MPA sample calcined at 400 °C as against 450 °C in the case of AMPA. This implies that the structure of the Keggin ion begins to decompose at a temperature between 350 and 400 °C in MPA with partial reduction and distortion of the Keggin unit, as also shown by Tsigdinos [46]. Increase in calcination temperature beyond 450 and 500 °C results in formation of MoO_3 phase. Thus, the thermal stability of MPA is lower compared to AMPA, closely agreeing with what has been reported by McMonagle and Moffat [47].

The FTIR spectra of AMPA and MPA samples subjected to thermal treatment at 300–500 °C are presented at Fig. 2.9. The characteristic peaks due to the Keggin ion, the presence of ammonium ion and a significant shift in the $Mo–O_b–Mo$ band to an extent of 15 cm^{-1} reveal the formation of AMPA (Fig. 2.9A), as also reported by Deltcheff and Fournier [48]. Calcination at 500 °C has resulted in total disappearance of AMPA and the domination of bands due to MoO_3. The FTIR spectra of the MPA catalysts calcined at lower temperatures (Fig. 2.9B), i.e., 300–350 °C, show the characteristic bands of the Keggin ion without any distortion. Upon increasing the pretreatment temperature to 400 °C, a broadband in the range of

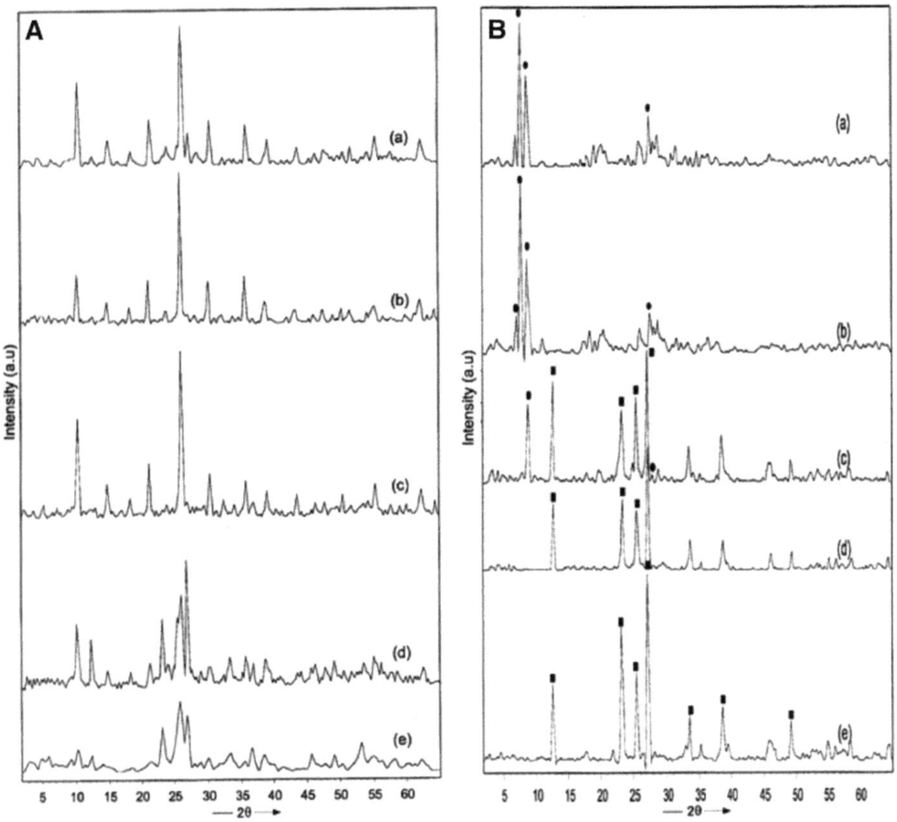

Fig. 2.8 XRD pattern of (**A**) AMPA and (**B**) MPA catalysts pretreated at (*a*) 300 °C, (*b*) 350 °C, (*c*) 400 °C, (*d*) 450 °C, and (*e*) 500 °C (Ref. [42])

600–1,000 cm^{-1} is observed; this might be due to formation of MoO_3. The peak at 790 cm^{-1} corresponding to as (Mo–O_c–Mo) also becomes very weak, confirming the distortion of the Keggin structure. After calcination at 450–500 °C, the MPA spectra clearly show characteristic bands of MoO_3, indicating complete destruction of the Keggin ion.

The ^{31}P NMR spectra of AMPA catalysts calcined at different temperatures are shown in Fig. 2.10A. The sample calcined at 300 °C shows two prominent peaks appearing at –5.98 and –12.3 ppm and two more at –5.27 and –7.75 ppm, respectively. The peak at –5.98 ppm corresponds to the $[(NH_4)_3PMo_{12}O_{40}\cdot 4H_2O]$ species, and the signals at –5.27 and –7.75 ppm may correspond to dehydrated species and that at –12.3 ppm is for the lacunary species. It is proposed that the signals in the lower field belong to the Keggin ion whether they are dehydrated or the lacunary species. The four different species are similar to those reported by vanVeen et al. [49]. These four signals appeared with very little variation in chemical shifts toward upfield after calcination at 350 °C. At 400 °C, the peak at –5.98 ppm becomes very small, and the

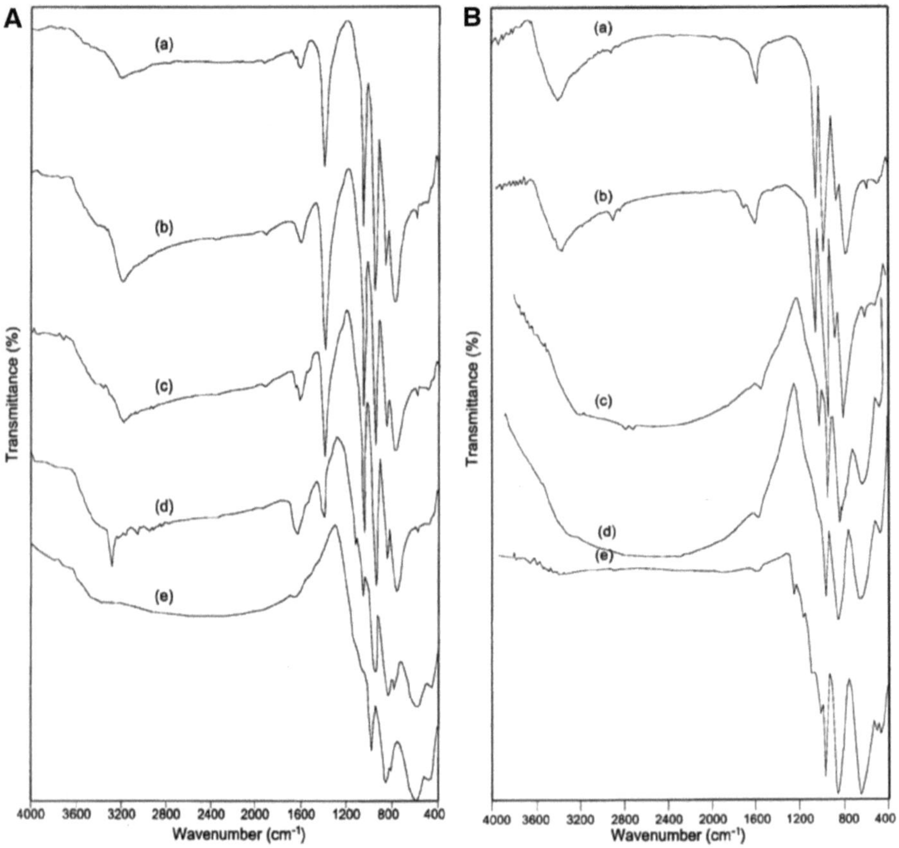

Fig. 2.9 FT-IR spectra of (**A**) AMPA and (**B**) MPA catalysts pretreated at (*a*) 300 °C, (*b*) 350 °C, (*c*) 400 °C, (*d*) 450 °C, and (*e*) 500 °C (Ref. [42])

peak at –12 ppm vanishes. A new peak appears at 0.56 ppm, which corresponds to monophosphate [50]. This observation clearly reveals the beginning of decomposition of lacunary species into the Mo and P oxides at this temperature. At 450 °C, the two signals at –6.0 and 0.56 ppm can be observed. The peak at 0.56 ppm has shifted to 1.94 ppm due to the decomposition of Keggin ion. At 500 °C, the peaks in the negative side vanish completely with the appearance of only one peak at 1.98 ppm. The peak at 1.94 ppm reflects the formation of P_2O_5 like structure or some unknown molybdenyl phosphate from decomposed Mo and P oxides. These results clearly reveal that the decomposition of Keggin ion completes at this temperature. The signal at 1 ppm appearing in all the samples corresponds to external reference (H_3PO_4).

The ^{31}P NMR spectra of MPA catalysts calcined at different temperatures are shown in Fig. 2.10B. The sample calcined at 300 °C shows two peaks at –1.132 and –2.67 ppm. The former, which is more intense than the latter, corresponds to [$H_3PMo_{12}O_{40}\cdot 9H_2O$], whereas the latter, which has shifted to up field, corresponds

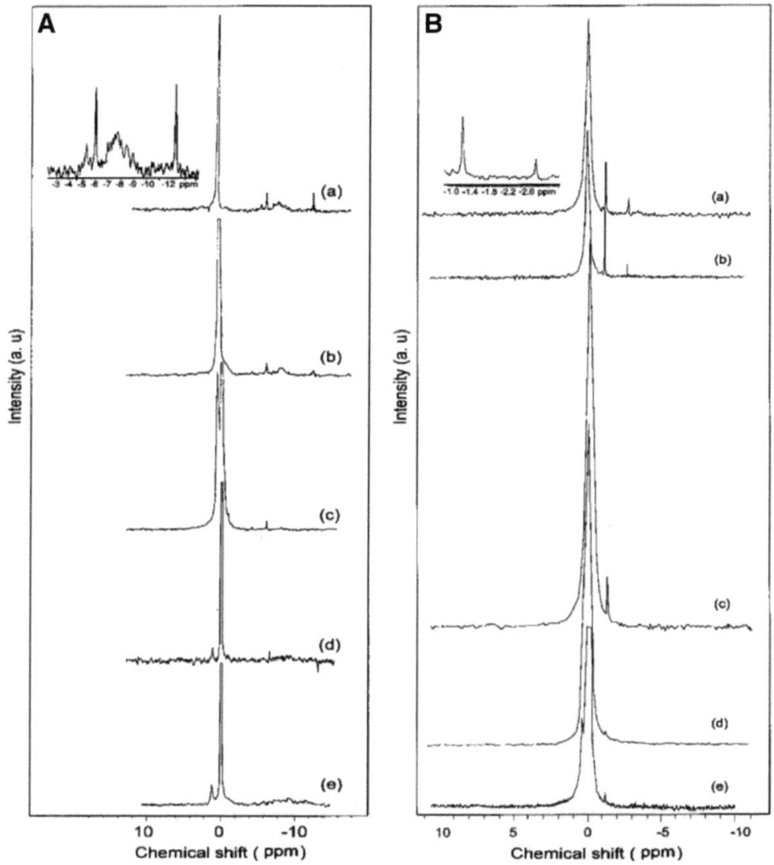

Fig. 2.10 ^{31}P NMR spectra of (**A**) AMPA and (**B**) MPA catalysts pretreated at (*a*) 300 °C, (*b*) 350 °C, (*c*) 400 °C, (*d*) 450 °C, and (*e*) 500 °C

to dehydrated Keggin ion species [51]. These two signals appear at the same positions in the sample calcined at 350 °C but with decrease in peak intensities. Only one clear peak at –1.15 ppm can be observed in the sample calcined at 400 °C with lesser intensity. The peak at –2.67 ppm disappears. However, a small peak at the positive side, corresponding to monophosphate can be seen.

This change demonstrates the start of the decomposition of Keggin ion at 300 °C. Upon increasing the calcination temperature to 450 °C, the peak which corresponds to Keggin ion becomes very small. A corresponding increase in the signal at 0.42 ppm gives a clear indication that the Keggin ion is almost decomposed to its oxides. The sample calcined at 550 °C shows only peak at 0.42 ppm, indicating complete decomposition. These results are in perfect agreement with the data obtained on XRD and FTIR.

The TPD profiles of AMPA, obtained after various pretreatments, are presented in Fig. 2.11A. The sample calcined at 300 °C exhibits three peaks: one centered

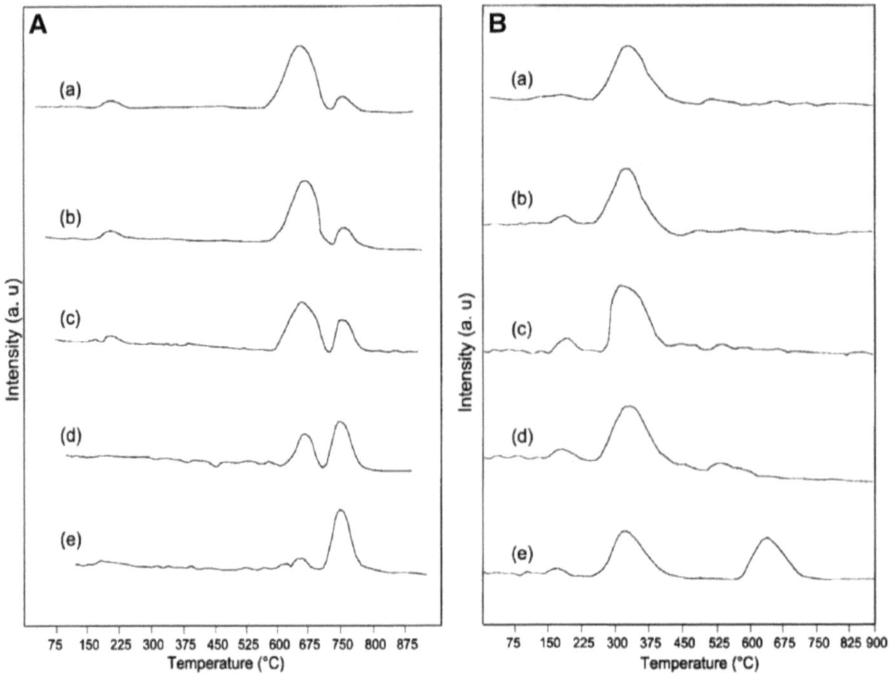

Fig. 2.11 TPDA patterns of (**A**) AMPA and (**B**) MPA catalysts pretreated at (*a*) 300 °C, (*b*) 350 °C, (*c*) 400 °C, (*d*) 450 °C, (*e*) 500 °C

around 200 °C, a second broad peak centered at 655 °C, and the third desorption peak at 730 °C. Hodnett and Moffat [45] reported that the peak at 200 °C is entirely due to water molecules. Essayem et al. [52] demonstrated two desorption peaks: one due to desorption of NH_3 from the ammonium salt and the other due to desorption of NH_3 from the products formed by the decomposition AMPA at higher temperatures. Upon increasing the pretreatment temperature, the intensity of the peak at 655 °C decreases, and the intensity of the peak at 730 °C increases. As already discussed, the peak due to desorption of NH_3 on the oxides of Mo and P represents the formation thermally decomposed products.

The TPD profiles of MPA pretreated at different temperatures are depicted in Fig. 2.11B. A single desorption peak with its maximum at 360 °C is observed in samples calcined at temperature in the range of 300–400 °C. It is because of desorption of adsorbed ammonia, as reported elsewhere [53]. Ammonia, as a polar molecule, can enter into the whole crystallites and neutralize all the bulk protons resulting in the formation of ammonium salt. Upon increasing the pretreatment temperature to 450–500 °C, a new desorption peak at 655 °C appears, which may correspond to adsorption of NH_3 by the oxides formed during the calcination at higher temperatures. TGA, XRD, FTIR, and ^{31}P NMR results reveal that the decomposition of Keggin ion starts beyond 350 °C in the case of MPA. The samples calcined at 450 and 500 °C have very little capability to adsorb NH_3 molecules because of complete destruction of Keggin structure.

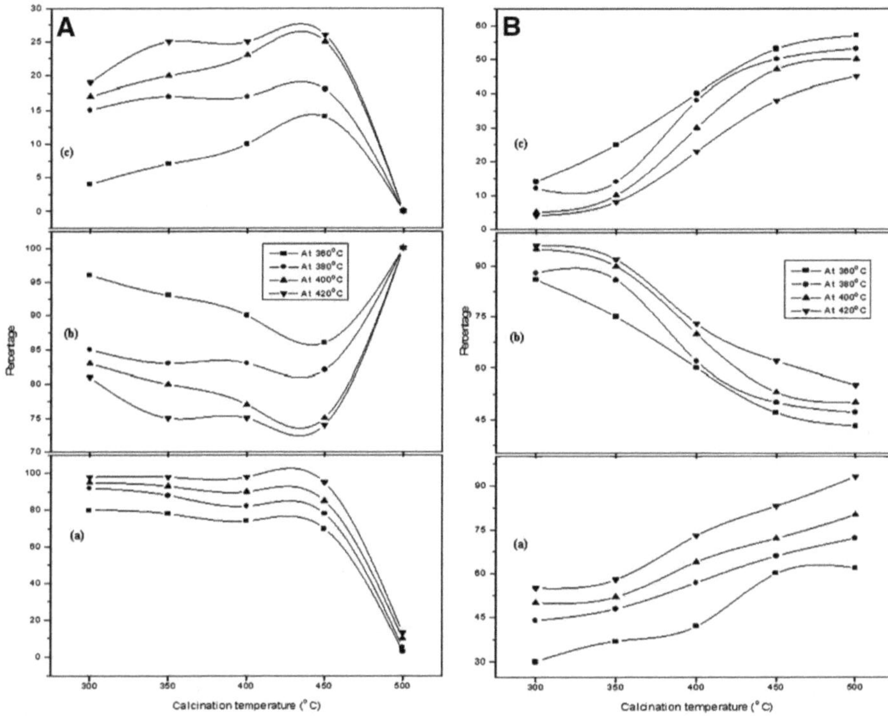

Fig. 2.12 Product distribution of (**A**) AMPA and (**B**) MPA catalysts pretreated at different temperatures: (*a*) conversion of 2-MP, (*b*) selectivity to CP, (*c*) selectivity to pyrazine

Variations in activity and selectivity of the catalysts against the temperature of pretreatment as well as reaction temperature are shown in Fig. 2.12A. The AMPA catalysts pretreated in the temperature region 300–450 °C show stable activity reflected in terms of overall conversion at all reaction temperatures in the region of 360–420 °C. Further, the conversion reaches a value of 80 % at the reaction temperature of 360 °C and increases further to a value close to 100 % at 420 °C. Obtaining very high and stable activity is a significant observation. The samples also exhibit very high CP selectivity falling in the region of 75–95 %. Pyrazine levels vary in the range of 5–25 %. Pyrazine formation is reported to be more facile on the acid form of the catalyst. Thus, the tendency to form more pyrazine with increasing calcination temperature can be attributed to the partial decomposition of the ammonium salt leading to the formation of the parent acid. The activity patterns reveal that the AMPA catalyst offers the flexibility of conducting the reaction either at maximum conversion with restricted selectivity or with high selectivity at restricted conversion, the latter being environmentally highly attractive.

The activity dropped to very low value when the pretreatment temperature reaches 500 °C. From the XRD, FTIR, ^{31}P NMR, and TPDA results, it can be understood that the catalyst retains the structure of AMPA up to 450 °C, beyond which there is complete destruction of Keggin unit with the release of ammonia and the formation of

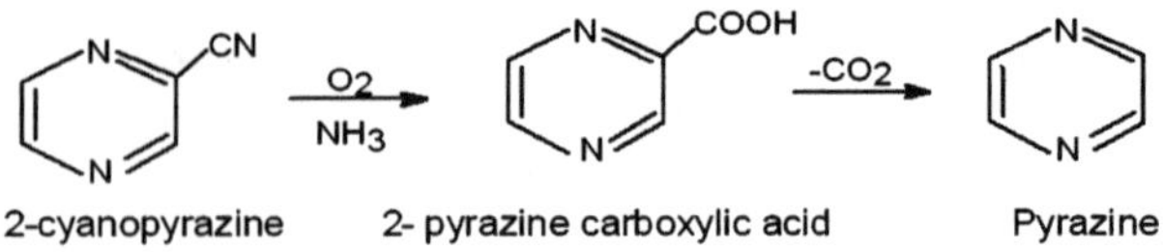

Scheme 2.3 Possible reaction pathways of pyrazine formation from cyanopyrazine

oxides of molybdenum and phosphorous. Thus, it may be expected that as long as the catalyst exists in the form of AMPA, higher activity can be realized. The partially decomposed salt can be regenerated during the reaction. Albonetti et al. [43] also have observed ammonium salt to be more reactive than the acid itself in the formation of methacrylic acid from isobutyric acid. The effect of the cation is related either to its electronegativity affecting the Mo oxidation state or to the modification of acidity.

Increase in reaction temperature from 360 to 420 °C is seen to increase the MP conversion, as a consequence of increase in reaction rate. However, the selectivity toward CP decreases and that of pyrazine increases. Partial dehydration of the catalyst and formation of the acid from the compound could be attributed to the increase in selectivity to pyrazine. Bondareva et al. [7] has also observed increase in pyrazine selectivity at higher reaction temperatures. They are of the opinion that as the MP conversion exceeds 80 %, the oxidation of CP occurs leading to pyrazine production. The reaction pattern is shown below in Scheme 2.3.

The product distribution in the ammoxidation reaction over MPA catalysts is shown in Fig. 2.12B. The conversion level gradually increases from 30 to 90 % when the pretreatment temperature is increased from 300 to 500 °C. Even though the MPA catalyst exhibits a facile decomposition at lower temperatures compared to that of AMPA, the increase in conversion with increase in calcination temperature can be attributed to the transformation of the oxides into the AMPA structure during the reaction, as also revealed by the XRD patterns of the used catalysts. This seems to be the main reason for higher conversion levels of the MPA catalysts pretreated at higher temperatures. The selectivity to CP, however, is decreasing with increase in pretreatment temperature. It is clear that the presence of AMPA is advantageous to get higher selectivity to CP. The samples calcined at lower temperatures have preserved the Keggin ion structure. The transformation of MPA into AMPA could be more facile during the reaction than the transformation of the oxides obtained at higher calcination temperatures. The transformation in the latter case may not be complete as revealed by the XRD of the used samples. Very high pyrazine selectivity exhibited by the MPA samples calcined at high temperatures also suggest that the AMPA form of the catalyst is much better both economically and environmentally.

3.3.2 Influence of Phosphate Precursor on Ammoxidation Activity

Preparation of AMPA by precipitation method, at controlled pH, using DAHP as the precursor is observed to be the best method than using ADHP in terms of obtaining relatively pure form of AMPA. The ^{31}P NMR spectra of the two catalysts (DAHP and ADHP) are shown in Fig. 2.13A, B. The DAHP sample calcined at 400 °C shows

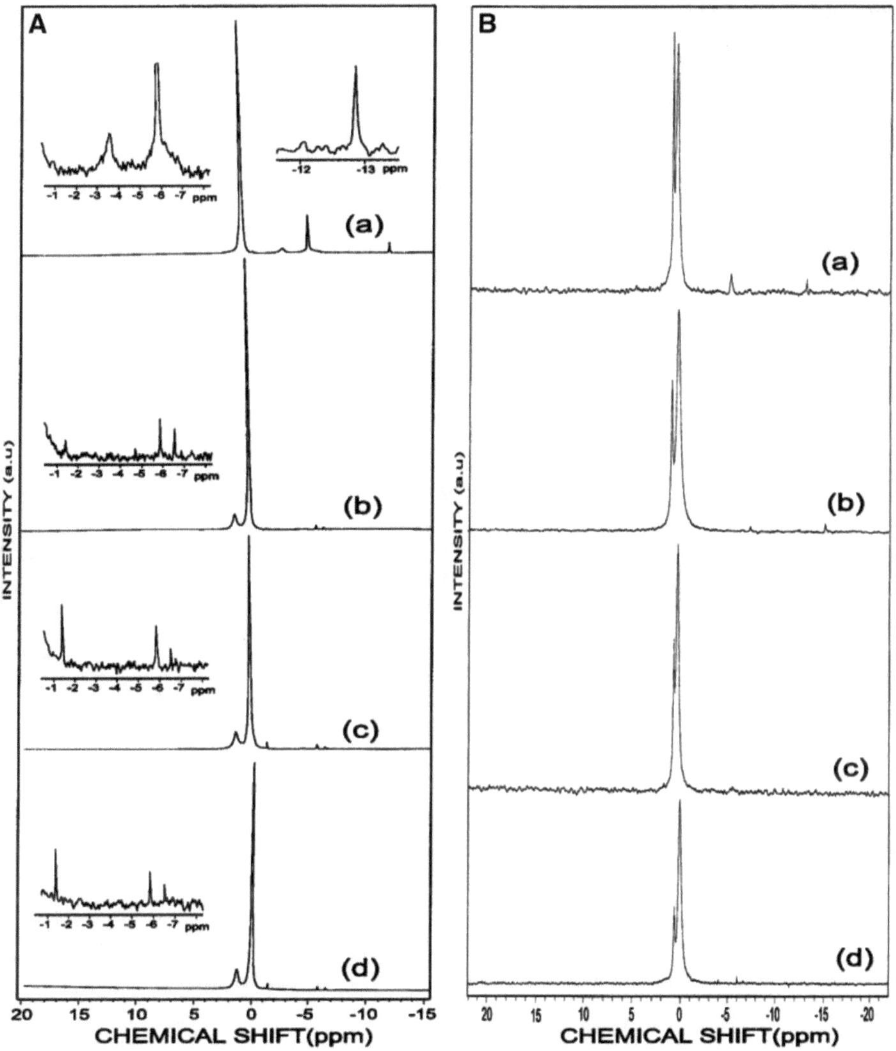

Fig. 2.13 ^{31}PNMR spectra of (**A**) DAHP catalysts and (**B**) ADHP catalysts at different calcination temperatures: (*a*) 400 °C, (*b*) 450 °C, (*c*) 500 °C, and (*d*) 550 °C (Unpublished data or PhD thesis of K.N. Rao, Jawaharlal Nehru Technological University, India 2004)

peaks at –3.5 and –5.8 ppm corresponding to Keggin ion $[PMo_{12}O_{40}]^{3-}$ and the dehydrated Keggin ion $[PMo_{12}O_{38}]^{3-}$ species [54]. The low-intensity peak seen at –13.0 ppm may be attributed to decomposed Keggin species [50]. The catalyst sample calcined at 450 °C shows reduction in the intensity of –3.5 and –6.0 ppm peaks (clearly shown in inset of the Fig. 2.13A). A new peak is observed at 1.4 ppm, which may be due to the basic molybdenum phosphate [55]. As the calcination temperature increases to 500 and 550 °C, the intensity of –6.0 ppm peak reduces very much and that of the 1.4 ppm

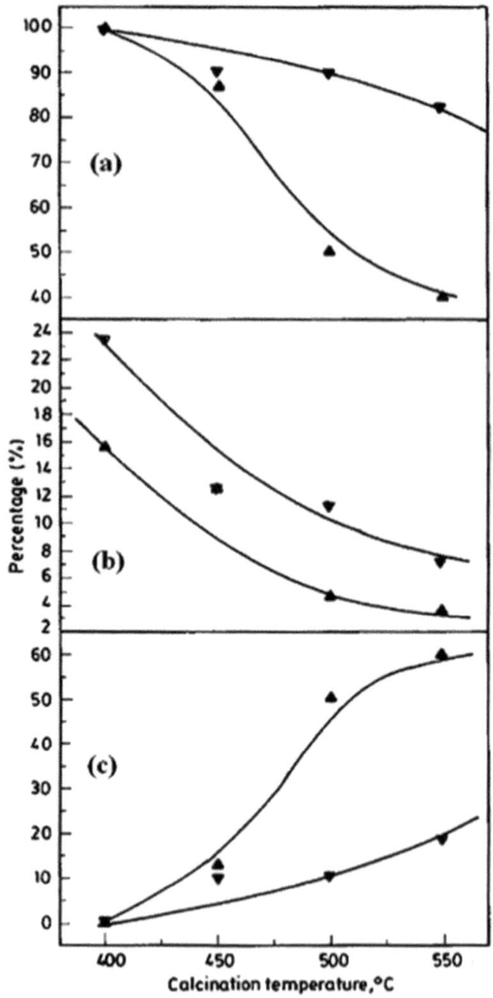

Fig. 2.14 Influence of calcination temperature on ammoxidation functionalities of catalysts at a reaction temperature of 380 °C: (**a**) selectivity to CP, (**b**) conversion of MP, and (**c**) selectivity to pyrazine (▲) DAHP and (▼) ADHP catalysts

peak increases presumably because of formation of stable basic molybdenum phosphate. The ADHP catalyst sample calcined at 400 °C shows a small peak at –4.0 ppm corresponding to Keggin ion along with a major peak at 0.56 ppm, which may be due to free phosphate or P_2O_5 like structure [50]. With increase in calcination temperature, the intensity of 0.56 ppm peak increases revealing the complete conversion of reactants to unknown compounds. The ^{31}P NMR results clearly corroborate the conclusions drawn from XRD and FTIR results (not shown).

The AMPA and not MPA as such, is found to be responsible for the activity and selectivity of the catalysts used in MP ammoxidation. This observation is confirmed by correlating the ammonium content of the catalyst with its activity during the ammoxidation. Optimization of the quantity of ammonium salt in the catalyst has resulted in identification of the maximum selectivity toward CP (Fig. 2.14).

Table 2.1 BET surface area and acidic strength values of the various heteropoly acid catalysts

Catalyst	BET surface area (m^2/g)	Acidic strength (Ei, mV)
$NbPO_4$	146	378
AMPA-$NbPO_4$-5[a]	85	530
AMPA-$NbPO_4$-10[a]	79	543
AMPA-$NbPO_4$-15[a]	41	581
AMPA-$NbPO_4$-20[a]	34	560
$FePO_4$	2.3	333
AMPA-$FePO_4$-5[a]	3.1	368
AMPA-$FePO_4$-10[a]	12	469
AMPA-$FePO_4$-15[a]	6.5	528
AMPA-$FePO_4$-20[a]	5.1	564
α-$VOPO_4$	2.2	798
β-$VOPO_4$	1.4	528
AMPA-α-$VOPO_4$	6.6	825
AMPA-β-$VOPO_4$	2.8	586
		Ammonia evolved by TPD (mmol/g)
DAHP-400[b]	4.5	5.04×10^{-2}
DAHP-450[b]	3.5	3.88×10^{-2}
DAHP-500[b]	2.0	3.00×10^{-2}
DAHP-550[b]	1.0	1.85×10^{-2}
ADHP-400[b]	3.4	4.40×10^{-2}
ADHP-450[b]	2.1	1.44×10^{-2}
ADHP-500[b]	1.6	1.12×10^{-2}
ADHP-550[b]	0.8	9.32×10^{-3}

[a]Mo-loading (wt%)
[b]Calcination temperature

Formation of AMPA is more favorable when the reactant salt contains more ammonia in the precursor salt, as in the case of catalysts prepared using diammonium hydrogen orthophosphate (Table 2.1). Formation of thermally stable basic molybdenum phosphate at high calcination temperatures and its facile transformation to AMPA during the reaction make it a better active and selective ammoxidation catalyst, compared to the less ammonia containing catalyst prepared by using ammonium dihydrogen orthophosphate.

3.4 *Studies on AMPA Supported on Nb_2O_5, SiO_2, TiO_2, ZrO_2, and Al_2O_3*

Even though it is proved that using AMPA in place of MPA is beneficial for getting enhanced activity and selectivity during ammoxidation of MP, there have been no detailed studies on the supported AMPA catalysts. Giving due consideration for this observation, it is found to be prudent to carry out a systematic study on AMPA

Table 2.2 BET surface area and acidic strength values of different supports and 20 % AMPA-supported catalysts

S. no.	Catalyst	BET surface area (m^2/g)	Amount of ammonia desorbed (mmol/g)
1	Hydrated Nb_2O_5	140	2.12×10^{-1}
2	SiO_2	300	1.96×10^{-1}
3	TiO_2	55	4.58×10^{-2}
4	ZrO_2	20	0.98×10^{-1}
5	Al_2O_3	196	1.10×10^{-1}
6	20 %AMPA/HNb_2O_5	43	6.92×10^{-2}
7	20 %AMPA/SiO_2	184	3.96×10^{-1}
8	20 %AMPA/TiO_2	1.4	5.57×10^{-2}
9	20 %AMPA/ZrO_2	2	8.54×10^{-2}
10	20 %AMPA/Al_2O_3	119	4.93×10^{-2}

supported on different supports like Nb_2O_5, SiO_2, TiO_2, ZrO_2, and Al_2O_3. The surface area, acidity, and nature of the support play an important role in ammoxidation to obtain high activity and selectivity. The BET surface area and acidity values of different supports and 20 % AMPA-supported catalysts are presented in Table 2.2. It is known that supports, which contain basic centers, bring about the decomposition of the polyanion, due to the interaction of the HPA with the support forming a compound between the two. In the case of acidic supports, the hydroxyls get protonated and interact with the negative heteropoly ion leading to electrostatic attraction which brings about better dispersion. Better HPA-support interaction improves the thermal stability of the HPA when compared with the bulk acid. The ammonium salt also contains strong acid sites. The super acidity is reported to be the manifestation of the presence of residual protons in the ammonium salt [33]. However, the ammonium salt is also expected to display variation in the nature of HPA-support interaction depending on the acid-base characteristics of the support.

XRD, FTIR, and ^{31}P NMR studies on AMPA/TiO_2 catalysts have shown (Figs. 2.15 and 2.16) that AMPA is well dispersed until 15 wt% loading due to strong electrostatic interaction between complex Keggin $[PMo_{12}O_{40}]^{3-}$ ion with the positively charged hydroxyls on the TiO_2 support. At higher loadings, the formation of crystalline AMPA $[(NH_4)_3PO_4(MoO_3)_{12}\cdot 4H_2O)]$ indicating the completion of interaction at 15 wt% is seen. The temperature-programmed desorption of ammonia studies (Fig. 2.17) on AMPA/TiO_2 catalysts reveal that with increase of AMPA loading there is a decrease in Lewis acidity and increase in Bronsted acidity of the supported catalysts. The increase in Bronsted acidity is associated with bulk AMPA formation. More details can be found in our previous publications [56, 57].

A comparison of ammoxidation activity obtained at 20 wt% AMPA supported on Nb_2O_5, SiO_2, TiO_2, ZrO_2, and Al_2O_3 obtained at 380 °C is shown in Fig. 2.18. The activity and selectivity obtained on the catalysts are in the following order: AMPA/Nb_2O_5 > AMPA/ZrO_2 > AMPA/TiO_2 > AMPA/SiO_2 > AMPA/Al_2O_3.

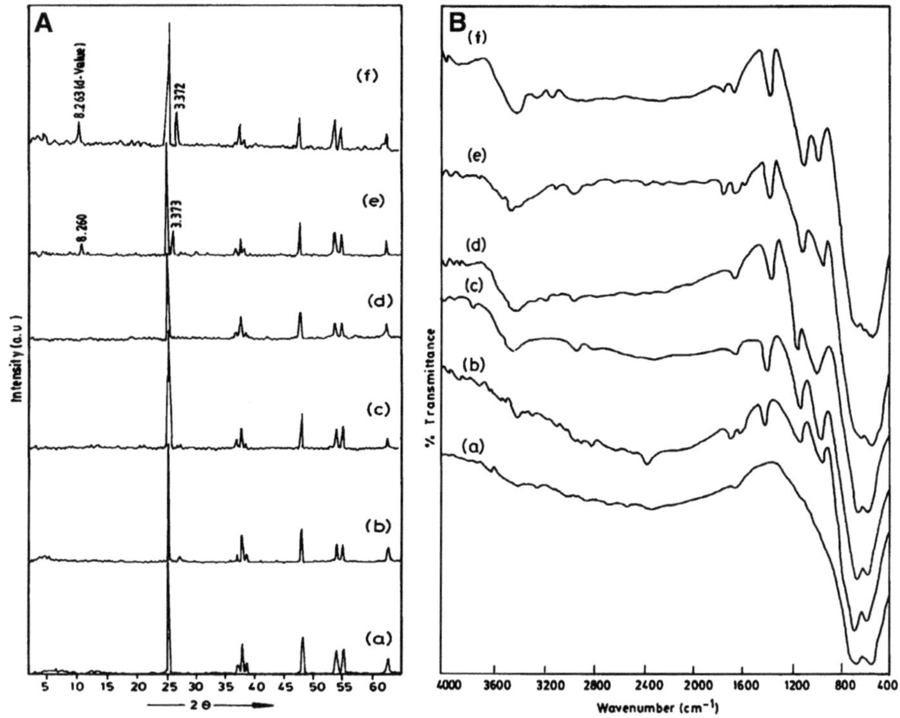

Fig. 2.15 (**A**) XRD and (**B**) FTIR patterns of TiO_2-supported AMPA catalysts: (*a*) support (*b*) AMPA-5, (*c*) AMPA-10, (*d*) AMPA-15, (*e*) AMPA- 20, (*f*) AMPA-25 (Ref. [56])

Nb_2O_5- and ZrO_2-supported catalysts show higher activity and selectivity because of high Bronsted acidity of AMPA, whereas silica-supported AMPA catalyst contains highest acidity than the other catalysts and hence shown very less selectivity toward CP because of its Lewis acidity generated by interaction of surface hydroxyl groups of silica with AMPA.

3.4.1 Half-Bandwidth Analysis: A Novel Method to Determine the Dispersion of HPA-Supported Catalysts [56]

The chemical interactions between the HPCs and the support are a matter of great interest because a strong interaction could fix the HPC to the carrier, avoiding the leaching of heteropoly compound in the reaction media or maintaining its high dispersion [32]. The determination of dispersion of active phase on the support is important to understand the role of active phase in reaction mechanism. It is not possible to determine the dispersion of HPC on the support by using conventional gas adsorption measurements. A novel approach called IR half-bandwidth analysis has been developed by our group to determine the dispersion of HPCs. By taking

Fig. 2.16 ^{31}P NMR spectra of TiO_2-supported AMPA catalysts: (*a*) AMPA-5, (*b*) AMPA-10, (*c*) AMPA-15, (*d*) AMPA-20, and (*e*) AMPA-25 (Ref. [56])

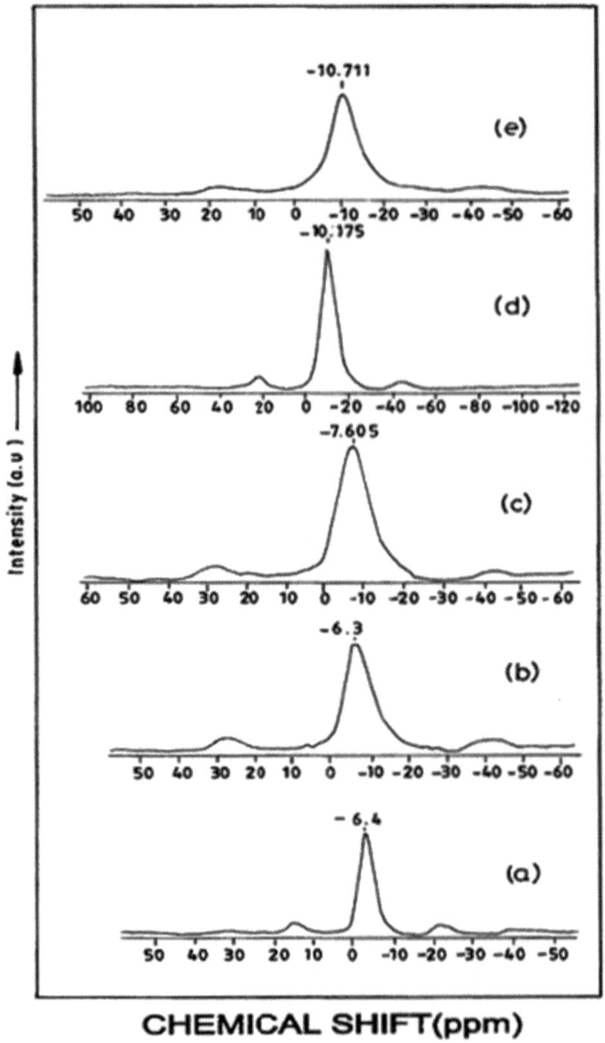

the P=O band at 1,060 cm^{-1} as reference, the half-bandwidth was measured for different AMPA loaded catalysts. The half-bandwidth of the catalysts increases linearly from 5 to 15 wt% AMPA loading, and beyond 15 wt%, it is almost equal for higher-loaded catalysts (Fig. 2.19). This indicates that the monolayer saturation occurs between 10 and 15 wt%. Above 15 wt% loading, the catalysts acquire bulk nature by forming multi-layers of AMPA.

A correlation between the half-bandwidth and ammoxidation functionality is shown in Fig. 2.19. The values of MP conversion on bulk AMPA are 5 and 23 % at 360 and 380 °C, respectively. The conversion has increased up to 15 wt% loading

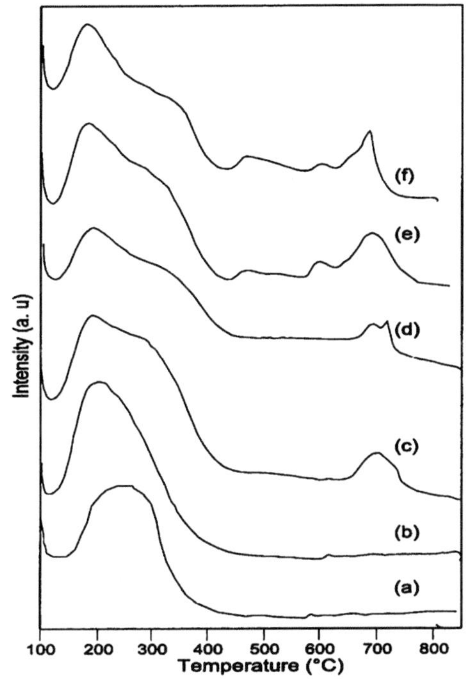

Fig. 2.17 TPDA spectra of TiO_2-supported AMPA catalysts: (*a*) support, (*b*) AMPA-5 (*c*) AMPA-10, (*d*) AMPA-15, (*e*) AMPA-20, (*f*) AMPA-25

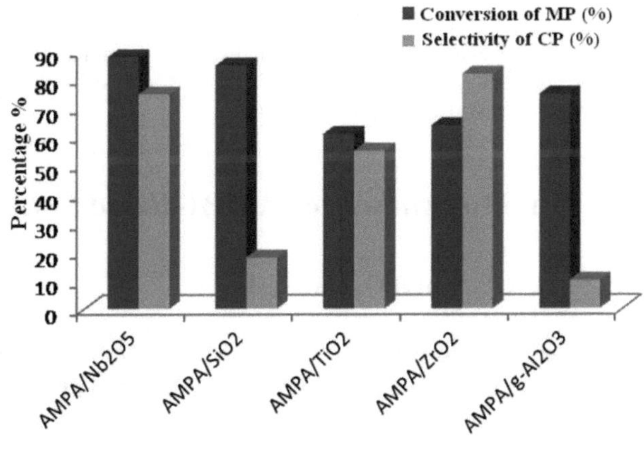

Fig. 2.18 Ammoxidation activity profiles of the supported AMPA (20 wt%) catalysts at 380 °C

and thereafter remains almost constant. The values of the half-width of the 1,060 cm^{-1} band also have displayed a similar trend. Hence, it appears that the extent of interaction of the salt with the support reaches a maximum at about 15 wt%. Further increase in loading (where crystallization of the salt is observed) does not increase the conversion. This observation may be valuable in the characterization of supported HPAs and their salts in determining the maximum extent of loading where the activity is also the maximum.

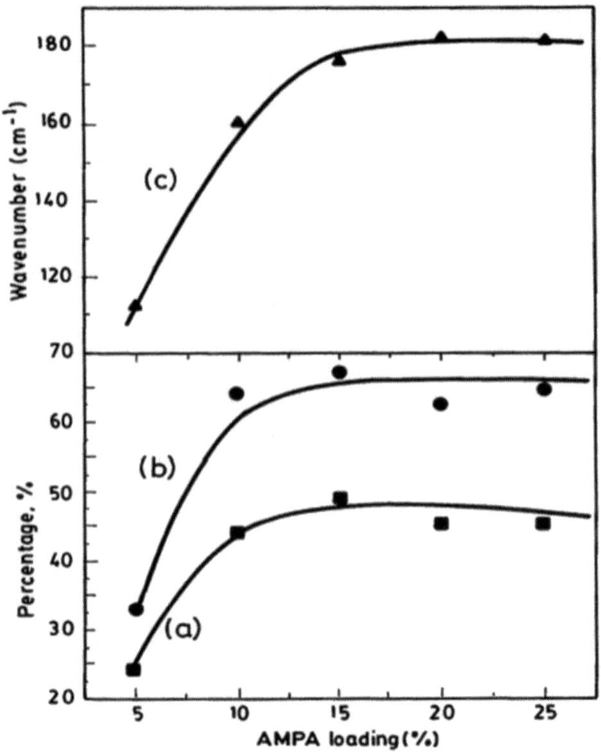

Fig. 2.19 Correlation between half-bandwidth and ammoxidation activity of the TiO_2-supported AMPA catalysts (*a*) conversion of MP at 360°C, (*b*) conversion of MP at 380°C, and (*c*) half-bandwidth of 1,060 cm^{-1} band (Ref. [56])

3.5 In Situ *Synthesized AMPA-Based Systems*

3.5.1 AMPA/$NbOPO_4$ Catalysts

It is interesting to verify whether in situ synthesis of AMPA on various supports is possible, so we have used $NbOPO_4$, $VOPO_4$, and $FePO_4$ to prepare AMPA by in situ technique, which markedly enhances the ammoxidation activity. The enhancement in the activity as correlated with the extent of salt formation has been discussed in our previous publications [58–62]. Figure 2.20A shows the XRD patterns of the AMPA generated using niobium phosphate as the support, as a function of MoO_3 content for the catalysts along with that of pure niobium phosphate. The formation of AMPA with the formula $(NH_4)_3PO_4(MoO_3)_{12}{\cdot}4H_2O$ (ASTM Card No-09-412) can be clearly observed in all the catalysts from 5 to 20 wt% of MoO_3. The present observation is very significant in the sense that the formation of AMPA exhibiting the Keggin structure occurs by participation of the surface phosphate of the support, contrary to the general methods of preparation of AMPA, wherein the phosphate

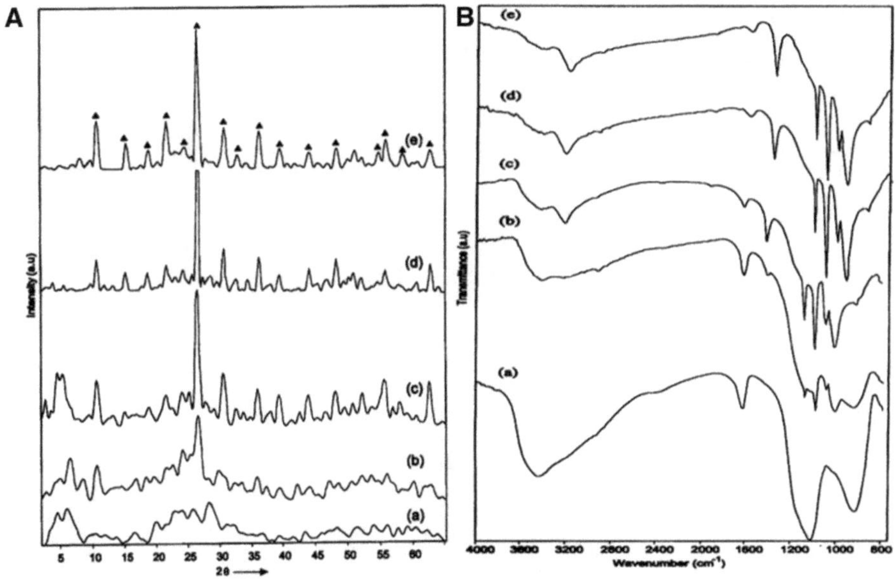

Fig. 2.20 (**A**) XRD and (**B**) FTIR spectra of AMPA-$NbPO_4$ catalysts: (*a*) $NbPO_4$, (*b*) AMPA-5, (*c*) AMPA-10, (*d*) AMPA-15, and (*e*) AMPA-20 (Ref. [58])

species are added in the form of phosphoric acid or its ammonium salts along with the Mo salt. Since powder X-ray diffraction patterns have not evidenced any discrete molybdenum containing phases, further characterization studies of AMPA/$NbOPO_4$ catalysts have been carried out by FTIR spectroscopy. FTIR spectra of catalysts prior to reaction are given in Fig. 2.20B. On the basis of previous literature, the features at 1,064, 970, and 870 cm^{-1} which are present in all the four catalysts have been assigned to Keggin structure in the moiety. Increase in the intensity of these bands with the AMPA content in the catalyst can be observed. In order to confirm whether the formation of AMPA is confined to the surface of the niobium phosphate or niobium enters into the primary structure of Keggin unit, a comparison of the positions of the bands with those reported in the literature is made. As there is no noticeable shift in the bands, it is considered that the formation of AMPA is confined to the surface of the niobium phosphate support, and niobium is not incorporated into the primary structure of the Keggin unit. FTIR studies on the catalysts correlate with the XRD results and do not show any formation of molybdenum trioxide species in the catalysts. Pronounced dependence of ammoxidation activity upon AMPA loading shown in Fig. 2.21 is due to increase in formation of highly dispersed AMPA content in the catalysts (5–10 wt% AMPA/$NbOPO_4$), whereas the decrease in activity in 15–20 wt% is due to formation of bulk crystallites of AMPA which resist diffusion of reactant species by blocking the pores. Details of the activity studies conducted at various temperatures can be found elsewhere [58]. The selectivity toward cyanopyrazine obtained a maximum value of 69 % on 20 wt% AMPA/$NbOPO_4$ at 380 °C.

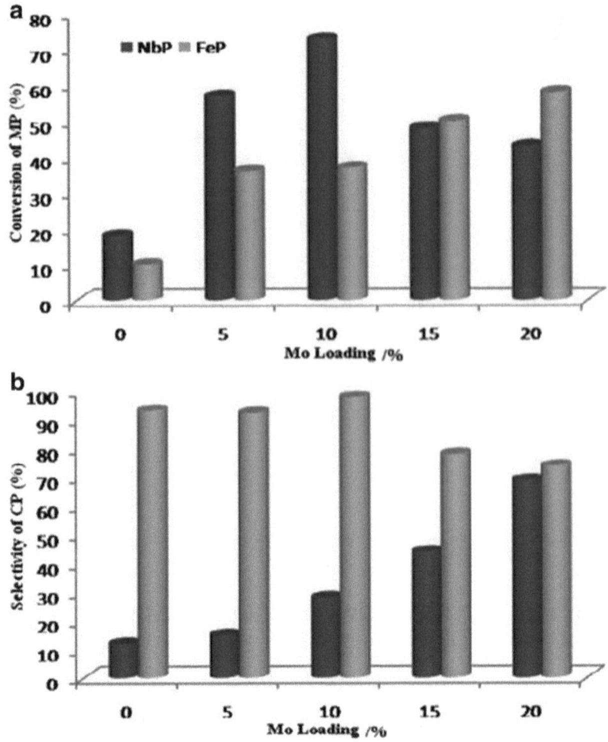

Fig. 2.21 Comparison of activity for ammoxidation of MP for AMPA-$NbPO_4$ and AMPA-$FePO_4$ catalysts: (**a**) conversion of MP, (**b**) selectivity to CP at 380 °C

3.5.2 Studies on AMPA/$FePO_4$ Catalysts

In order to design a highly selective catalyst for MP ammoxidation, attention has been paid on iron phosphate [59]. Iron phosphate is found to be less active and highly selective and shows 98 % CP selectivity compared to AMPA/$NbOPO_4$. The AMPA/$FePO_4$ catalysts have also showed enhancement in the activity while retaining the CP selectivity of iron phosphate to a large extent as seen in Fig. 2.21. The high selectivity of iron phosphate can be explained as due to the absence of labile M=O bonds which otherwise contributes to oxygen insertion. It could also be due to the ease of formation of an ammonium complex of the type $NH_4FeP_2O_7$. Formation of Mo oxide phase is observed at higher loadings of ≥15 wt% in the XRD spectra. This observation is further supported by FTIR and Raman spectroscopy [60] analysis. The formation of Mo oxide could be the reason for the decline in selectivity at higher loadings.

3.5.3 *Studies on* AMPA/$VOPO_4$ Catalysts

Two $VOPO_4$ (α- and β) and two 20 wt% AMPA-$VOPO_4$ (α- and β) catalysts are selected and tested for their activity in ammoxidation. The influence of reaction temperature on the conversion of MP is shown in Fig. 2.22. Phase composition is

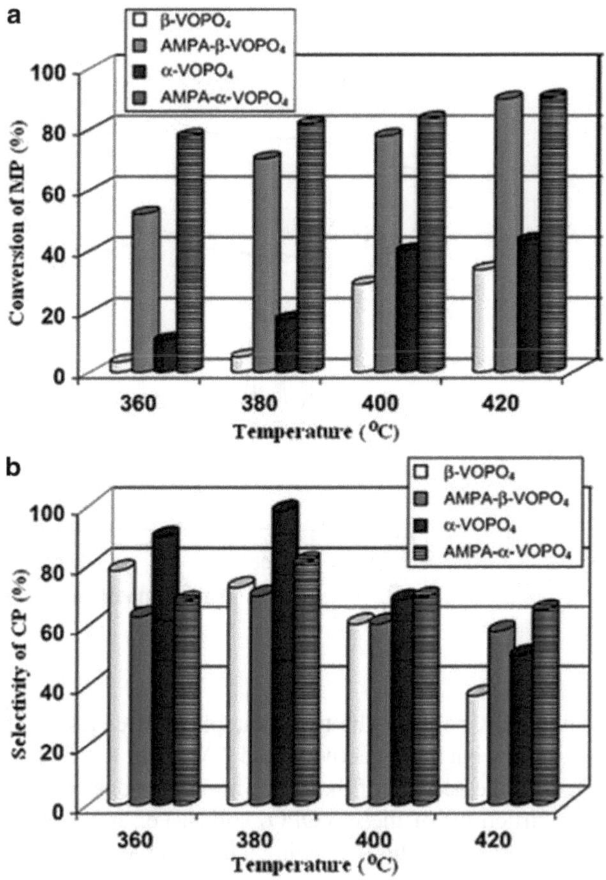

Fig. 2.22 (**a**) Influence of reaction temperature on the catalytic performance of $VOPO_4$ and AMPA-$VOPO_4$ catalysts. (**b**) Variation of CP selectivity as a function of reaction temperature over $VOPO_4$ and AMPA-$VOPO_4$ catalysts (Ref. [61])

identified by Raman spectroscopy and the data are shown in Fig. 2.23. Detailed description of all the spectra can be found elsewhere [61, 62]. The main characteristic features of the Keggin structure in the spectra c and d can be observed at 988 cm^{-1} (νs Mo–Od), 919–877 cm^{-1} (νas Mo–O_b–Mo), and 598–615 cm^{-1} (νas Mo–O_c–Mo) with an important bridge stretching character. The spectra agree well with those values reported in the literature [63]. Thus, the information obtained from the Raman data confirms the formation of AMPA on the $VOPO_4$ materials. Figure 2.22 clearly demonstrates that the reaction temperature has a promotional effect on the conversion of MP irrespective of the nature of catalyst applied. It is also obvious that there is a distinct influence of in situ synthesized AMPA samples on the catalytic performance compared to their corresponding analogues. Both α- and β-$VOPO_4$ compose of AMPA exhibited better performance in comparison with their parent α- and β-$VOPO_4$ samples. Another interesting observation is that between these AMPA-containing solids, the α-form of AMPA displayed better performance compared to its

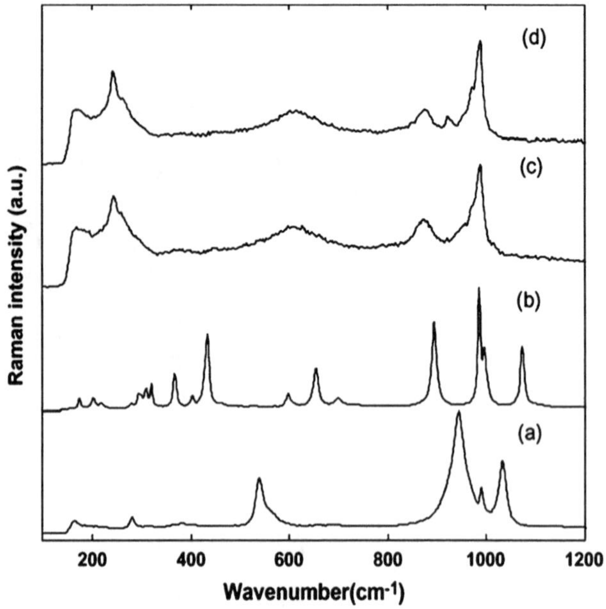

Fig. 2.23 Raman spectra of catalysts: (*a*) α-$VOPO_4$, (*b*) β-$VOPO_4$, (*c*) AMPA-α-VOPO4, (*d*) AMPA-β-$VOPO_4$ (Ref. [61])

β-form. In a similar way, between the two monophosphates tested, the α-$VOPO_4$ is observed to show somewhat higher activity than β-$VOPO_4$. It is reasonable to assume that the differences in the activity of these two monophosphates could be due to the differences in the structures of these $VOPO_4$ solids, i.e., α-$VOPO_4$ consisting of a layered structure, whereas β-$VOPO_4$ showing a three-dimensional one.

The acidity characteristics of the catalysts are another possible reason for such deviations in their catalytic performance. For instance, the highly acidic samples of this study such as AMPA-α-$VOPO_4$ and α-$VOPO_4$ (see Table 2.1) exhibit higher activity and selectivity and thus lend support to the above assumption that acidity plays a critical role on the catalytic performance. It can be deduced that the presence of AMPA enhances the Brønsted acidity of the catalysts. Such enhancement in the Bronsted acidity in turn increases the ammonia adsorption capacity of the catalysts. This seems to be the more probable reason for the increased activity of the AMPA/α-VOPO4 catalysts. As shown in the Fig. 2.22, temperature has profound influence on the selectivity of CP, α-$VOPO_4$ exhibiting the highest selectivity ≥95 % but only a low degree of conversion ≤5 %. However, with rise in conversion levels, the CP selectivity is significantly decreased. The AMPA catalysts disclose a slightly lower selectivity (60–70 %) but remarkably higher conversions (55–90 %) than the pure $VOPO_4$ catalysts (CP selectivity = 60–95 % and conversion of MP = 5–30 %). Overall, better yields have been obtained over in situ synthesized AMPA-containing $VOPO_4$ solids compared to their corresponding analogues.

XRD and the FTIR data obtained on the spent samples [62] reveal that the $VOPO_4$ catalysts partly transformed into the ammonium-containing vanadium phosphates,

Table 2.3 Phase identification by XRD of V-, Bi-, and Sb-modified AMPA catalysts

Type of catalyst	Temperature region: Appearance of AMPA (°C)	Temperature region: Keggin ion decomposition observed (°C)	Remarks from powder X-ray diffraction patterns
V_1	300–400	450–500	Cubic structure of AMPA stable up to 400 °C and decomposition starts at 450–500 °C
$AMPV_2$	300–350	400–500	Cubic structure of AMPA stable up to 350 °C and decomposition started between 400 and 450 °C
$AMPBi_1$	300–400	400–500	Cubic structure of AMPA stable up to 400 °C and decomposition starts between 400 and 450 °C
$AMPBi_2$	300–350	400–500	Diffraction lines due to AMPA disappeared at 400 °C, and the lines due to MoO_3 along with $Bi_9PMo_{12}O_{52}$ were observed
$AMPSb_1$	300–350	400–500	Decomposition started at 400 °C and complete decomposition observed in 450–500 °C
$AMPSb_2$	300–350	400–500	Decomposition started at 400 °C and complete decomposition observed in 450–500 °C
$AMPSb_3$	300–500	400–500	Small diffraction lines of AMPA were observed even at 450 °C

such as $(NH_4)_2(VO)_3(P_2O_7)_2$, during the course of the reaction. From the comparison of the absorption band appearing at 1,420 cm^{-1} (corresponding to NH^{4+} ion) in both the $VOPO_4$ catalysts, it seems plausible that the aforesaid ammonium complex formation is more facile in the case of α- than in the β-isomorph.

3.6 V, Sb, and Bi Modified AMPA-Based Systems

The addition of transition metals to heteropoly compounds is an important approach to control their redox properties and improve their thermal stability. Metal can be coordinated with a heteropoly ion in three different ways to form metal coordinated polyions, which show unique catalytic activities for various reactions. A simple combination of metal salts with heteropoly ion is the most commonly used method to prepare the transition metal modified heteropoly compounds. V, Bi, and Sb substituted AMPA catalysts are prepared and the influence of their physicochemical properties on ammoxidation functionality determined. The active phases identified by XRD in all the catalysts at calcination temperatures of 300–500 °C are shown in Table 2.3.

Table 2.4 MP conversion on transition metal modified AMPA catalysts at 380 °C

	Calcination temperature of catalysts				
Types of catalyst	300 °C	350 °C	400 °C	450 °C	500 °C
$AMPV_1$	62	60	53	50	45
$AMPV_2$	50	48	46.2	43.5	35
$AMPBi_1$	54	56	60	63.2	66
$AMPBi_2$	60	59.6	55.3	51	51.8
$AMPSb_1$	50.4	47	45.3	43	40
$AMPSb_2$	30	37	43	43.2	72
$AMPSb_3$	21	20	25	53	58

Table 2.5 Selectivity to CP (S–CP %) on transition metal modified AMPA catalysts at 380 °C

	Calcination temperature of catalysts				
Types of catalyst	300 °C	350 °C	400 °C	450 °C	500 °C
$AMPV_1$	93	90	89	87.3	86.2
$AMPV_2$	95	94.5	93.4	91	91.3
$AMPBi_1$	87	85	83	81	77
$AMPBi_2$	90	85.4	64	80.8	78.9
$AMPSb_1$	96	92.5	85.4	84.8	76
$AMPSb_2$	96.6	96.8	91.6	87	83
$AMPSb_3$	98	98.2	97	94.2	90.6

3.6.1 Influence of Calcination Temperature on the Modified AMPA Catalysts

From the activity results (Tables 2.4 and 2.5), it can be observed that the decrease in conversion at higher calcination temperatures from 450 to 500 °C in the case of V and Bi modified catalysts is due to low thermal stability and decomposition of AMPA at these temperature and could also be due to formation of mixed oxide as observed by the formation of bismuth phosphomolybdate phase. Reverse is the case with AMPSb catalysts which show greater stability to AMPA even at higher calcination temperatures. With increase in calcination temperature from 300 to 500 °C, all AMPSb samples show increase in conversion and a slight decrease in selectivity to CP. This fact can be explained by taking the hypothesis proposed by Cavani et al. [64] that with increase in calcination temperature, the amount of ammonia expelled out from the secondary structure is more and the amount of antimony going into the secondary structure will increase. It is known that, due to the redox reaction between the $[PMo_{12}O_{40}]^{3-}$ and the residual Sb^{3+} ions, the oxidation state of Mo reduces from 6 to 5. The extent of reduction is directly proportional to the antimony content. From this discussion, we can say that the sample $AMPSb_3$ has more amount of Mo^{5+} species. The presence of reduced molybdenum species (Mo^{5+}), which are known to be more active in redox-type reactions, is the reason for high conversions and yield of CP.

3.6.2 Influence of the Number of Transition Metal Atoms Incorporated to AMPA

Incorporation of one V metal atom, which goes into the primary structure of Keggin ion of AMPA catalyst leads to enhancement of the redox properties of the catalysts. However, addition of further V may lead to the formation of amorphous material like $[VO]^{2+}$ species in the secondary structure, which also helps enhance the redox behavior of the catalysts. However, the addition of two V atoms leads to lower thermal stability of the cubic secondary structure of AMPA catalysts possibly due to replacement of ammonium cations by other vanadium oxo species.

Addition of one Bi atom enhances the thermal stability of the AMPA catalysts contrary to the general observation that the addition of guest metal reduces the thermal stability of the parent compound by formation of lacunary species. But in the present case, better thermal stability can be attributed to interaction between Mo-Bi species. Further, the addition of Bi in place of two molybdenum atoms in the Keggin ion leads to formation of bismuth phosphomolybdate during the preparation of catalyst with a consequent decrease in the formation of AMPA. The formation of this new species reduces the selectivity to CP considerably.

Addition of one Sb atom leads to a dramatic increase in thermal stability in AMPA catalysts. In general, the AMPA decomposes into its oxides at temperature between 400 and 450 °C, but the addition of three Sb atoms leads to existence of AMPA even after calcination at 450 °C revealing the enhanced thermal stability. This effect can be attributed to a partial replacement of (NH_4^+) by Sb metal atoms. When compared with AMPA, the Sb-modified catalysts show lower activity. Reduced ammonium content in the secondary structure, with addition of Sb, seems to the reason for the decreased activity. But there is considerable improvement in the CP selectivity. This can be attributed to the redox reaction between the $[PMo_{12}O_{40}]^{3-}$ and the residual Sb^{3+} ions. The oxidation state of Mo reduces from 6 to 5. The extent of reduction is directly proportional to the antimony content. The presence of reduced molybdenum species (Mo^{5+}) may be responsible for higher selectivity.

3.7 Supported Vanadium Incorporated AMPA Systems

Bulk catalysts have low surface area and thermal stability. Therefore, in order to increase thermal stability and surface area, these catalysts are to be deposited on different supports. Diffusion resistance observed in bulk catalysts can be overcome by dispersed on different supports. An enhancement in the reduction of molybdenum species can be achieved in AMPV because of vanadium incorporation. The thermal stability of supported catalysts is higher than that of the bulk catalysts.

The powder X-ray diffraction patterns (XRD) of fresh AMPV/TiO_2 catalysts calcined at 350 °C are shown in Fig. 2.24A. Details of the characterization and catalytic activity of all the catalysts can be found in our previous publications [65–67]. The low weight percent catalysts did not reveal any diffraction peaks of

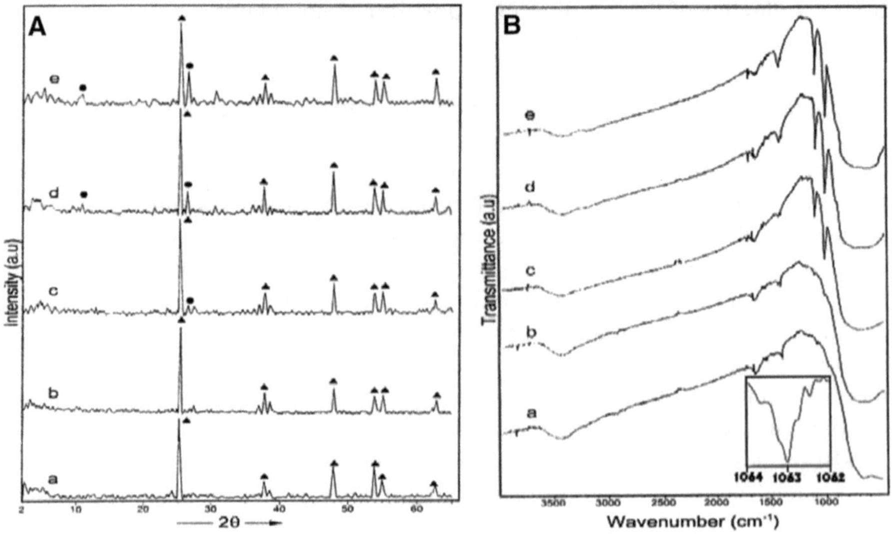

Fig. 2.24 (**A**) XRD and (**B**) FTIR spectra of AMPV/TiO_2 catalysts: (*a*) AMPV-5, (*b*) AMPV-10, (*c*) AMPV-15, (*d*) AMPV-20, and (*e*) AMPV-25; (●) Keggin ion (▲) TiO_2 (Ref. [65])

crystalline AMPV. They exhibited mainly the patterns of titania support. This might be due to well dispersion of Keggin units on titania surface. The hydroxyl groups bonded to the titania surface are protonated in the acidic solution thereby creating positively charged surface hydroxyl groups. These groups can bind a complex anion like that of the Keggin $[PMo_{12}O_{40}]^{3-}$ ion by electrostatic attraction leading to strong interaction of the AMPV with the support surface. Such an opinion is also expressed by Bruckman et al. [68]. The XRD peaks of the Keggin ion [69] are visible for the catalysts with 20 wt% of AMPV or higher. These catalysts show crystalline patterns of the salt. The intensities of the two main peaks corresponding to the salt have increased with increase in AMPV loading. It is also further confirmed with the FTIR spectra of AMPV supported on titania samples shown in Fig. 2.24B. The IR spectra show bands at 1,410, 1,065, 960, 873, and 786 cm^{-1} assigned to stretching vibrations of NH_4^+ ion, (P-O_d), (Mo-O_t), (Mo-O_b-Mo), and (Mo-O_c-Mo), respectively [7]. The peak obtained at 1,065 cm^{-1} is resolved using a resolution of 0.1 cm^{-1}, and it is shown in the inset of Fig. 2.24B; the splitting of P–O_a band at 1,065 cm^{-1} can be observed. It is known that the introduction of a metal other than Mo in the Keggin ion induces a decrease in the Mo–Od stretching frequencies and a possible splitting of the P–O_a band [48]. This splitting suggests the incorporation of V into the Keggin ion. These data suggest that the Keggin structure has been formed on titania during the synthesis. The intensity of the peaks related to Keggin ion are weak at lower loading but show an increase trend with increasing loading, similar to the observations made in the case of XRD.

TPR measurements on the catalysts have shown (Fig. 2.25) that the main reduction peak of HPCs are within the temperature range of 650–730 °C. The reduction peaks

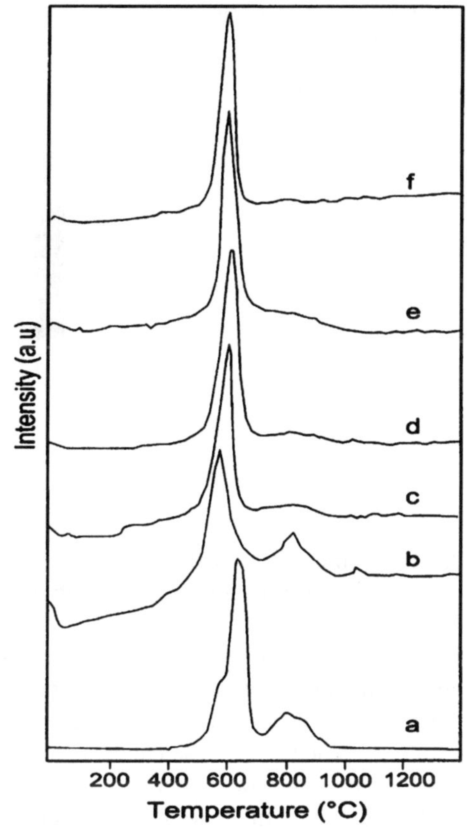

Fig. 2.25 TPR profiles of AMPV/TiO_2 catalysts: (*a*) support, (*b*) AMPA-5, (*c*) AMPA-10, (*d*) AMPA-15, (*e*) AMPA-20, (*f*) AMPA-25

in the 650–730 °C range may be ascribed to the reduction of free metal oxides originating from the decomposition of Keggin oxoanion, whereas the reduction peaks below 540 °C can be ascribed to the reduction of transition metal cations in the framework structure of HPCs [48, 70–73]. The TPR pattern of the unsupported bulk AMPV catalyst is also included for comparison. The bulk AMPV catalyst shows three distinct reduction temperatures at 613, 658, and 821 °C. The first two peaks may be due to reduction of more than one oxomolybdenum species. The high-temperature reduction peak, however, can be ascribed to the reduction of bulk MoO_3 formed at high temperatures and/or reduction of new phase containing V and P [72]. In contrast, the titania-supported AMPV catalysts exhibit only one reduction temperature maximum around 575–595 °C, which can be attributed to the reduction of octahedrally coordinated Mo^{6+} polymolybdates to a lower valence state. In the supported catalysts, the shift in the principal TPR peak indicates that the reduction of molybdenum species is enhanced due to more interaction of AMPV with titania. Sainero et al. [73] also observed a low-temperature shift in TPR peak when an interacting support like zirconia is added to silica in the case of supported MPA catalysts. As loading increases, the reduction temperature shifts to higher temperature.

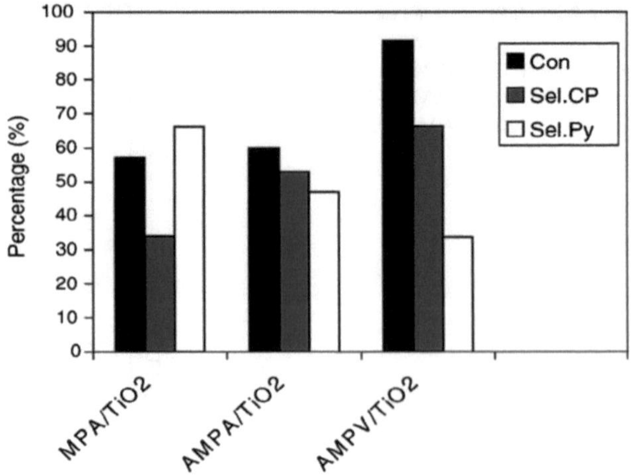

Fig. 2.26 Activity patterns of the various catalysts during the ammoxidation of MP: MPA/TiO_2, AMPA/TiO_2, and AMPV/TiO_2 catalysts (Ref. [65])

3.7.1 Comparison of the Catalytic Activity of Titania-Supported Bulk, Ammonium Salt, and Modified Ammonium Salt

In order to determine the effect of vanadium substitution for molybdenum in the ammoxidation reaction, ammoxidation of MP was studied on titania-supported MPA, the unsubstituted AMPA (of 20 wt% loading), and the AMPV catalysts. The results obtained over these catalysts are shown in Fig. 2.26. The titania-supported AMPV catalyst shows better activity than the other two catalysts. The performance of the catalysts is in the following order: AMPV > AMPA > MPA. The MPA catalyst leads to formation of an undesired dealkylation product, pyrazine. The conversion of acid (MPA) to its ammonium salt (AMPA) improves the formation of CP, a desired product, by reducing the pyrazine selectivity to some extent. However, incorporation of vanadium into AMPA leads to substantial improvement in the selectivity toward CP, by minimizing the formation of pyrazine with high conversion of MP.

The conversion of MP and acidity of the catalyst are in good agreement. As can be seen from an example shown in Fig. 2.27, the conversion obtained at 420 °C is proportional to the acidity on AMPV/CeO_2 catalysts. Conversion of MP at 380 °C and acidity values are tabulated and shown in the Table 2.6. The conversion and acidity values increase as the AMPV loading increases up to 20 wt%. Further increase to 25 wt% decreases the conversion of MP and the acidity, except in AlF_3-supported catalysts. All the supported systems have shown 10–15 % difference in their conversion and negligible selectivity to CP and high selectivity to pyrazine. The conversion of MP to CP is dependent upon the acid strength of the catalyst. The conversion of MP is not proportional to specific surface area of the supported catalysts, possibly due to the contribution from the support. The observation that increasing AMPV loading increases the performance of the catalyst indicates the active role of AMPV. Highly acidic supports like AlF_3 are not favorable, because they react with AMPV and form unwanted salt, which drastically reduce the catalyst

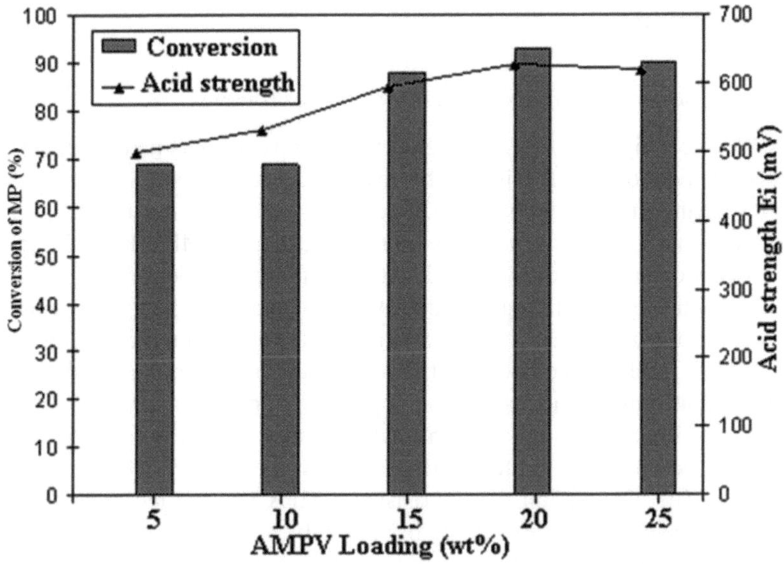

Fig. 2.27 Conversion and acid strength values of AMPV/CeO_2 catalyst at 420 °C (Ref. [67])

Table 2.6 Conversion and selectivity values of 20 % AMPV on different supports at 380 °C (Ref. [67])

Catalyst	Conversion of MP (%)	Selectivity to CP (%)	Acid strength (Ei)
20 %AMPV/AlF_3	58	50	610
20 %AMPV/ZrO_2	62	48	458
20 %AMPV/TiO_2	82	66	591
20 %AMPV/SiO_2	96	54	730
20 %AMPV/CeO_2	72	80	628

performance and selectivity to desired product. These supports need higher loading to get the optimum conversion and selectivity. Supports having moderate acidity seem to be best suited for the activity and selectivity of the catalysts. The selectivity of the catalysts obtained at 380 °C is given in Table 2.6. A comparison of the activity and selectivity data obtained on the AMPV/MO (MO-Metal Oxide) catalysts at various temperatures from 360 to 420 °C (not shown here) reveals that the activity and selectivity of different supports vary in the following order; AMPV/CeO_2 > AMPV/SiO_2 > AMPV/TiO_2 > AMPV/ZrO_2 > AMPV/AlF_3. The high activity of ceria-supported catalysts is due to its reducible nature and high oxygen storage capacity of ceria. Therefore, this system is expected to show significant salt–support interaction. All the catalysts have shown considerable amounts of by-product pyrazine at lower loadings, and it is reduced gradually with increased amount of AMPV content in the catalysts.

4 Conclusions

The conclusions are drawn from the above study as follows:

1. AMPA offers higher activity and selectivity compared to its parent acid, MPA, in the ammoxidation of MP. It is possible to tune the catalytic performance of AMPA; by restricting the conversion, the CP selectivity could be maximized to achieve atom-economy. AMPA is thermally more stable than the MPA. Even though these heteropoly compounds decompose into their component oxides during their preparation when subjected to higher calcination temperatures, they get regenerated in the course of reaction. However, at very high calcination temperatures, of the order of 500 °C, AMPA decomposes, at least partially, into non-regenerable, inactive species, thus limiting the maximum temperature of pretreatment.
2. The catalytic activity is not proportional to specific surface area of the supported catalysts, possibly due to the contribution from the support, which is difficult to quantify. The observation that increasing AMPA loading increases the performance of the catalyst dramatically indicates the active role of AMPA. Supports with more number of basic sites like γ-Al_2O_3 are not favorable, because they react with AMPA and form unwanted salt, which drastically reduce the catalyst performance. These supports need higher loading to get the optimum conversion and selectivity. Supports having moderate acidity such as Nb_2O_5, TiO_2, and ZrO_2 seem to be best suited for the activity and selectivity of the catalysts. Supported catalysts with optimum loading of AMPA (between 15 and 20 wt%) on TiO_2 and ZrO_2 offered better MP conversion and CP selectivity than bulk AMPA. The amount of AMPA loading to obtain optimum activity varied with acidity of the supports. Interacted species seems to be responsible for the activity and selectivity as revealed by the ^{31}P MAS NMR studies.
3. Addition of one V metal atom, which goes into the primary structure of Keggin ion of AMPA catalyst leads to enhancement of the redox properties of the catalysts. Addition of further V may lead to the formation of amorphous material like $[VO]^{2+}$ species in the secondary structure, which also helps enhance the redox behavior of the catalysts. Addition of two V atoms leads to lower thermal stability of the cubic secondary structure of AMPA catalysts; it may be due to the replacement of ammonium cations by other vanadium oxo species.
4. Addition of one Bi atom enhanced the thermal stability of the AMPA catalysts, contrary to the general observation that the addition of guest metal reduces the thermal stability of the parent compound by formation of lacunary species. But in the present case, better thermal stability could be attributed to interaction between Mo-Bi species. Further addition of Bi corresponding to two molybdenum atoms in the Keggin ion leads to formation of bismuth phosphomolybdate during the preparation of catalyst with a consequent decrease in the formation of AMPA. The formation of the new species reduced the selectivity to CP considerably.
5. Addition of Sb metal atom leads to a dramatic increase in thermal stability in AMPA catalysts. In general, the AMPA decomposes into its oxides at temperature between 400 and 450 °C, but the addition of three Sb atoms leads to existence of

AMPA even after calcination at 450 °C revealing the enhanced thermal stability. This effect can be attributed to a partial replacement of (NH_4^+) by Sb metal atoms. When compared with AMPA, the Sb-modified catalysts show lower activity. Reduced ammonium content in the secondary structure, with addition of Sb, seems to the reason for reduced activity. But there is considerable improvement in the CP selectivity. This can be attributed to the redox reaction between the $[PMo_{12}O_{40}]^{3-}$ and the residual Sb^{3+} ions. The oxidation state of Mo reduces from 6 to 5. The extent of reduction is directly proportional to the antimony content. The presence of reduced molybdenum species (Mo^{5+}) may be responsible for higher selectivity. The catalytic performance of the transition metal incorporated AMPA catalysts is in the order of AMPSb ≥ AMPV > AMPBi.

6. V incorporated MPA catalysts offered better catalytic performance than the bulk MPA and supported MPA, probably due to increase in redox properties of the VMPA catalyst. We obtained superior catalytic activity for the VMPA supported on TiO_2 catalysts, which could be related to fine dispersion of catalytically active HPA clusters on the support.
7. In situ synthesis of AMPA on the surface of metal phosphates is advantageous by several means. This protocol has been established by synthesizing AMPA supported on $NbOPO_4$, $FePO_4$, and $VOPO_4$ (α and β forms). At optimum MoO_3 loading (between 15 and 20 wt%) of polyatom (Mo), the catalysts offered more catalytic activity than the simple impregnated supported AMPA catalysts. The conversion rate and yield of CP are varied with physicochemical properties of the metal phosphate. $NbOPO_4$ and $VOPO_4$ offered high conversion of MP. $FePO_4$ offered less conversion but high selectivity to CP than $NbOPO_4$ and $VOPO_4$.
8. FTIR technique can be used to determine the dispersion of AMPA on supports.

Acknowledgments Thanks are due to Director, CSIR-Indian Institute of Chemical Technology, Hyderabad, for permitting to carry out the work. The authors are thankful to Dr. I Suryanarayana for his help in the interpretation of NMR results.

References

1. Wiberg KB (ed) (1995) Oxidation in organic chemistry. Academic, New York
2. Moffat JB (2001) Metal-oxygen clusters. The surface and catalytic properties of heteropoly oxometalates. Kluwer Publications, New York
3. Misono M (1987) Catal Rev Sci Eng 29:269
4. (a) Hill CL (ed) (1998) Chem Rev 98:1; (b) Okuhara T, Mizuno N, Misono M (1996) Adv Catal 41:113
5. Inamaru K, Ono A, Kubo H, Misono M (1998) J Chem Soc Faraday Trans 97:1765
6. Martin A, Lucke B (2000) Catal Today 57:61
7. Bondareva VM, Andrushkevich TV, Detushera LG, Latvak GS (1996) Catal Lett 42:113
8. (a) Keggin JF (1933) Nature 131:908; (b) Wells AF (1945) Structural inorganic chemistry. Oxford University Press, Oxford, p 344; (c) Dawson B (1953) Acta Crystallogr 6:113; (d) Anderson JS (1937) Nature 140:850
9. Pope MT (1983) Heteropoly and isopoly oxometalates. Springer, Berlin/New York
10. (a) Mc Garvey GB, Moffat JB (1991) J Catal 130:483; (b) Hu J, Burns RC (2000) J Catal 195:360
11. Knoth WH, Harlow RL (1981) J Am Chem Soc 103:1856

12. Misono M, Nojiri N (1990) Appl Catal 64:1
13. Ahmed S, Moffat JB (1988) Appl Catal 40:101
14. Faraj M, Hill CL (1987) J Chem Soc Chem Commun 1487
15. Kozhevnikov IV, Matveev KI (1983) Appl Catal 5:135–150
16. Keana JFW (1986) J Am Chem Soc 108:7951
17. (a) Buzt T, Vogdt C, Lerf H, Knozinger H (1989) J Catal 116:31; (b) Smit JVR (1958) Nature 181:1530; (c) Guilbault GG, Brignac PJ (1971) Anal Chim Acta 56:139; (d) Seidle AR, Newmark RA, Gleason WB, Skarjune RP, Hodgson KO, Rol RA, Day VB (1988) Solid Sate Ionics 26:109
18. Mizuno N, Misono M (1998) Chem Rev 98:199
19. Marchal-Roch C, Bayer R, Moison FF, Teze A, Herve G (1996) Top Catal 3:407
20. (a) Cavani F, Etienne E, Favaro M, Falli A, Trifiro F, Hecquet G (1995) Catal Lett 32:215; (b) Knapp C, Ui T, Nagai K, Mizuno N (2001) Catal Today 71:111
21. Centi G, Perathoner S (1998) Catal Rev Sci Eng 40:175
22. Kozhevnikov IV (1997) J Mol Catal A Chem 111:109
23. Ressler T, Timpe O, Girgsdies F, Wienold J, Neisius T (2005) J Catal 231:279
24. Liu H, Iglesia E (2003) J Phys Chem B 107:10840
25. Liu H, Iglesia E (2004) J Catal 223:161
26. Mestl G, Ilkenhans T, Spielbaur D, Dieterle M, Timpe O, Krohnert J, Jentoft F, Knozinger H, Schlogl R (2001) Appl Catal A Gen 210:13
27. Kozhevnikov IV (1997) J Mol Catal A 117:151
28. Berzelius J (1826) Pogg Ann 6:369
29. McGarvey GB, Moffat JB (1991) J Catal 132:100
30. Bielanski A, Malecka A, Kubelkova L (1989) J Chem Soc Faraday Trans 85(9):2847
31. Rao KM, Gobetto R, Innibello A, Zacchina A (1989) J Catal 119:512
32. Kozhevnikov IV (1995) Catal Rev Sci Eng 37(2):311
33. Nowinska K, Fiedorow R, Adamiec J (1991) J Chem Soc Faraday Trans 87:749
34. Lapham D, Moffat JB (1991) Langmuir 7:2273
35. Ito T, Irumaru K, Misono M (2001) Chem Mater 13:824
36. Lingaiah N, Mohan Reddy K, Nagaraju P, Sai Prasad PS, Wachs IE (2008) J Phys Chem C 112:8294
37. Li X-K, Zhao J, Ji W-j, Zhang Z-B, Chen Y, Chak-Tong A, Han S, Hibst H (2006) J Catal 237:58
38. Sopa M, Waclaw-Held A, Grossy M, Pijanka J, Nowinska K (2005) Appl Catal A Gen 285:119
39. Garte JH, Hamm DR, Mahajan S (1994) In: Pope MT, Muller A (eds) Polyoxometalates: from platonic solids to anti-retroviral activity. Kluwer Academic Publisher, Dordrecht/Boston, p 281
40. Narasimha Rao K, Gopinath R, Sai Prasad PS (2001) Green Chem 3:20
41. Marchal-Roch C, Laronze N, Guillou N, Teze A, Herve G (2000) Appl Catal A Gen 199:33
42. Rao KN, Gopinath R, Hussain A, Lingaiah N, Sai Prasad PS (2000) Catal Lett 68:223
43. Albonetti S, Cavani F, Triffiro F, Gazzano M, Koutyrev M, Aissi FC, Aboukais A, Guelton M (1994) J Catal 146:491
44. Damyanova S, Cubeiro ML, Fierro JLG (1999) J Mol Catal A Chem 142:85; Damyanova S, Fierro JLG (1998) Chem Mater 10:876
45. Hodnett BK, Moffat JB (1984) J Catal 88:253
46. Tsigdinos GA (1974) Ind Eng Chem Prod Res Dev 13:267
47. McMonagle JB, Moffat JB (1985) J Catal 91:132
48. Rocchiccioli-Deltcheff C, Fournier M (1991) J Chem Soc Faraday Trans 87:3913
49. Van Veen JAR, Sudmeijer O, Emeis CA, de Wit H (1986) J Chem Soc Dalton Trans 1825–1831
50. Iwamoto R, Fernandez C, Amoureux JP, Grimblot J (1998) J Phys Chem B 102(22):4343
51. Damyanova S, Fierro JLG, Sobrados I, Sanz J (1999) Langmuir 15:469
52. Essayem N, Frety R, Coudurier G, Vedrine JC (1997) J Chem Soc Faraday Trans 93(17):3243
53. Nowinska K, Kaleta W (2000) Appl Catal A Gen 203:91
54. Black JB, Clayden NJ (1984) J Chem Soc Dalton Trans 2765
55. Kraus H, Prins R (1996) J Catal 164:251
56. Narasimha Rao K, Gopinath R, Santhosh Kumar M, Suryanarayana I, Sai Prasad PS (2001) Chem Commun (2088)

57. Narasimha Rao K, Mohan Reddy K, Lingaiah N, Suryanarayana I, Sai Prasad PS (2006) Appl Catal A Gen 300:139
58. Srilakshmi Ch, Narasimha Rao K, Lingaiah N, Suryanarayana I, Sai Prasad PS (2002) Catal Lett 83:3
59. Srilakshmi Ch, Lingaiah N, Suryanarayana I, Sai Prasad PS, Ramesh K, Anderson BG, Niemantsverdriet JW (2005) Appl Catal 296:54
60. Srilakshmi Ch, Lingaiah N, Nagaraju P, Sai Prasad PS, Kalevaru V, Narayana A, Martin A, Lucke B (2006) Appl Catal 309:247
61. Srilakshmi Ch, Nagaraju P, Sreedhar B, Sai Prasad PS, Narayana Kalevaru V, Lucke B, Martin A (2009) Catal Today 141:337
62. Srilaxmi C, Lingaiah N, Hussain A, Sai Prasad PS, Narayana KV, Martin A, Lucke B (2004) Catal Commun 5:199
63. Rocchiccioli-Deltcheff C, Aouissi A, Bettahar MM, Launay S, Fournier M (1996) J Catal 164:16
64. Albonetti S, Cavani F, Trifiro F, Koutrev M (1995) Catal Lett 30:253
65. Mohan Reddy K, Lingaiah N, Rao KN, Nilofer R, Sai Prasad PS, Suryanarayana I (2005) Appl Catal A Gen 296:108
66. Mohan Reddy K, Lingaiah N, Nagaraju P, Sai Prasad PS, Suryanarayana I (2008) Catal Lett 122:314
67. Mohan Reddy K, Lingaiah N, Rao PSN, Nagaraju P, Sai Prasad PS, Suryanarayana I (2009) Catal Lett 130:154
68. Bruckman K, Che M, Haber J, Tatibouet JM (1994) Catal Lett 25:225
69. Marchal-Roch C, Laronze N, Villanneau R, Guillou N, Teze A, Herve G (2000) J Catal 190:173
70. Mizuno N, Sun DJ, Han W, Kudo T (1996) J Mol Catal A Chem 114:309
71. Dimitratos N, Védrine JC (2003) Appl Catal A Gen 256:251
72. Spojakina AA, Kostova NG, Sow B, Stamenova MW, Jiratova K (2001) Catal Today 62:315
73. Gomez Sainero LM, Damyanova S, Fierro JLG (2001) Appl Catal A Gen 208:63

Chapter 3
Transition Metal-Substituted Salt of Tungsten-Based Polyoxometalate-Supported Mesoporous Silica as a Catalyst for Organic Transformation Reactions

Surjyakanta Rana and Kulamani Parida

Contents

1 Introduction 58
  1.1 Solid Acid Catalyst 58
  1.2 Polyoxometalates 59
  1.3 Heteropoly Acid 59
  1.4 Synthesis 59
  1.5 Structure of Heteropoly Acids 60
  1.6 Acidic Properties of Heteropoly Acid 62
  1.7 Redox Properties of Heteropoly Acid 63
  1.8 Heteropoly Acid as Homogeneous Catalysis 63
  1.9 Heteropoly Acid as Heterogeneous Catalysis 64
2 Experimental 66
  2.1 Cesium Salt of Phosphotungstic Acid-Supported MCM-41 Toward Acylation of Anisole 66
3 Results and Discussion 67
  3.1 Characterization 67
  3.2 Catalytic Activity Toward Acylation of Anisole 68
  3.3 Influence of Various Substrates 70
  3.4 Cu Salt of Phosphotungstic Acid-Supported MCM-41 Toward Heck Vinylation Reaction 72
  3.5 Fe-Modified Lacunary Phosphotungstate-Supported MCM-41 as an Excellent Catalyst for Acid-Catalyzed as well as Oxidation Reaction 76
  3.6 Cs Salt of Pd Substituted Lacunary Phosphotungstate-Supported MCM-41 Toward Hydrogenation of *p*-Nitrophenol to *p*-Aminophenol 83
4 Conclusions 89
References 89

S. Rana • K. Parida (✉)
Colloids and Materials Chemistry Department, CSIR-Institute of Minerals and Materials Technology, Bhubaneswar 751 013, Orissa, India
e-mail: kmparida@immt.res.in

A. Patel (ed.), *Environmentally Benign Catalysts: For Clean Organic Reactions*,
DOI 10.1007/978-94-007-6710-2_3, © Panch Tattva Publishers, Pune, India 2013

Abstract Solid acids and super acids are attracting substantial interest due to their applicability in many chemical industries, especially in the petroleum industry for alkylation, isomerization, and cracking reactions, as well as in the synthesis of fine chemicals. The understanding of acid-catalyzed reactions is very much important since it covers a wide field of applications, ranging from large-scale industrial processes to enzyme-controlled reactions in the living cell. This chapter deals with synthesis and characterization of different types of salts as well as transition metal-modified phosphotungstates and their applications as catalyst for acylation, Heck vinylation, bromination of phenol, oxidation of *trans*-stilbene, and hydrogenation of *ortho*-nitrophenol, respectively.

1 Introduction

1.1 Solid Acid Catalyst

A solid acid catalyst should possess high stability, numerous strong acid sites, large pores, and a hydrophobic surface providing a favorable condition for reaction and should also be economically viable. In general, a catalyst that is to be used for synthesis of biodiesel should be selective and specific and result in esterification/transesterification with high conversion and yield of biodiesel.

Traditionally almost all the chemical reactions were catalyzed by liquid acids. But, the enforcement of stringent environmental regulations has demanded the replacement of these environmentally hazardous materials, which left with no other alternative than using solid acid catalysts. More than three hundreds of solid acids have been developed in the last four decades such as natural clay minerals, cation exchange resins, zirconia, alumina, silica, mixed metal oxides, heteropoly acids (polyoxometalates), lacunary, and zeolites [1]. The surface properties and the structures have been clarified by the newly developed sophisticated analytical techniques. The characterized solid acids are applied as catalysts to various reactions, and the role of acid–base properties has been studied extensively. Now, the use of solid acid catalyst is one of the most economically and ecologically important fields in catalysis. The solid acid catalysts have many advantages over liquid Brønsted and Lewis acid catalysts such as the following: they are noncorrosive, the nature of acid sites are known, their acid strength can be modulated, and they are environmentally benign, presenting less disposal problems, and safe to handle. Their repeated use is possible and the catalyst separation from the product stream is easier. Furthermore, they can be designed and their textural properties can be engineered to develop a catalyst with higher activity, selectivity, and longer lifetime. Therefore, heterogeneous catalysis has gained so much attention and becoming more and more popular in chemical, petrochemical, and life science industries. Tanabe and Holderich [2] made a statistical survey on various types of catalysts that are used in the industrial processes. It shows that metal oxide is the second largest group of catalysts employed in industrial processes next to zeolites.

Recent report showed that heteropoly acids and heteropoly anions, like phosphotungstic acids, are efficient "super acid" catalysts which can be used both in the homogeneous and heterogeneous phases [3]. Bulk heteropoly acids generally exhibit low catalytic performances due to their low surface areas. For this reason, the main criterion, which can lead to highly active heterogeneous acid catalysts, is an improvement of the heteropoly acid dispersion with respect to its primary acidic features. For this purpose, the main two approaches are the impregnation of the HPA on classical porous materials and the direct preparation of the acidic porous salts of HPA.

1.2 Polyoxometalates

Polyoxometalates (POMs) are self-assembled anionic metal oxide clusters. They are typically synthesized under acidic aqueous conditions. There are two broad classes of POMs, isopoly and heteropoly. In the heteropoly case, X is the heteroatom and is located in the center of the cluster. The element (M) that composes the framework is usually molybdenum or tungsten. The heteroatom is often P=3 or Si=4, but there are numerous examples for over 70 different elements.

1.3 Heteropoly Acid

Heteropoly acids (HPAs) are the acidic forms of polyoxometalates. They are widely used as homogeneous and heterogeneous acids and oxidation catalysts due to their unique physicochemical properties. Generally, all heteropoly acids are strong acids. This strong acidity can be attributed to the delocalization of surface charge density throughout the large sized polyanion, leading a weak interaction between the proton and the anion.

Solid HPAs in the acid form are more efficient catalyst than conventional solid acids. The ability to absorb large amount of polar molecules in the bulk, coupled with high proton mobility, leads to a high catalytic efficiency for liquid phase reaction. They have been widely used as acid and oxidation catalyst for organic synthesis, and they are found in several industrial applications. HPA is used as catalyst due to the following properties:

1. Solubility in aqueous and organic media
2. Intrinsic multifunctional, strong acidity (Brønsted sites), and redox properties

They are also of great interest as model systems for studying fundamental problems of catalysis.

1.4 Synthesis

Preparation of heteropoly acids is getting more and more importance for their applications. Such heteropoly acids as $H_3PW_{12}O_{40}$, $H_4SiW_{12}O_{40}$, and $H_3PMo_{12}O_{40}$

are commercially available as crystalline hydrates. The simplest way to prepare heteropoly acids involves the acidification of an aqueous solution containing the oxoanions and the heteroatom:

$$12WO_4^{2-} + HPO_4^{2-} + 23H^+ \rightarrow PW_{12}O_{40}^{3-} + 12H_2O$$

Control of the pH and the ratio of central atom to metal atom (X/M) is necessary to obtain the desired structure. Acidification is achieved by direct addition of a mineral acid.

1.5 Structure of Heteropoly Acids

Heteropoly acids are polymeric oxoanions with well-defined primary structure. They are made by the combination of hydrogen and oxygen with certain metals and nonmetals. Various types of heteropoly acid are known each having its own characteristic structure like:

1. Keggin structure
2. Silverton structure
3. Dawson structure
4. Waugh structure
5. Anderson structure

1.5.1 Keggin Structure

Keggin heteropoly acids have general formula $X^{n+}M_{12}O_{40}^{n-8}$, where X is the central atom (Si^{4+}, Ge^{4+}, P^{5+}, As^{5+}, etc.), *n* the degree of its oxidation, and M is molybdenum or tungsten, which can be partly replaced by other metals [4]. The structure of Keggin compounds comprises four trigonal groups of edge-sharing MO_6 octahedral, each group sharing corners with neighboring groups and with the central tetrahedron. In each octahedron, the metal is displaced toward the terminal oxygen atoms. This structural arrangement leads to the formation of a spherical polyanion. The first characterized well-known heteropoly compound is the one having Keggin structure which is fairly stable and easily available.

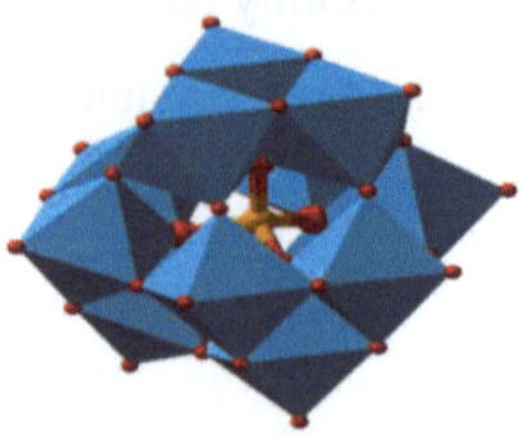

1.5.2 Wells–Dawson Structure

Well–Dawson structure heteropoly acids have general formula $[X_2M_{18}O_{62}]$, where X is P^{5+}, S^{6+}, As^{5+} and M may be either W^{6+} or Mo^{6+}. These HPCs are formed by dimerization of PM_9O_{34} moieties, thus under suitable conditions of pH. It consists of two half units, each of which is derived from the Keggin structure by removal of three adjacent corner linked MO_6 octahedron anions, leaving three octahedra over a ring of six.

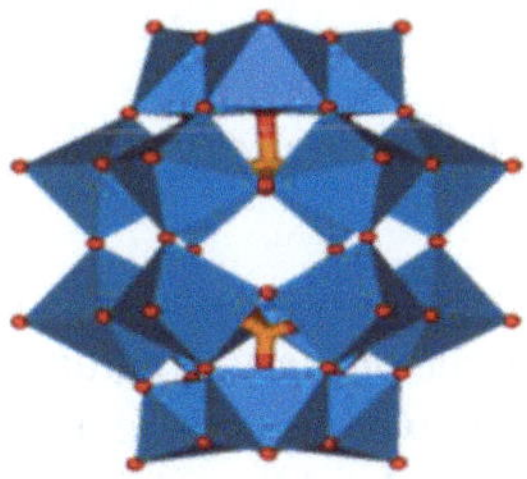

1.5.3 Anderson Structure

Anderson structure of heteropoly acid has general formula $[XM_6O_{24}]$, where X is Mn^{4+}, Ni^{4+}, Pt^{4+}, Te^{6+} and M may be either W^{6+} or Mo^{6+}. The structure is planar in which each MoO_6 octahedron has two terminal oxygen and the heteroatom X adopts the octahedral coordination. Anderson anions are usually obtained from aqueous solutions at a pH of 4–5.

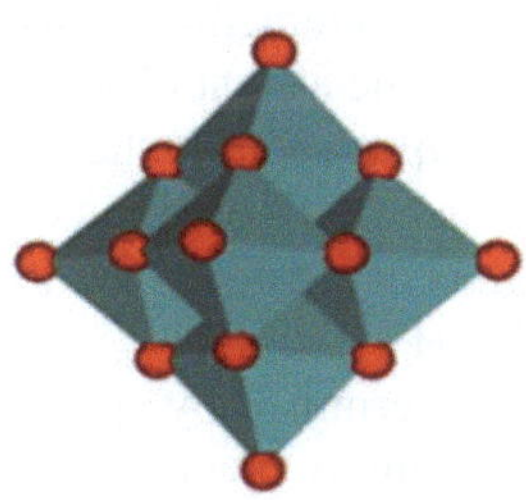

1.5.4 Silverton Structure

Silverton structure of heteropoly acid has general formula $[XM_{12}O_{42}]$, where X is Ce^{4+}, Th^{4+} and M may be either W^{6+} or Mo^{6+}. The central atom is surrounded by 12 oxygen atoms that form an icosahedron (12 vertices, 20 faces, and 30 sides) as a central polyhedron around which MO_6 octahedra are arranged in a face sharing pairs.

1.5.5 Waugh Structure

Waugh structure of heteropoly acid has general formula $[XM_9O_{32}]$, where X is Mn^{4+}, Ni^{4+} and M may be either W^{6+} or Mo^{6+}. It is built around an octahedral coordinated heteroatom. Three octahedral addenda atoms are arranged at the vertices of a triangle, which is coplanar with the central XO_6 octahedron and another two groups of edge-shared M_3O_{13} triplets, are placed above and below the middle layer of the four octahedral.

1.6 Acidic Properties of Heteropoly Acid

The acid–base properties of heteropoly compounds can be modified by the choice of the hetero atom, oxometal in the primary structure, and the cation [5]. The acidity can be generated by the protons which act as counter cations in heteropoly acids (i.e., in $H_3PMo_{12}O_{40}$) and in mixed acidic salts (for instance, in $K_xH_{3-x}PW_{12}O_{40}$). There are two types of protons in crystalline HPA: (1) nonlocalized hydrated protons bound to one metal cation as a whole and rapidly exchanging with the protons of the water molecules in the hydration shell of the acid and (2) nonhydrated protons localized at the peripheral oxygen atoms of the polyanion [4].

All heteropoly acids are strong acids, much stronger than conventional solid acids such as SiO_2–Al_2O_3, H_3PO_4/SiO_2, and HY and HX zeolites and mineral acids as H_2SO_4, HCl, and *p*-toluenesulfonic acid. They are completely dissociated in aqueous solutions and partly dissociated in organic solvents. This strong acidity can be attributed to the delocalization of surface charge density throughout the large sized polyanion, leading to a weak interaction between the protons and the anion. The order of acid strength is the same as that observed in solution.

The acid strength can be expressed by the Hammett acidity function H_0: $H_0 = pK_{BH+} - \log\left([BH^+]/[B]\right)$ where [B] is the concentration of the indicator B, $[BH^+]$ is the concentration of the conjugated acid, and K_{BH+} is the equilibrium constant for the reaction: $BH^+ \rightarrow B + H^+$.

The $[BH^+]/[B]$ ratio can be determined by spectroscopy in the UV and visible bands or can be measured more roughly by visually observing the point in the titration at which the color changes.

The H_0 value of 100 % H_2SO_4 (−11.94) is taken as a reference number. The acids with values >12 are classified as superacids [6]. Superacids with H_0 values of −20 (i.e., 108 times stronger than 100 % H_2SO_4), such as HSO_3F and SbF_5, are able to protonate methane.

According to this scale heteropoly compounds can be classified as super acid compounds. This extraordinarily high acidity has led to increasing interest in the possibility of using them as alternative catalysts for acid-catalyzed transformations which employ environmentally homogeneous liquid acids like HF, $AlCl_3$, or H_2SO_4. On the other hand, a very high acidity may be responsible for undesired side reactions and for quick deactivation phenomena due to the formation of heavy by-products. The possible way to control the acid properties is through partial neutralization of the protons, which can be achieved by exchanging the acid form with a suitable metal ion.

1.7 Redox Properties of Heteropoly Acid

The redox properties of HPA are a function of the nature of the metal atoms in the primary structure (addenda atoms) and of both the heteroatom and the counterions. In solutions, the reduction potentials of heteropoly anions containing Mo and V are high as these ions are easily reduced. Oxidative ability decreases generally in the order V > Mo > W containing heteropoly anions, which means that the vanadium-containing heteropoly compounds are the strongest oxidants [7]. The replacement of one or more molybdenum atoms in the primary structure of Keggin heteropoly compound of molybdenum for vanadium leads to an enhancement of the oxidation potential of the heteropoly compound due to the vanadium reducibility.

The nature of the heteroatom affects the overall charge of the polyanion. An increase in the charge leads to a decrease in the oxidation potential for the W^{6+}/W^{5+} couple in $P^{V}W_{12}O_{40} > Si^{IV}W_{12}O_{40} \approx Ge^{IV}W_{12}O_{40}{}^{4-} > B^{III}W_{12}O_{40}{}^{5-} \approx Fe^{III}W_{12}O_{40}{}^{5-} > H_2W_{12}O_{40} \approx Co^{II}W_{12}O_{40}{}^{6-} > Cu^{I}W_{12}O_{40}{}^{7-}$.

The effect of the counter cation may be significant. When the cation is easily reducible, the redox properties of the heteropoly compound are parallel to those of the cation. When the cation is not reducible, like alkali metals, the reducibility of the metal in the primary structure is nevertheless affected by the nature of the cation [5]. The oxidizing ability has been estimated from the rate of reduction of heteropoly compounds by H_2, CO, and organic compounds, but sometimes the data appear inconsistent, due to the difference in the kind of reductant, homogeneity, nonstoichiometry, and decomposition of the catalysts [3].

1.8 Heteropoly Acid as Homogeneous Catalysis

Heteropoly acids catalyze a wide variety of reactions in homogeneous liquid phase offering strong options for more efficient and cleaner processing compared to conventional mineral acids [8]. In principle, mechanisms of homogeneous catalysis by heteropoly acids and by ordinary mineral acids are of the same origin. Both HPAs and mineral acids function as proton donors [9]. There are, however, some specific features in the HPA catalysis. First, being stronger acids, HPAs have significantly higher catalytic activity than mineral acids. In organic media, the molar catalytic activity of heteropoly acid is usually 100–1,000 times higher than that of H_2SO_4 [4]. This makes it possible to carry out the catalytic process at a lower catalyst concentration and/or at a lower temperature. Further, heteropoly acid catalysis lacks side reactions such as sulfonation and chlorination, which occur with mineral acids.

As stable, relatively nontoxic crystalline substances, HPAs are also preferable as regard to safety and ease of handling [8].

The relative activity of Keggin heteropoly acids depends primarily on their acid strength. Other properties such as the oxidation potential, the thermal, and hydrolytic stability are also important factors. These properties for the most common heteropoly acids are summarized below:

Acid strength: $H_3PW_{12}O_{40} > H_4SiW_{12}O_{40} > H_3PMo_{12}O_{40} > H_4SiMo_{12}O_{40}$
Oxidation potential: $H_3PMo_{12}O_{40} > H_4SiMo_{12}O_{40} \gg H_3PW_{12}O_{40} > H_4SiW_{12}O_{40}$
Thermal stability: $H_3PW_{12}O_{40} > H_4SiW_{12}O_{40} > H_3PMo_{12}O_{40} > H_4SiMo_{12}O_{40}$
Hydrolytic stability: $H_4SiW_{12}O_{40} > H_3PW_{12}O_{40} > H_4SiMo_{12}O_{40} > H_3PMo_{12}O_{40}$

Generally, tungsten heteropoly acids are the catalysts of choice because of their stronger acidity, higher thermal stability, and lower oxidation potential compared to molybdenum heteropoly acids. Usually, if the reaction rate is controlled by the catalyst acid strength, $H_3PW_{12}O_{40}$ shows the highest catalytic activity in the Keggin series. However, in the case of less demanding reactions as well as in reactions at higher temperatures in the presence of water, $H_4SiW_{12}O_{40}$, having lower oxidation potential and higher hydrolytic stability, can be superior to $H_3PW_{12}O_{40}$. Some homogeneous reactions catalyzed by HPA are hydration of olefins, esterification reaction, condensation of acetone to mesityl oxide, and alkyl benzenes.

The major problem, limiting the utility of homogeneously catalyzed processes, is the well-known difficulty in catalyst recovery and recycling. As the cost of the heteropoly acids is higher than mineral acids, the recycling of heteropoly acid catalyst is the key issue to their application. Only a few homogeneous reactions allow for easy heteropoly acid recycling like hydration of olefins. In some cases, HPA can be recovered from polar organic solution without neutralization by precipitating with a hydrocarbon solvent. HPA can also be extracted from an acidified aqueous solution of its salt with a polar organic solvent. Even though the neutralization of HPA is necessary, the amount of alkali needed and the amount of waste formed there upon is much less than that of mineral acids. A more efficient way to overcome the separation problem is the use of biphasic systems or solid heteropoly acid catalysts [8, 9].

1.9 Heteropoly Acid as Heterogeneous Catalysis

Solid heteropoly acids are more efficient than conventional solid acids, like zeolites, SiO_2, and Al_2O_3, due to their ability to absorb large amount of polar molecules in the bulk, coupled with high proton mobility, which leads to a high catalytic efficiency for liquid phase reactions. This behavior favors the reaction kinetics and the participation of all structural protons in the reaction [5]. This high activity allows the operation to be carried out at milder conditions than that of other solid acids. Obvious advantage of heterogeneous systems over homogeneous is easy separation of a catalyst from reaction products. Furthermore, since heteropoly acids are soluble only in wet polar solvents, their strong acidity cannot be utilized in homogeneous systems [8]. Therefore, heteropoly acids must be used as solid acid catalyst for catalyzing highly demanding reactions like Friedel–Crafts reaction. In order to enhance the acid strength, solid heteropoly acid catalysts are usually dehydrated by evacuating at 150–300 °C for 1–2 h [10].

A serious problem with the solid heteropoly acid catalysts is their deactivation during organic reactions due to the coke formation on the catalyst surface, which remains to be solved to put heterogeneous heteropoly acid catalysis in practice. Instead of burning the coke as in the case of alumina silicates and zeolites, supporting heteropoly acids on a carrier inhibits the formation of coke on the catalyst surface.

1.9.1 Supported Heteropoly Acids

The drawback of heteropoly acids is their low surface area (1–10 m^2/g) that limits their application in many reactions. This disadvantage can be overcome by dispersing the heteropoly acid on solid supports with high surface area. The catalytic activity of supported heteropoly acid depends mainly on the heteropoly acid loading, the pretreatment conditions, and the type of carrier. Generally, heteropoly acids strongly interact with supports at low loading levels, while the bulk properties of heteropoly acids prevail at high loadings. Acidic or neutral substances like SiO_2, active carbon, acidic ion-exchange resin, and TiO_2 are suitable supports, but on the other hand, solids like Al_2O_3 and MgO having basicity tend to decompose heteropoly acids [8], causing a significant decrease in catalytic activity.

1.9.2 HPA on Silica

SiO_2 is relatively inert toward heteropoly acids above a certain loading level. The thermal stability of silica-supported heteropoly anions decreases with respect to that of the bulk acid, particularly at low concentrations. On the other hand, tungstic acid keeps its Keggin structure when impregnated on silica, provided tungsten content is greater than 10 % w/w but at lower contents the acid may experience partial degradation.

Moffat and Kasztelan [11] have concluded that the Keggin structure of phosphotungstic acid dispersed on silica is maintained even after calcination at 550 °C. At low loadings, $H_3PW_{12}O_{40}$ and $H_4SiW_{12}O_{40}$ form finely dispersed species on the surface of silica and a heteropoly acid crystal phase is developed at heteropoly acid loading above 50 wt%. Silica-supported tungsten HPAs like $H_3PW_{12}O_{40}$ and $H_4SiW_{12}O_{40}$ maintain the Keggin structure at high loadings but decompose at very low loadings because of the strong interactions with surface silanol groups.

A study based on the synthesis of chemicals catalyzed by transition metal/ salts of tungsten-based polyoxometalate-modified mesoporous silica (MCM-41) in the liquid phase has been focused in this article. We will also comment briefly on the preparation of the modified catalyst by various methods and its characterization by different techniques, as well as on the structure and the active sites. It is out of the scope of this article to cover comprehensively all reactions that have ever been reported using tungsten-based polyoxometalate-modified MCM-41-based solid acids.

2 Experimental

2.1 *Cesium Salt of Phosphotungstic Acid-Supported MCM-41 Toward Acylation of Anisole*

Salts of HPA were prepared by partially exchanging protons of the parent HPA with large cations, such as Cs^+, K^+, Rb^+, and NH_4^+, which could be water insoluble and present a rather high surface area (>100 m^2/g) [10, 12]. Extensive studies have been carried out over Cs salt of HPA supported on metal oxides such as Al_2O_3, MgO, silica, titania, and zirconia, MCM-41, MCM-48, and SBA-15 [13–15]. The catalytic activities of these materials have been investigated thoroughly for various alkylation, acylation, esterification reactions, etc. [16]. Recently, the effect of acylating agent on the acylation of anisole using mesoporous silica-supported HPA and Cs salt of HPA has been studied by various research groups [17–19]. Cardoso et al. reported 90 % of conversion for acylation of anisole taking phosphotungstic acid-supported silica as catalyst [17]. Kaur and Kozhevnikov [18] reported the acylation of anisole with carboxylic acids on $Cs_{2.5}$-HPA as catalyst having 51 % yields. Kamala and Pandurangan reported butylation of anisole taking phosphotungstic acid-supported Al MCM-41 as catalyst [20]. Acylation of aromatic compounds is a widely used reaction for the production of fine chemicals, pharmaceutical, and cosmetics. So many papers are published in Cs salts of PTA supported on different supports, but our catalysts are different from others and characterized by different techniques and also focus on various organic transformation reactions.

(a) Synthesis of MCM-41
In a typical synthesis process, 1.988 g of cetyl trimethyl ammonium bromide (CTAB, 98 %, S.D. fine chem.) was dissolved in 120 g of water at room temperature. After complete dissolution, 8 ml of aqueous NH_3 (32 % in water, Merck) was added to the above solution. Then 10 ml of tetraethyl orthosilicate (TEOS, 99 %, Aldrich) was added to the solution under vigorous stirring (300 rpm). The hydrolysis of TEOS takes place during the first 2 min at room temperature (the solution becomes milky and slurry forms), whereas the condensation of the mesostructured hybrid material is achieved after 1 h of reaction. The material was then filtered and allowed to dry under static air at 80 °C for 12 h. The surfactant removed mesoporous material was obtained by acid treatment method. For acid extraction, the as-obtained materials (1 g) were treated with a mixture of ethanol (100 ml) and concentrated HCl (1 ml, 38 % in weight) at 80 °C for 6 h.

(b) Synthesis of Cs Salt of Phosphotungstic Acid
Cs of phosphotungstic acid ($Cs_{2.5}H_{0.5}PW_{12}O_{40}$) was prepared by adding the Cs_2CO_3 solution dropwise to the PTA solution according to the literature method [10]. The resulting precipitate was dried at 110 °C overnight in vacuum and calcined at 300 °C for 3 h. The chemical composition of the samples was determined using ICP atomic emission spectrometer (PerkinElmer).
(c) Synthesis of Cs Salt of Phosphotungstic Acid Supported on MCM-41
MCM-41 was first impregnated with aqueous solution of the Cs^+ precursor (Cs_2CO_3), dried at 110 °C for 12 h. Following this, a methanolic solution of PTA was impregnated, dried at 110 °C for 12 h, and calcined at 200. The catalysts are designated as x Cs-PTA/MCM-41(x = 10–60 wt%).

3 Results and Discussion

3.1 Characterization

3.1.1 Surface Area and Pore Size Distribution

Figure 3.1 shows the nitrogen adsorption–desorption isotherms of the parent and modified samples. MCM-41 shows a type IV isotherm with a sharp inflection capillary condensation step at $P/P_0 = 0.35$, corresponding to the pore size of about 2.7 nm. Apart from the narrow pore size distribution, MCM-41 exhibits an H1-type hysteresis loop at P/P_0 between 0.9 and 1, reflecting secondary mesoporosity due to inter particle condensation [21]. In case of the modified sample, the sharp capillary condensation step shifts to low relative pressure of $P/P_0 = 0.1$, indicating the decrease in pore volume and surface area after modification. A reasonable explanation for this

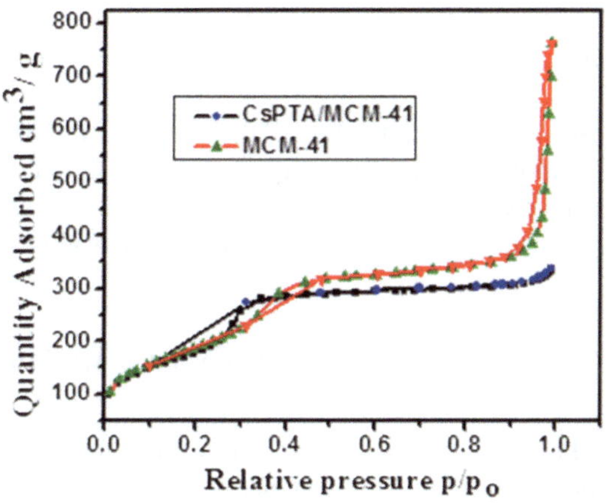

Fig. 3.1 N_2 adsorption–desorption isotherms of MCM-41 and 50 wt% Cs-PTA/MCM-41

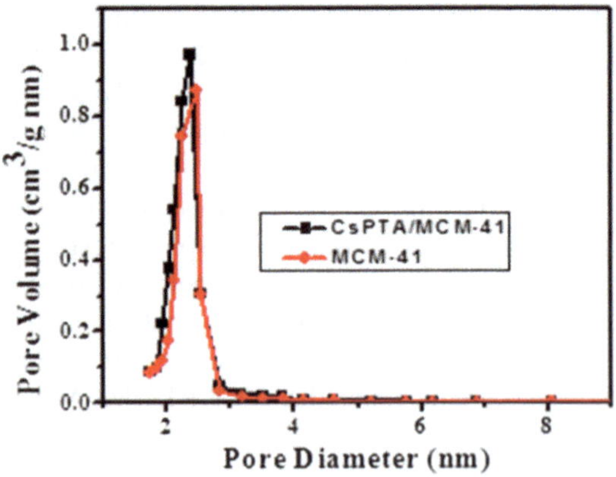

Fig. 3.2 Pore size distribution of MCM-41 and 50 wt% Cs-PTA/MCM-41 sample

observation is that most of the Cs-PTA clusters were introduced into the pore and the partial blockage of one-dimensional mesopores of MCM-41 by small aggregates of Cs-PTA [21].

The pore size distribution curves of the samples are shown in Fig. 3.2. The pore size distribution curves show that there is slight decrease in pore diameter after impregnation of Cs salt of PTA.

3.1.2 FTIR Studies

The FTIR spectra of Cs-PTA and Cs-PTA loaded MCM-41 are shown in Fig. 3.3. The spectra show a broad band around 3,100–3,600 cm^{-1}, which is due to adsorbed water molecules. The absorption band due to H–O–H bending vibration in water is at 1,620–1,640 cm^{-1}. The absorption band around 1,087–1,092 cm^{-1} is due to Si–O asymmetric stretching vibrations of Si–O–Si bridges. The characteristic signature of the Keggin structure was seen in each case, with dips at 1,080, 985, 890, and 800 cm^{-1}. The band assignments are available in the literature [22, 23].

3.1.3 SEM Studies

SEM micrographs of Cs-PTA/MCM-41 microcrystals recorded at different magnifications are shown in Fig. 3.4a and b. Each crystal has a cylindrical form having diameter ranging from 0.4 to 0.6 μm. The total length of the crystals is 1–2 μm and consists of packages of cylindrical fibers (Fig. 3.4a).

3.2 Catalytic Activity Toward Acylation of Anisole

The acylation reaction was carried out in the liquid phase in a 50 ml three-necked round-bottom flask fitted with a thermometer, reflux condenser with $CaCl_2$ guard tube, and a magnetic stirrer. A mixture of 100 mmol anisole and 10 mmol of acetic anhydride

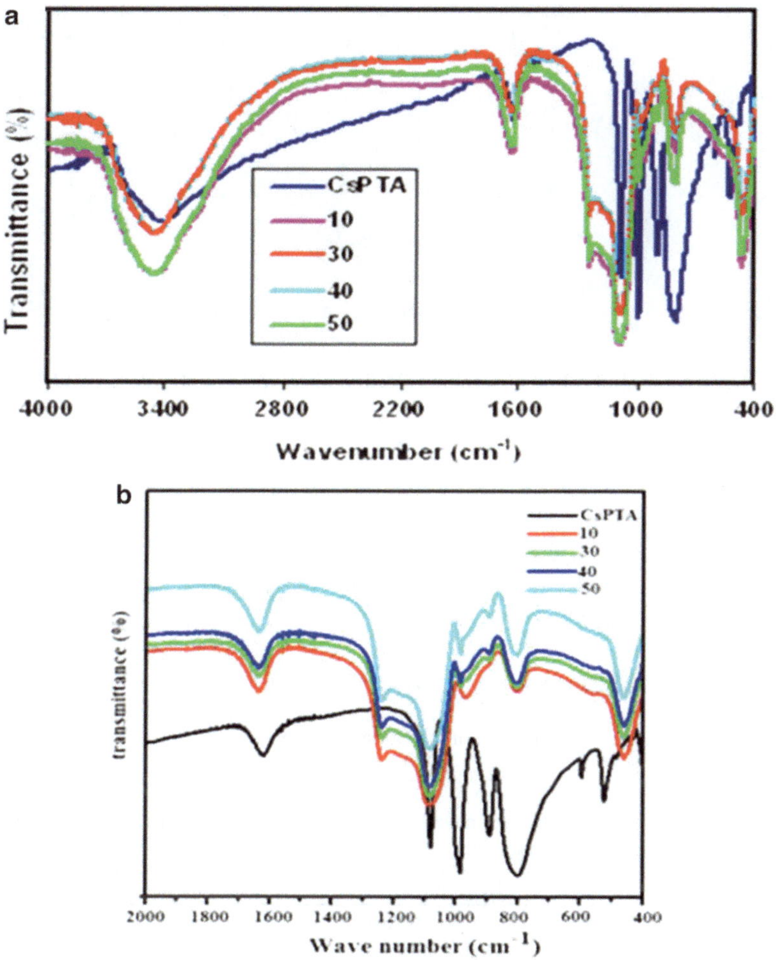

Fig. 3.3 FTIR spectra of Cs-PTA/MCM-41 samples (**a**) 400–4,000 cm^{-1}, (**b**) 400–2,000 cm^{-1}

was added to the flask along with 0.1 g of *n*-tridecane, which was used as an internal standard for GC analysis. The catalyst was added after adjusting the temperature to 70 °C. The reaction mixture was separated from the catalyst after 1 h and analyzed by off-line gas chromatography (GC) as shown in Scheme 3.1 (Table 3.1).

Among the catalysts with different Cs-PTA loadings, the 50 wt% Cs-PTA/MCM-41 catalyst gave the highest conversion (98 %). Further, increase in the Cs-PTA loading decreases the acetic anhydride conversion. The sample containing 60 wt% Cs-PTA shows 95 % conversion. The activity of the catalysts has been found to be related to the number of Brønsted acid sites. Since no metal ion was present in the reaction filtrate, it is presumed that the true active species is the solid acid, which developed due to chemical interaction between the Cs-PTA and MCM-41. This also supports that the catalyst is stable under reaction conditions.

Fig. 3.4 Different magnification of scanning electron micrograph of 50 wt% Cs-PTA/MCM-41

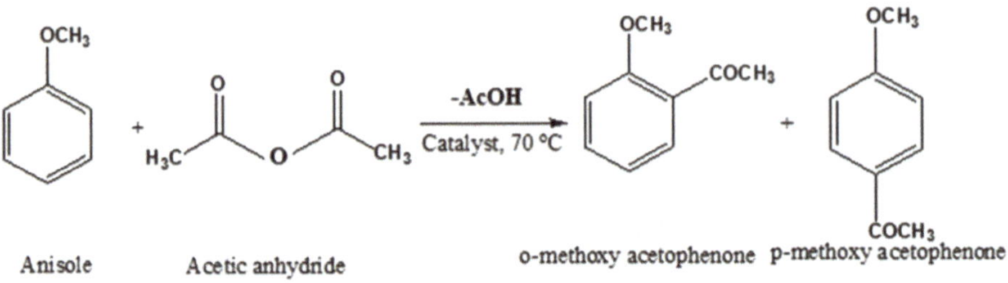

Scheme 3.1 Schematic representation of acylation of anisole

Table 3.1 Liquid phase acylation of anisole over 50 wt. % Cs-PTA/MCM-41 and comparison of the result with other reported methods

Catalyst used	Reactant	Conversion (%)	References
Phosphotungstic acid/silica	Acetic anhydride	90	[24a]
$Cs_{2.5}H_{0.5}PW_{12}O_{40}$	Dodecanoic acid	51	[24b]
ZSM-5	Acetic anhydride	90	[24c]
Aluminosilicates	Octanoyl chloride	90	[24d]
$Cs_{2.5}H_{0.5}PW_{12}O_{40}$/silica	Acetic anhydride	50	[24e]
$Cs_{2.5}H_{0.5}PW_{12}O_{40}$/K-10	Benzoyl chloride	37	[24f]
$H_3PW_6Mo_6O_{40}$/Zirconia	Acetic anhydride	89	[24g]
Cs-PTA/MCM-41	Acetic anhydride	98	Our work

3.3 *Influence of Various Substrates*

The reaction procedure was applied to anisole as well as activated aromatic compounds such as toluene, aniline, and deactivated compounds such as chlorobenzene, and the results were summarized in Table 3.2. The results show that introduction of an electron withdrawing group (e.g., nitro group) on the aromatic ring substantially decreases the conversion of acylation, while an electron-donating group (e.g., CH_3

Table 3.2 Effect of substrate on acylation reaction

Substrate	Conversion (%)	Selectivity	
		Para	Ortho
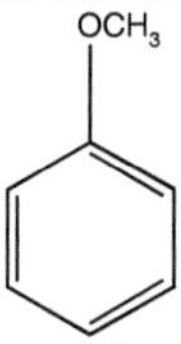	98	97	3
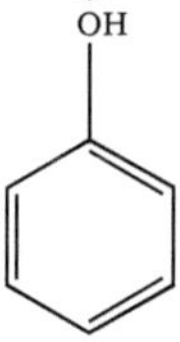	94	96	4
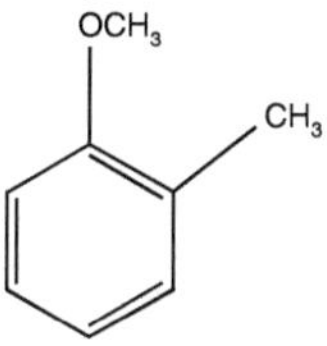	99	100	–
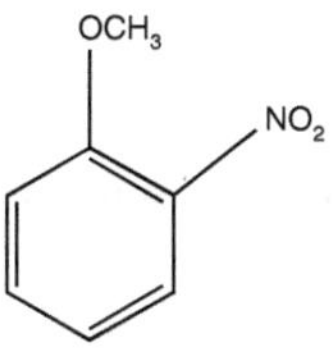	65	100	–
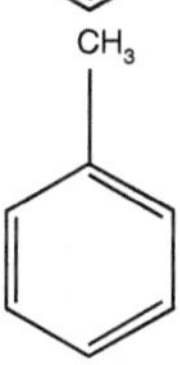	75	90	10
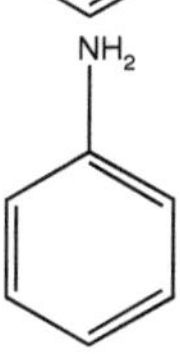	90	92	8
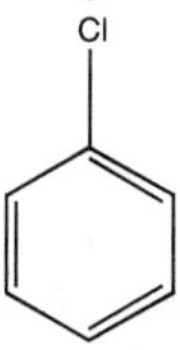	35	75	25

Substrate = 100 mmol, acetic anhydride = 10 mmol, catalyst amount = 0.1 g, temperature = 70 °C, time = 1 h

group) increases it. Like anisole, aniline and phenol give preferentially para product in high yield. Activated aromatic compounds such as toluene gives around 72 % conversion having 61 % of para selectivity. Chlorobenzene was sluggish in the acylation reaction and gave very low conversion.

3.4 Cu Salt of Phosphotungstic Acid-Supported MCM-41 Toward Heck Vinylation Reaction

A lot of papers have already been published on Cs, K, and Na salts of PTA toward various reactions [25a, b]. A few publications are also there on the salts of PTA supported on MCM-41 [25c]. However, there has been no example on the use of Cu salt of HPA supported on MCM-41 for the Heck vinylation of aryl halides with olefins. Hence for the first time, we have performed the coupling reaction using $Cu_xH_{3-2x}PW_{12}O_{40}$/MCM-41 as catalyst in aqueous medium.

3.4.1 Preparation of Catalyst

The Cu salt of the $H_3PW_{12}O_{40}$ was prepared [25d] as precipitate by adding 0.18 g of $Ba(OH)_2$ (to neutralize protons) to the aqueous solution containing 2 g of $H_3PW_{12}O_{40}$. Later 0.16 g of $CuSO_4{\cdot}5H_2O$ was added to replace Ba with Cu by eliminating Ba as $BaSO_4$. The $Cu_{1.5}PW_{12}O_{40}$ salt was recovered from the solution by recrystallization. Varying the amount of $Ba(OH)_2$ and $CuSO_4{\cdot}5H_2O$, we prepared $Cu_1HPW_{12}O_{40}$ and $Cu_{0.5}H_2PW_{12}O_{40}$. The catalyst mass was dried at 120 °C for 12 h in an oven and finally calcined at 300 °C for 2 h. The materials are further termed as Cu_xH_{3-2x} $PW_{12}O_{40}$ (x=0.5–1.5).

A series of catalysts having different Cu/HPA loading 10–60 wt% were synthesized by impregnating 2 g of MCM-41 with an aqueous solution of Cu/HPA under constant stirring followed by heating till complete evaporation of water takes place (4 h). Then the sample was dried in an oven at 110 °C for 24 h and calcined at 500 °C. The catalysts are herein after referred to as $y$$Cu_xH_{3-2x}PW_{12}O_{40}$/MCM-41 ($y$=10–60 wt%).

3.4.2 Characterization

X-Ray Diffraction

The PXRD patterns of $H_3PW_{12}O_{40}$/MCM-41 and 50$Cu_{1.5}PW_{12}O_{40}$-promoted MCM-41 samples are shown in Fig. 3.5. It can be observed that both the materials exhibit a strong peak at $2\theta=2.2°$ due to (100) plane. Also, small peaks due to higher order (110), (200), and (210) plane reflections within 5° indicate the formation of well-ordered mesoporous materials. Thus, the mesoporosity remains intact after the modification of the silica network with $Cu_xH_{3-2x}PW_{12}O_{40}$. There is a little bit reduction and broadening of the (100) peak of $H_3PW_{12}O_{40}$/MCM-41 after modification of $Cu_{1.5}PW_{12}O_{40}$ indicating a slight disturbance in hexagonal symmetry [26].

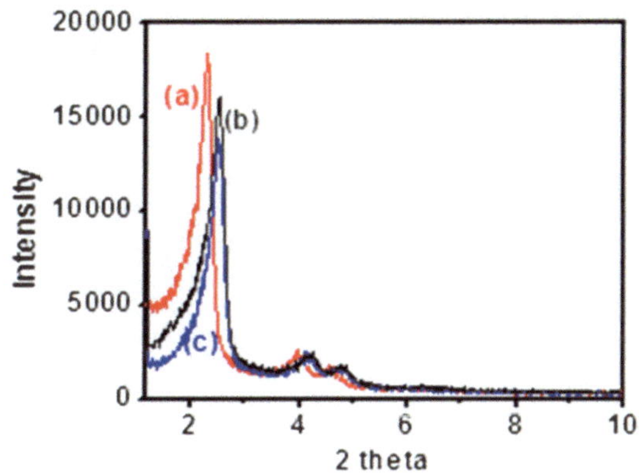

Fig. 3.5 Low angle (0–10°) XRD patterns of MCM-41 (*a*), HPA/MCM-41 (*b*), $Cu_{1.5}$ PA/MCM-41 (*c*)

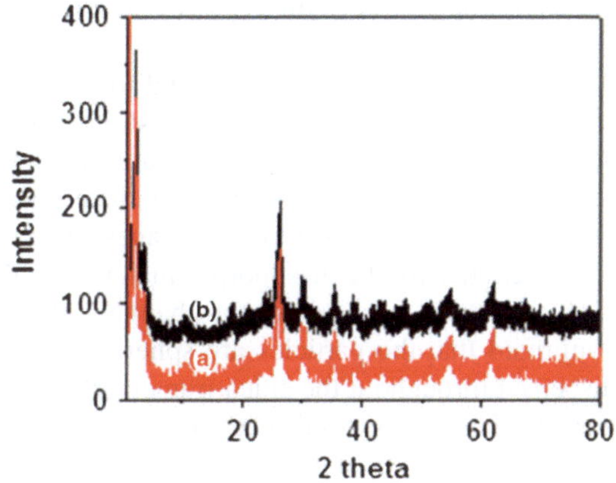

Fig. 3.6 Wide angle (20–80°) XRD patterns of HPA/MCM-41 (*a*), $Cu_{1.5}$PA/MCM-41 (*b*)

The characteristic peaks for HPA [27] are observed in both cases (Fig. 3.6), which indicated that the structure of HPA remains intact after metal modification.

Temperature-Programmed Reduction

The temperature-programmed reduction profiles of the 50 wt% Cu-/HPA-modified MCM-41 are shown in Fig. 3.7. All the samples show reduction peaks in mainly two regions: one is around 250–400 °C and another is 650–800 °C. The first peak corresponds to reduction of Cu^{+2} to Cu metal. Reduction may take place in one step ($Cu^{+2} \rightarrow Cu$) or in two subsequent steps ($Cu^{+2} \rightarrow Cu^{+1} \rightarrow Cu$). The high-temperature reduction peak corresponds to reduction of CuHPA species adsorbed on the MCM-41 support. As a result of their high dispersion, these species interacting with the support are more easily reducible than the species of the bulk CuHPA. The peak intensity increases with increase in Cu content in the sample.

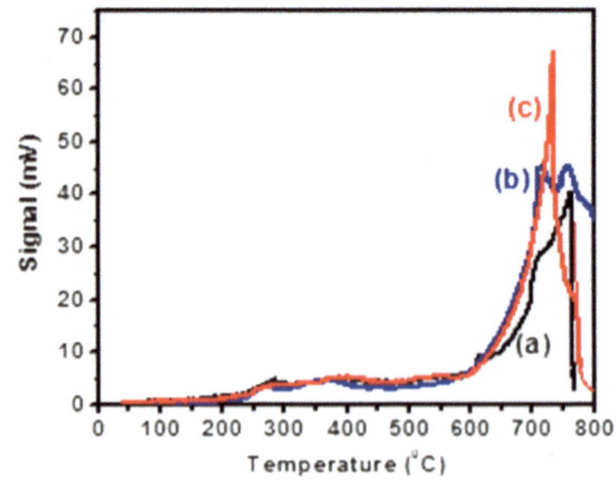

Fig. 3.7 TPR spectra of $Cu_{0.5}H_2PA$/MCM-41 (*a*), Cu_1HPA/MCM-41 (*b*) and $Cu_{1.5}PA$/MCM-41 (*c*)

Transmission Electron Microscopy (TEM)

To study the surface topography and to assess the surface dispersion of the active components over the MCM-41 substrate, TEM investigation was performed on $50Cu_{1.5}PA$/MCM-41 and the images are shown in Fig. 3.8. It has been found that the catalysts are well-ordered spherical particles. The particles are uniformly distributed over the support surface. TEM images of the mesoporous molecular sieves are characteristics for the mesoporous materials with hexagonal channel array, showing high quality in organization of channels of these catalysts. From the figure it is confirmed that the particles are spherical in nature. The metal particles are well dispersed throughout the silica framework, which is clearly seen in the figure. The particle size can be confirmed from the TEM images. The average particle size is calculated to be 100 nm.

3.4.3 Catalytic Evaluation Toward Heck Vinylation

Phenyl iodide (1 mmol), water (2 ml), and the catalyst (0.02 g) were taken in a round- bottom flask. The reaction mixture was stirred at 100 °C followed by the addition of acrylic acid (2 mmol) and K_2CO_3 (1.5 mmol). After 8 h, the solution was allowed to cool down and filtered. The filtrate was analyzed by off-line GC and the solid residue was subjected to H^1NMR. The conversion was determined from the amount of phenyl iodide consumed in the reaction and is shown in Scheme 3.2. Leaching experiments for Cu metal were performed for each catalyst by using AAS analysis of the filtrates.

Two types of mechanisms are suggested for the Heck vinylation reaction. One is through the neutral pathway and another is the cationic pathway [10]. Yang et al. [28] suggested the mechanism for Heck vinylation using silica-supported poly-γ-aminopropylsilane transition metal (Ni^{+2}, Cu^{+2}, Co^{+2}) complexes similar to that of homogeneous catalyst for Heck reaction. In the case of homogeneous Pd catalyst, the Pd (IV) is reduced to Pd (II) during the reaction procedure.

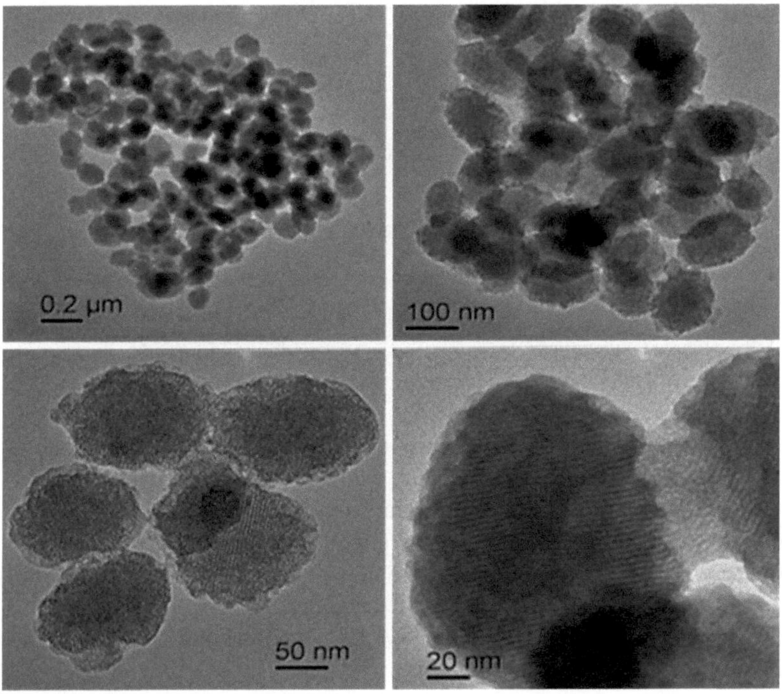

Fig. 3.8 TEM images of $50Cu_{1.5}PA/MCM\text{-}41$

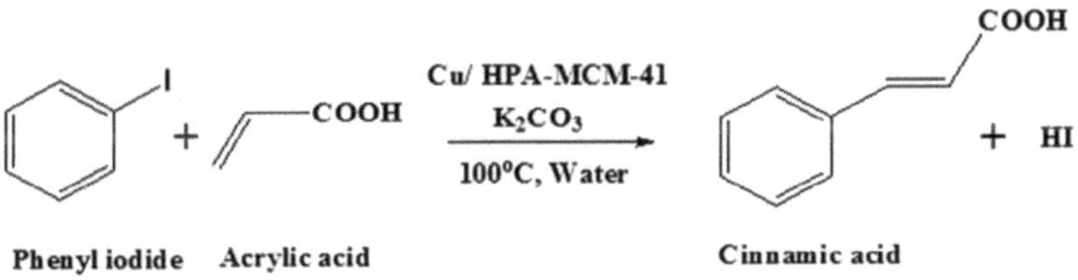

Scheme 3.2 Schematic presentation of Heck vinylation reaction

Similarly we can assume the reaction will proceed by the reduction of supported Cu^{+2} to the active Cu metal in case of CuHPA/MCM-41 (Scheme 3.3). First of all there is oxidative addition of phenyl iodide to CuHPA/MCM-41. With subsequent addition of HNu (acrylic acid) and base (K_2CO_3), there is reductive elimination of product (cinnamic acid) and the catalyst can be regenerated in subsequent steps.

Parent MCM-41 is less active toward Heck vinylation reaction and gave only 5 % conversion. In order to investigate the effect of Cu content, $Cu_{0.5}H_2PW_{12}O_{40}$, Cu_1H_1 $PW_{12}O_{40}$, and $Cu_{1.5}PW_{12}O_{40}$ catalysts were used in Heck vinylation. From the experimental results we can confirm that the percentage of phenyl iodide conversion

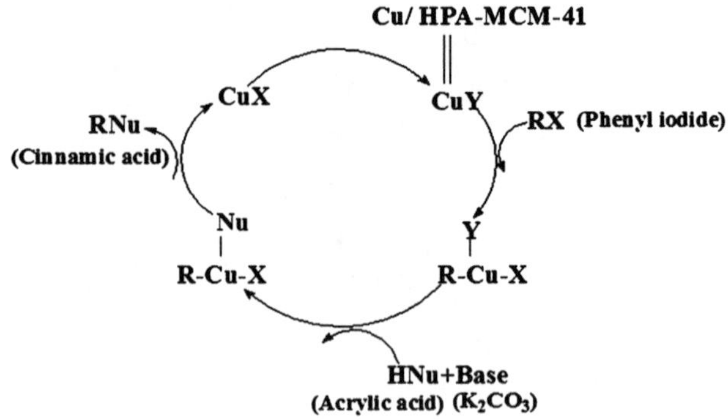

Scheme 3.3 Proposed mechanism for Heck vinylation reaction using Cu/HPA-MCM-41 as catalysts

increased from 88 to 98 % with increase in Cu content from 0.5 to 1.5 mol. Also the effect of CuHPA loadings was studied by using 10–60 wt% CuHPA/MCM-41 catalysts in Heck vinylation. Among the catalysts with different CuHPA loadings, the 50 wt% $Cu_{1.5}PW_{12}O_{40}$/MCM-41 catalyst gave the highest conversion (98 %). With further increase in CuHPA loadings, there may be possibility of formation polylayer of material on the support surface. So the catalytic conversion decreased in a small extent (94 %) in case of 60$Cu_{1.5}PW_{12}O_{40}$/MCM-41and the materials are characterized by different technique.

3.4.4 Influence of Various Substrates

Various aryl halides are tested toward Heck vinylation using acrylic acid as the olefin (Table 3.3). Electron-donating groups at the para position on the phenyl substrate enhanced the percentage of conversion, whereas electron-subtracting groups decreased it. Electron-rich phenyl iodide is more appreciable to form complex with Cu (II) and attack the double bond of acrylic acid to form the product.

3.5 Fe-Modified Lacunary Phosphotungstate-Supported MCM-41 as an Excellent Catalyst for Acid-Catalyzed as well as Oxidation Reaction

The most investigated Keggin-type heteropoly acids are represented by the general formula $[X^{n+}M_{12}O_{40}]^{(8-n)-}$. Removal of one or two MO units from the fully occupied polyoxometalates $[XM_{12}O_{40}]^{n-}$ gives rise to mono-lacunary $[XM^{VI}{}_{11}O_{39}]^{(n+4)-}$ and di-lacunary $[XM^{VI}{}_{10}O_{36}]^{(n+5)-}$ polyoxometalates, respectively.

Table 3.3 Effect of substrates on the conversion of aryl halide

Substrate	Product	Conversion (%)
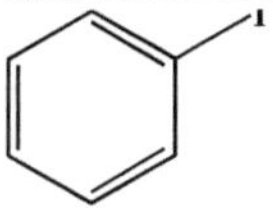	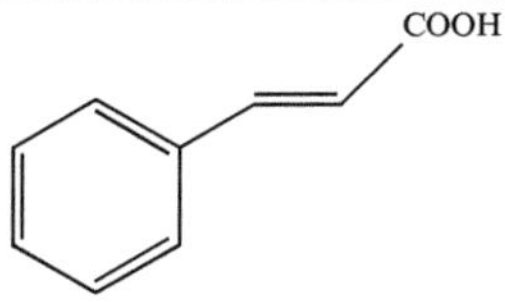	98
	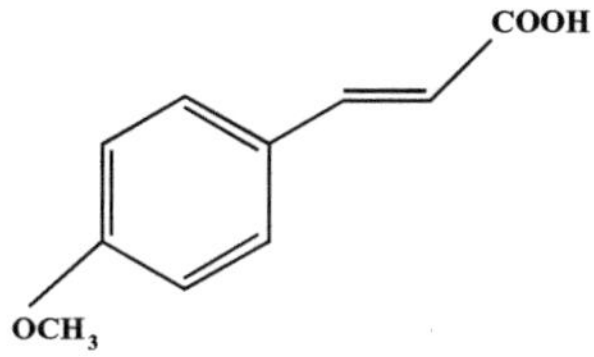	100
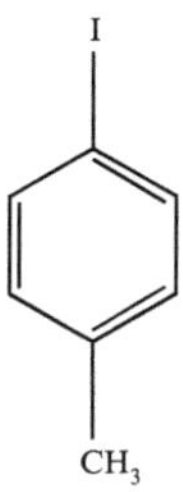	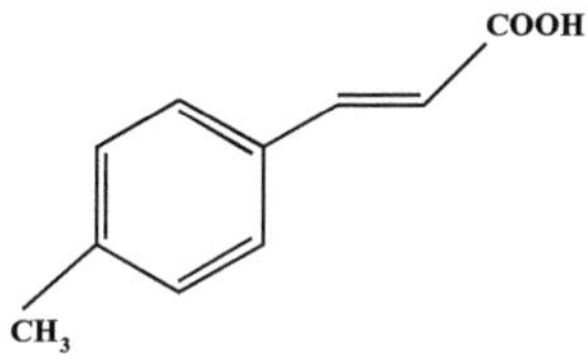	100
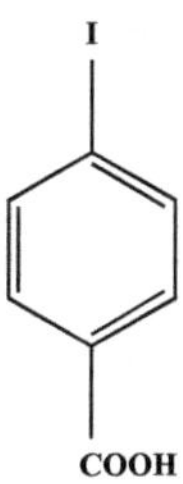	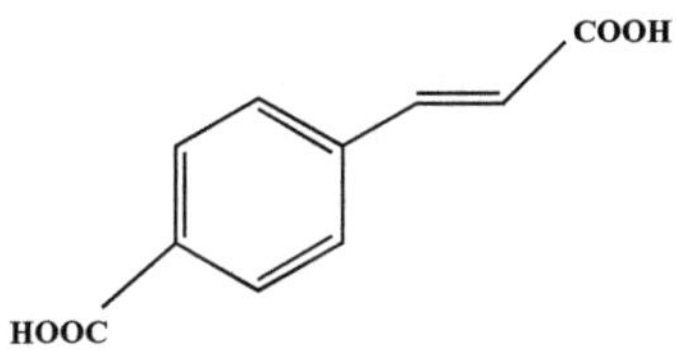	87
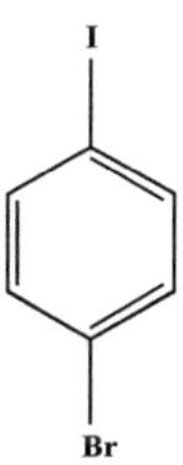	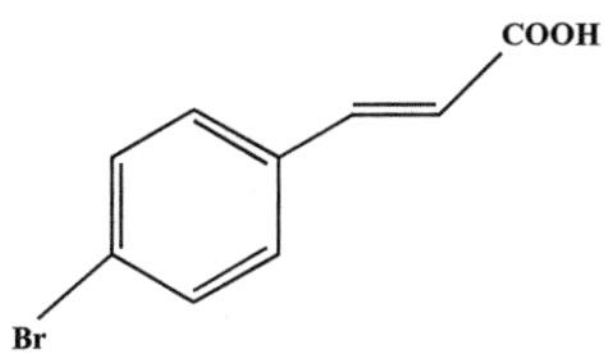	72
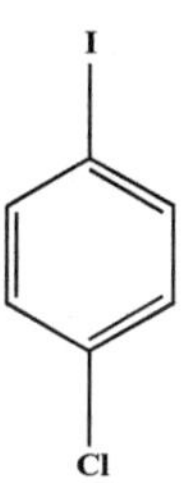	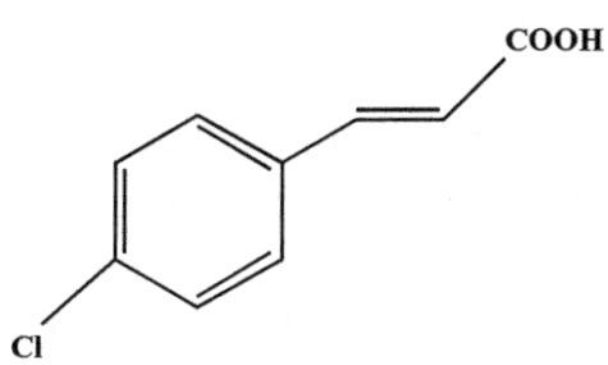	69

Aryl iodide = 1 mmol, acrylic acid = 2 mmol, K_2CO_3 = 1.5 mmol, water = 2 ml, catalyst = 0.02 g, temperature = 100 °C, time = 8 h

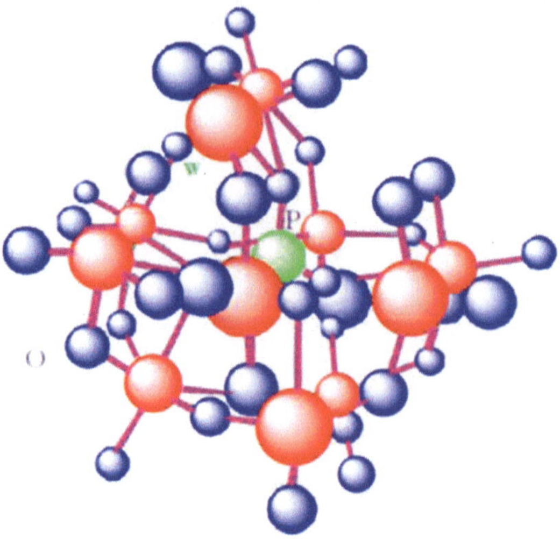

(Mono-lacunary)

Lacunary and di-lacunary polyoxometalates are gaining more importance because of their unique structural properties. It is well known that when the lacunary of Keggin anions $[XW_{11}O_{39}]^{(n+4)-}$ is substituted by other transition metal cations, it gives rise to transition metal-modified lacunary heteropoly compounds having the general formula $[XW_{11}O_{39}M]^{n-}$ (where M = first row transition metal). These species have recently attracted considerable attention [29], because of their thermal and chemical stability and the range of possibilities for their modification of electrocatalytic property without affecting the primary Keggin structure [30].

Recently, Patel et al. reported the detailed synthesis and characterization of Keggin-type manganese (II)-substituted phosphotungstate and its activity toward liquid phase oxidation of styrene [29]. Various works have already been carried out on Fe metal-substituted heteropoly acids. Mizuno et al. [31] reported synthesis of Fe, Ni-substituted Keggin-type heteropoly anion and evaluated its catalytic activity toward oxidation reaction. Nagai et al. reported iron in the Keggin anion of heteropoly acid catalysts for selective oxidation of isobutene [32]. But so far there is no literature available on the catalytic aspects of supported Cs salt of iron-substituted lacunary anions.

Many a study was already carried out on oxidation of *trans*-stilbene as well as bromination of phenol using various catalysts. Maurya and Amit Kumar [33] reported oxidation of *trans*-stilbene. But the inherent disadvantages associated were higher temperature and longer reaction time. Our group [34] reported bromination of phenol over heteropoly acid (HPA)-impregnated zirconium phosphate (ZrP), with 86 % conversion. But the most enchanting part of the present study is that this single catalyst is showing its superlative catalytic activity toward both the reactions.

We have investigated the use of various LFeW/MCM-41 as catalysts in the acid-catalyzed bromination of phenol as well as in the oxidation of *trans*-stilbene. The bromination of phenol was carried out in acetic acid medium with KBr and hydrogen peroxide at room temperature.

(a) Synthesis of sodium salt of iron-substituted lacunary phosphotungstate ($Na_5FePW_{11}O_{39}$)

The sodium salt of lacunary heteropoly compound modified with iron ion was prepared by the alkalization of a solution of dodecatungstophosphoric acid with an aqueous solution of $NaHCO_3$. First $H_3PW_{12}O_{40}{\cdot}nH_2O$ (2.88 g) was dissolved in water (10 ml) and the pH of the solution was adjusted to 4.8 using $NaHCO_3$ solution. This resulted in the formation of lacunary heteropoly anion $[PW_{11}O_{39}]^{7-}$. The solution having pH 4.8 was heated to 90 °C with constant stirring. A solution of $FeCl_2$ (0.197 g, 1 mmol) in water (10 ml) was added to this hot solution. The $Na_5FePW_{11}O_{39}$ was obtained by solvent evaporation and recrystallization from water, followed by subsequent drying at 110 °C for 12 h.

(b) Synthesis of Cs salt of iron-substituted lacunary phosphotungstate supported onto MCM-41 (*x* LFeW/MCM-41)

A series of catalysts having different loading of Cs salt of Fe-modified lacunary phosphotungstate (30–60 wt%)-promoted MCM-41 were synthesized by incipient weight impregnation method by adopting the following procedure.

MCM-41 was first impregnated with aqueous solution of the Cs^+ precursor (Cs_2CO_3), dried at 110 °C for 12 h. Following this, an aqueous solution of $Na_5FePW_{11}O_{39}$ was impregnated, dried at 110 °C for 12 h, and calcined at 200°C for 3 h. The catalysts are designated as *x* LFeW/MCM-41 ($x = 30$–60 wt%).

(c) Synthesis of cesium salt of iron-substituted lacunary phosphotungstate (LFeW) and Cs salt of lacunary phosphotungstate (LW)

The preparation procedure for LFeW was same as described in the section for the preparation of $Na_5FePW_{11}O_{39}$ up to the addition of $FeCl_2$. After which a saturated solution of the Cs_2CO_3 was added to the hot filtrate and the resulting mixture was allowed to stand overnight at room temperature. The mixture was filtered and the residue was dried at 110 °C for 12 h, which resulted in LFeW. The filtrate was used for the estimation of W and Fe, in order to see the loss during synthesis.

3.5.1 Characterization

Surface Area and Pore Size Distribution

The nitrogen adsorption–desorption isotherms for MCM-41 and 50 LFeW/MCM-41 are shown in Fig. 3.9. N_2 sorption resulted typical type IV isotherm which is defined by Brunauer et al. [35]. It is observed that there are three different well-defined stages in the isotherm of MCM-41. The initial increase in nitrogen uptake at low P/P_0 may be due to monolayer adsorption on the pore walls; a sharp steep increase at intermediate P/P_0 indicates the capillary condensation in the mesopores and a plateau portion at higher P/P_0 is associated with multilayer adsorption on the external surface of the materials [35].

Parent MCM-41 sample exhibits N_2 uptake at a relative pressure of 0.32 which corresponds to the pre-condensation loop. The isotherm shows a H4-type hysteresis

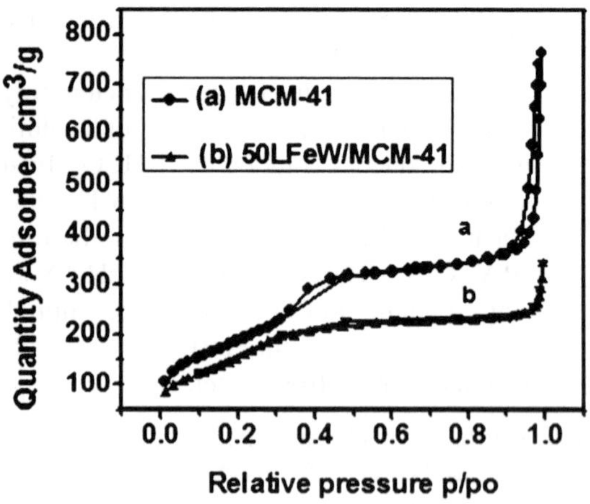

Fig. 3.9 N_2 ads–des isotherm of MCM-41 (*a*) and 50 LFeW/MCM-41 (*b*)

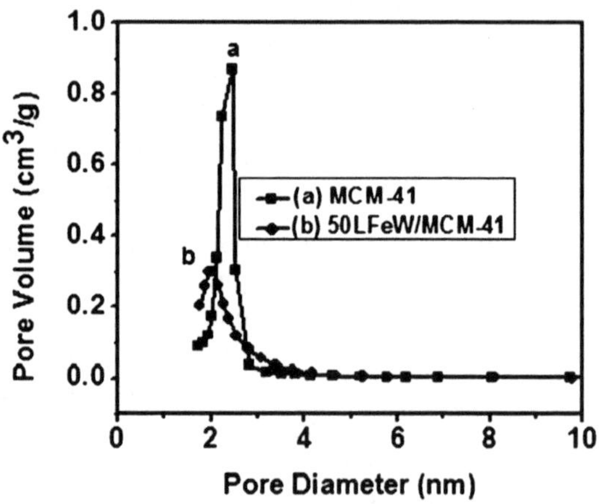

Fig. 3.10 Pore size distribution curve of MCM-41 (*a*) and 50 LFeW/MCM-41 (*b*)

loop (according to IUPAC nomenclature) with well-developed step in the relative pressure range ≈0.9. The incorporation of LFeW in the MCM-41 framework is found to lower the P/P_0 value for capillary condensation step, indicating the shift in pore size to lower value due to incorporation of lacunary acid. The pore diameter is found to decrease with increasing loading of LFeW content over the MCM-41 surface (Fig. 3.10).

The textural properties such as BET surface area, pore diameter, and pore volume derived from the N_2 adsorption–desorption measurements. The parent MCM-41 has a surface area of 1,250 m^2/g. But there is a gradual decrease in the value with increasing LFeW content in MCM-41. This may be due to the fact that increasing loading may result in the formation of polylayer on the pores of the silica surface. Pore size and pore volume exhibit similar trend as that of surface area. There is a gradual decrease in the values with increasing LFeW content on MCM-41 samples.

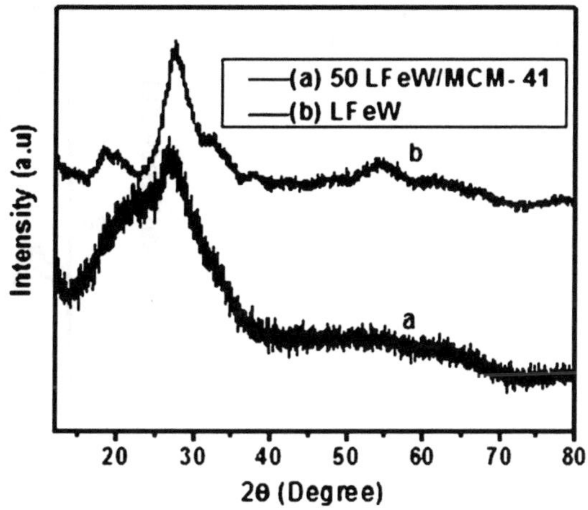

Fig. 3.11 Wide angle (10–80°) XRD patterns of (*a*) 50 LFeW/MCM-41 and (*b*) LFeW

X-Ray Powder Diffraction Studies

The wide angle XRD of LFeW and LFeW/MCM-41 is shown in Fig. 3.11. The XRD pattern of LFeW shows that it is crystalline in nature. But in the case of 50 LFeW/MCM-41 sample, the XRD pattern shows a broad peak showing no characteristic peaks of LFeW, indicating a very high dispersion of LFeW in a noncrystalline form on the surface of MCM-41.

FTIR Studies

The FTIR spectra of various samples are shown in Fig. 3.12. In the case of LFeW/MCM-41, the broad band around 3,500 cm^{-1} may be attributed to surface silanols and adsorbed water molecules, while deformational vibrations of adsorbed molecules cause the absorption bands at 1,623–1,640 cm^{-1} [36]. The spectrum of the Keggin anion $[PW_{12}O_{40}]^{3-}$ shows prominent bands at 1,080, 985, 890, and 800 cm^{-1} which are characteristic of Keggin structure and are assigned to $\nu_{(P\text{–}O)}$, $\nu_{(W=O)}$, corner-sharing $\nu_{(W\text{–}O\text{–}W)}$, and edge-sharing $\nu_{(W\text{–}O\text{–}W)}$, respectively [37]. In the case of LW, the 1,080 cm^{-1} band has been split into two components (1,084–1,044 cm^{-1}), due to the symmetry decrease of the PO_4 tetrahedron. The other bands found are 953 ($\nu as_{(W\text{–}Od)}$), 860 ($\nu as_{(W\text{–}Ob\text{–}W)}$), 809, and 742 cm^{-1}($\nu as_{(W\text{–}Oc\text{–}W)}$) and differ from those of $[PW_{12}O_{40}]^{3-}$ [37]. The spectra for LFeW showed characteristic splitting for the P–O bond frequency at 1,074 and 1,052 cm^{-1}, which faced a slight shifting toward lower frequency compared to bulk lacunary unit. This clearly indicated that Fe was introduced into the octahedral lacuna. The slight shifting of bands in FTIR spectra of LFeW sample compared to bulk LW may be due to formation of pseudosymmetric environment that resulted from the replacement of a W atom with a Fe atom. It can be observed that 50 LFeW/MCM-41 sample has similar vibration bands to those of the corresponding

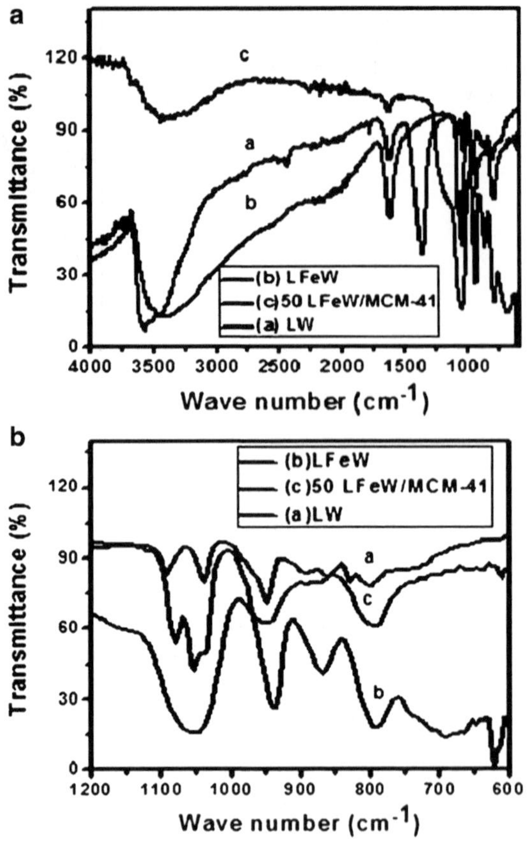

Fig. 3.12 FTIR spectra of LW (*a*), LFeW (*b*) and 50 LFeW/MCM-41 (*c*)

pure LFeW, which suggests that the LFeW structures remained intact regardless of their functionality. The shifting in the positions of the IR absorption peaks is due to both hydrogen bonding and chemical interactions that exist between the surface of LFeW and the MCM-41.

3.5.2 Catalytic Activity Toward Bromination of Phenol and Oxidation of *trans*-Stilbene

Bromination of phenol was carried out in a 50 ml two-necked round-bottom flask, charged with 0.2 g catalyst, phenol (2 mmol) in acetic acid (4 ml), and KBr (2.2 mmol). Then 30 % H_2O_2 (2.2 mmol) was added dropwise to the reaction mixture and the contents in the flask was stirred continuously at room temperature for 5 h [38]. After 5 h of the reaction, the catalyst was filtered and the solid was washed with ether. The combined filtrates were washed with saturated sodium bicarbonate solution and then shaken with ether in a separating funnel. The organic extract was dried over anhydrous sodium sulfate. The products were analyzed by GC, through capillary column (Scheme 3.4).

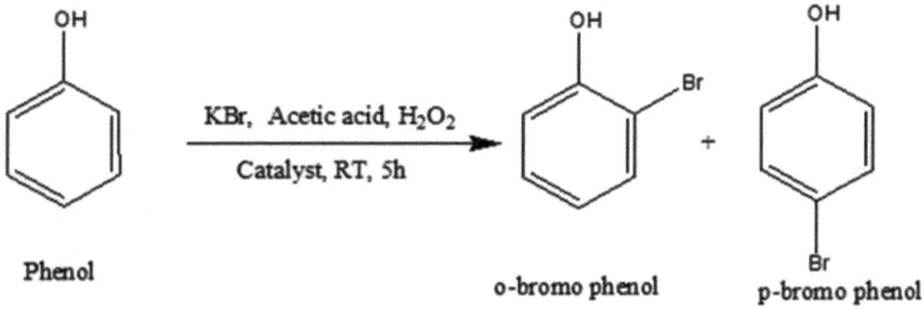

Scheme 3.4 Schematic representation of bromination of phenol

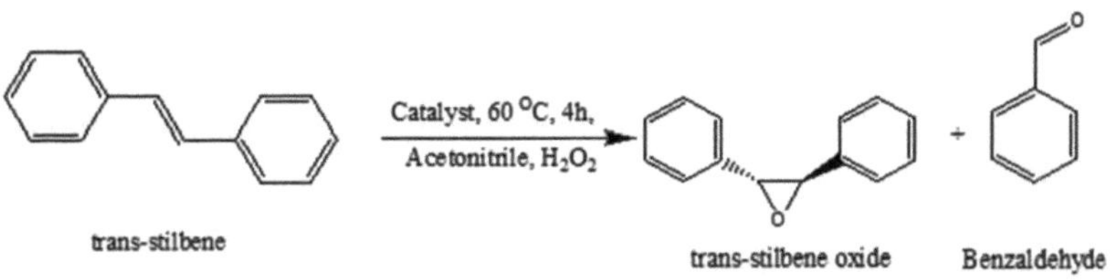

Scheme 3.5 Schematic representation of *trans*-stilbene oxidation

The oxidation of *trans*-stilbene was carried out in a 50 ml two-necked round-bottom flask, provided with a mercury thermometer for measuring the reaction temperature and a reflux condenser. The reaction mixture containing 0.015 g catalyst, 1.82 g *trans*-stilbene (10 mmol) and 30 % H_2O_2 (20 mmol), 20 ml of acetonitrile was heated at 60 °C for 4 h in an oil bath with stirring. The reaction products were analyzed by gas chromatograph using capillary column (ZB MAX) (Scheme 3.5).

Fe metal-modified Keggin-type lacunary phosphotungstate-supported mesoporous silica acts as an excellent catalyst for acid-catalyzed bromination of phenol as well as oxidation reaction of *trans*-stilbene. The 50 wt% Cs_5 $[PFeW_{11}O_{39}]$/MCM-41 showed remarkable catalytic performance, obtaining mono bromophenol having 95 % conversion with 99 % selectivity and *trans*-stilbene oxide having 52 % conversion with 99 % selectivity. The materials are characterized by different techniques.

3.6 Cs Salt of Pd Substituted Lacunary Phosphotungstate-Supported MCM-41 Toward Hydrogenation of p*-Nitrophenol to* p*-Aminophenol*

Considering enormous interest in palladium chemistry, it must be realized that homogeneous palladium catalysis has gained great success in Suzuki cross-coupling as well as other coupling reactions for the formation of C–C bonds [39]. To address the disadvantages associated with the homogeneous catalysis, one of the effective strategies is to support palladium on various inorganic supports including mesoporous silica, mesoporous alumina, mesoporous tin, and mesoporous carbon [40–44] and polymers [45, 46]. Palladium constitutes a very promising catalyst toward hydrogenation reactions [47].

Various methods have been reported earlier to synthesize *p*-aminophenol, such as multistep iron–acid reduction of *p*-nitrochlorobenzene or *p*-nitrophenol [48], catalytic hydrogenation of nitrobenzene [49], and electrochemical synthesizing method. However, to meet the growing commercial demand of *p*-aminophenol, the direct catalytic hydrogenation of *p*-nitrophenol is presently becoming important. So in this study we tried to explore the reducing ability of Keggin-type Cs salt of palladium metal-substituted mono-lacunary phosphotungstate-supported mesoporous silica (Cs_5 [$PPdW_{11}O_{39}$]/MCM-41) toward catalytic hydrogenation of *p*-nitrophenol to *p*-aminophenol. The catalyst showed eye-catching activity toward the desired reaction.

(a) Synthesis of sodium salt of palladium substituted lacunary phosphotungstate ($Na_5PdPW_{11}O_{39}$)

The sodium salt of lacunary heteropoly compound modified with palladium ion was prepared by the alkalization of a solution of dodecatungstophosphoric acid with an aqueous solution of $NaHCO_3$. First $H_3PW_{12}O_{40}{\cdot}nH_2O$ (2.88 g) was dissolved in water (10 ml) and the pH of the solution was adjusted to 4.8 using $NaHCO_3$ solution. This resulted in the formation of lacunary heteropoly anion $[PW_{11}O_{39}]^{7-}$. The solution having pH 4.8 was heated to 90 °C with constant stirring. A solution of $PdCl_2$ (0.177 g, 1 mmol) in water (10 ml) was added to this hot solution. The $Na_5PdPW_{11}O_{39}$ was obtained by solvent evaporation and recrystallization from water, followed by subsequent drying at 110 °C for 12 h.

(b) Synthesis of Cs salt of palladium-substituted lacunary phosphotungstate supported onto MCM-41 (*x* LPdW/MCM-41)

MCM-41 was first impregnated with aqueous solution of the Cs^+ precursor (Cs_2CO_3), dried at 110 °C for 12 h. Following this, a methanolic solution of $Na_5PdPW_{11}O_{39}$ was impregnated, dried at 110 °C for 12 h, and calcined at 200. The catalysts are designated as *x* LPdW/MCM-41(*x*=30–60 wt%). The materials are characterized by different technique.

3.6.1 Characterization

X-Ray Powder Diffraction Studies

The PXRD patterns of MCM-41 and 50 LPdW/MCM-41 samples are shown in Fig. 3.13. It can be observed that both materials exhibit a strong peak at about $2\theta = 2.2°$ due to reflection at (100) plane. A little bit reduction and broadening of the (100) peak of 50 LPdW/MCM-41 which can be seen with a slight shift toward a higher 2θ indicates a slight disturbance in its hexagonal symmetry. Also small peaks

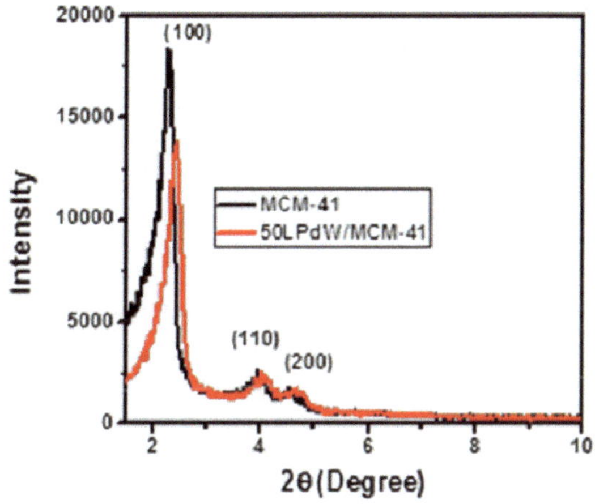

Fig. 3.13 Low angle (0–10°) XRD patterns of MCM-41 and 50 LPdW/MCM-41

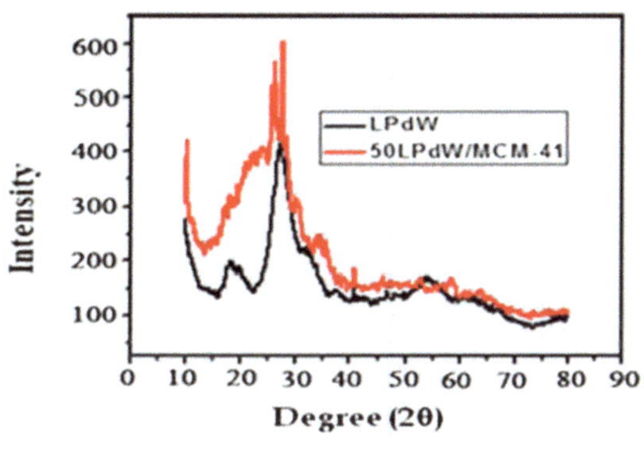

Fig. 3.14 PXRD of LPdW and 50 LPdW/MCM-41

due to higher order reflections at (110), (200), and (210) planes within 5° indicate that both the materials possess well-ordered mesoporosity [50].

The wide angle PXRD of LPdW and 50LPdW/MCM-41 is shown in Fig. 3.14. The XRD pattern of palladium metal has major diffraction peaks at $2\theta = 40.1°$ (111) and 46.7° (200), which is in agreement with the reported literature [51]. The characteristic peaks for LPdW are observed in both cases, which indicate that the structure of LPdW remains intact after metal modification.

^{31}P MAS NMR Spectrum Studies

The ^{31}P MAS NMR spectrum for 50LPdW/MCM-41 catalyst is shown in Fig. 3.15. The spectra show two peaks, the main peak at −15.4 ppm and small peaks at −13.5 ppm. From the literature [52], the spectra of LPdW shows two peaks, the major peak at −15.17 ppm (95 %) attributable to $[PPdW_{11}O_{39}]^{5-}$ and a small peak at −13.32 ppm (5 %) attributable to an impurity of the starting material, $[PW_{11}O_{39}]^{7-}$. The 50 LPdW/MCM-41 catalysts with high LPdW content exhibit a sharp resonance at −15.4 ppm,

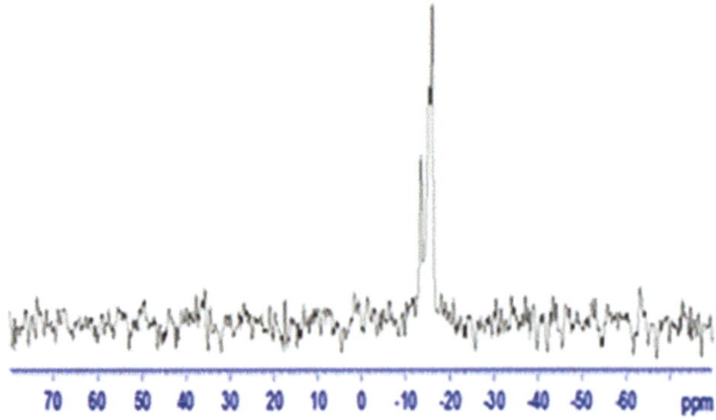

Fig. 3.15 ^{31}P MAS NMR spectra of 50 LPdW/MCM-41

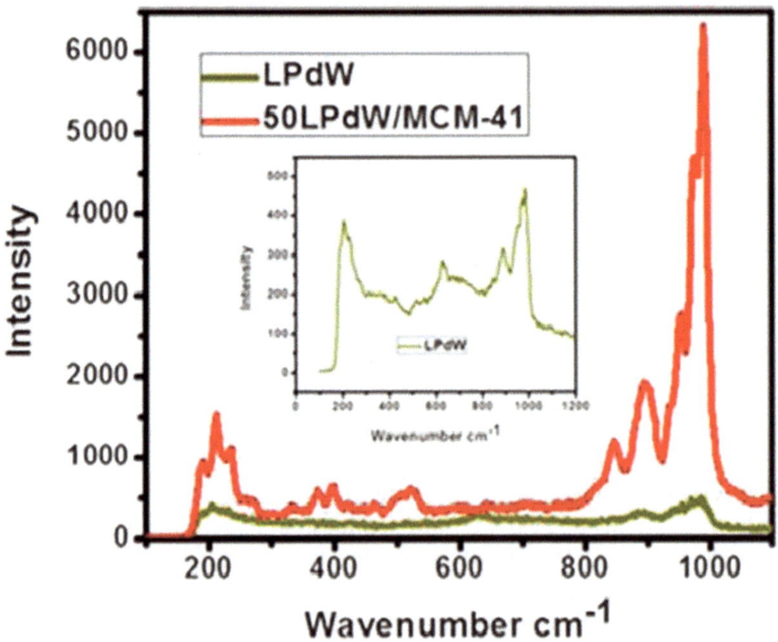

Fig. 3.16 Raman spectroscopy of LPdW and 50LPdW/MCM-41

which is close to that of bulk LPdW. This indicates unambiguously that the Keggin structure is retained even after LPdW is loaded on MCM-41, and it is slightly shifted toward right due to chemical interaction, hydrogen bonding, and covalent bonding between support and LPdW.

Raman Spectroscopy

Figure 3.16 shows the Raman scattering spectroscopy of LPdW and 50 LPdW/MCM-41. The bulk LPdW gives peaks at 984, 888, 847, and 799 cm^{-1} which are

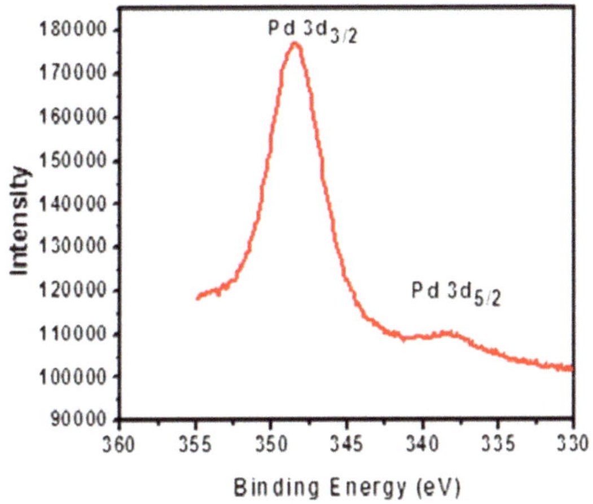

Fig. 3.17 XPS spectrum of Pd $3d_{3/2}$ and Pd $3d_{5/2}$ of 50 LPdW/MCM-41

attributed to the stretching vibrations of P–O, W–Ob–W, W–Oc–W, and W–Ot bonds of metal-modified mono-lacunary Keggin unit, respectively. The 50 LPdW/MCM-41 sample shows all the above-described bands of LPdW, but the intensities of the bands are low and slightly shifted toward higher wave number due to strong interactions between MCM-41 support and lacunary Keggin unit [53].

XPS Spectrum Studies

The incorporation of Pd ions into lacunary phosphotungstate, its oxidation states, and its interaction with the support were confirmed from XPS measurements. The Pd 3d XPS spectrum of 50 LPdW/MCM-41 sample are shown in Fig. 3.17.

From this figure two distinct palladium peaks were observed at binding energies 338.7 and 347.9 eV. The binding energy of about 337.9 eV for the Pd 3d 5/2 peak is reported in the literature [53]. However, shifting of 3d 5/2 peak toward a higher binding energy suggests the interaction of Pd (II) with the support surface [53].

3.6.2 Catalytic Activity Toward Hydrogenation of *p*-Nitrophenol to *p*-Aminophenol

The hydrogenation of *p*-nitrophenol was carried out in a two-necked round-bottom flask with a reflux condenser at atmospheric pressure, which was charged with 1.0 g of *p*-nitrophenol dissolved in 50 ml anhydrous ethanol and 0.05 g of catalyst. The flask was purged with nitrogen for 10 min. The reaction was started by vigorous stirring of the reaction mixture under H_2 gas-flow (10 ml min^{-1}) at room temperature for 1 h. The reaction products were analyzed by off-line gas chromatography (Scheme 3.6).

A green and effective method is reported for the hydrogenation of *p*-nitrophenol to *p*-aminophenol using a Cs salt of Pd-substituted Keggin-type mono-lacunary

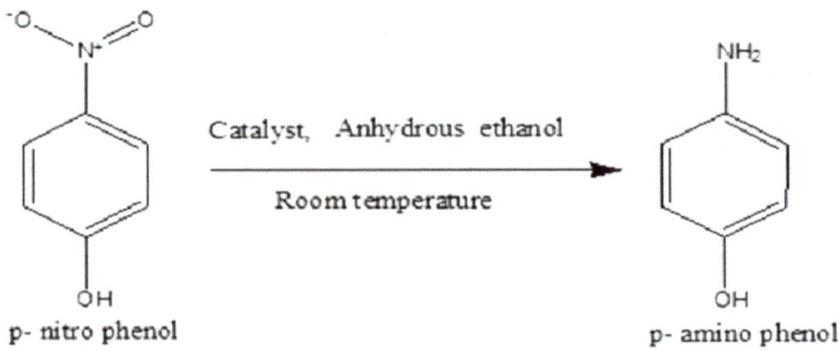

Scheme 3.6 Schematic representation of *hydrogenation of p-nitrophenol to p-aminophenol*

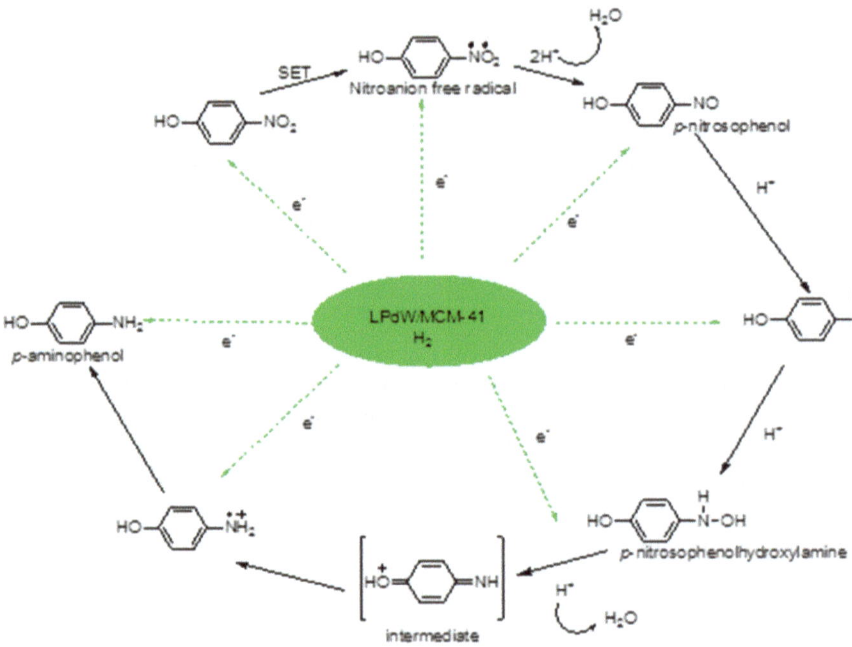

Scheme 3.7 Mechanism of hydrogenation of *p*-nitrophenol to *p*-amino phenol

phosphotungstate-supported mesoporous silica (LPdW/MCM-41). This economical and environmentally friendly method carried out at room temperature gives 99 % conversion and 100 % selectivity.

As shown in Scheme 3.7, a mechanism is proposed for this reaction. The key step in this mechanism involves the single electron transfer (SET) to form a nitro anion free radical. The subsequent step involves the electron transfer and also hydrogen ion transform to form an intermediate. The intermediate then gets converted to *p*-aminophenol.

4 Conclusions

We have successfully synthesized Cs salt of phosphotungstic acid, Cu salt of phosphotungstic acid, Fe-modified lacunary phosphotungstate, and Cs salt of Pd-substituted lacunary phosphotungstate supported on MCM-41 and unambiguously characterized by XRD, UV-vis DRS, FTIR, BET surface area, acid sites by ammonia TPD, and morphology by SEM. The catalytic activity of the transition metal salts of polyoxometalate-promoted MCM-41 was evaluated for various organic transition reactions. Among the catalysts with different Cs-PTA loadings, the 50 wt% Cs-PTA/MCM-41 catalyst gave the highest conversion (98 %) toward acylation of anisole with acetic anhydride. 50 wt% $Cu_{1.5}PW_{12}O_{40}$/MCM-41 catalyst gave the highest conversion (98 %) of phenyl iodide to cinnamic acid (Heck vinylation reaction). The 50 wt% Cs_5 [$PFeW_{11}O_{39}$]/MCM-41 showed remarkable catalytic performance, obtaining mono bromophenol having 95 % conversion with 99 % selectivity and *trans*-stilbene oxide having 52 % conversion with 99 % selectivity. Out of various wt% of the lacunary salt loadings, 50LPdW/MCM-41 was found to be an excellent catalyst toward hydrogenation of *p*-nitrophenol to *p*-aminophenol at room temperature. The catalyst could be reused several times without a significant degradation in catalytic activity. This suggests the commercial exploitation of this polyoxometalate-based heterogeneous catalyst for the synthesis of additional organic target molecules.

Acknowledgments The authors are thankful to Prof. B. K. Mishra, Director, IMMT (CSIR), Bhubaneswar, for his keen interest, encouragement, and kind permission to publish this work. Mr. Surjyakanta Rana is obliged to CSIR for a senior research fellowship. We are thankful to DST for the financial support.

References

1. Tanabe K (1970) Solid acid and bases. Academic, Tokyo
2. Tanabe K, Holderich WF (1999) Appl Catal A 181:399
3. Mizuno N, Misono M (1994) J Mol Catal 86:319
4. Kozhevnikov IV (1987) Russ Chem Rev 56:811
5. Cavani F (1998) Catal Today 41:73
6. Misono M, Okuhara T (1993) CHEMTECH 23:23
7. Kozhevnikov IV, Matveev KI (1983) Appl Catal 5:135
8. Pizzio LR, Caceres CV, Blanco MN (1999) Appl Surf Sci 151:91
9. Kozhevnikov IV (1995) Catal Rev Sci Eng 37:311
10. Misono M (1987) Cat Rev Sci Eng 29:269
11. Moffat JB, Kasztelan S (1988) J Catal 109:206
12. Izumi Y, Ogawa M, Ohara WN, Urabe K (1992) Chem Lett 39:1987
13. Pizzio LR, Caceres CV, Blanco MN (1998) Appl Catal A 167:283
14. Pizzio LR, Vazquez PG, Caceres CV, Blanco MN (2003) Appl Catal A 256:125
15. Knifton JF, Edwards JC (1999) Appl Catal A 183:1
16. Rao PM, Wolfson A, Landau MV, Herskowitza M (2004) Catal Commun 5:327

17. Cardoso LAM, Alves W Jr, Gonzaga ARE, Aguiar LMG, Andrade HMC (2004) J Mol Catal A Chem 209:189
18. Kaur J, Kozhevnikov IV (2002) Chem Commun 21:2508
19. Sarsani VR, Lyon CJ, Hutchenson KW, Harmer MA, Subramaniam B (2007) J Catal 245:184
20. Kamala P, Pandurangan A (2008) Catal Commun 9:2231
21. Luque R, Campelo JM, Luna D, Marinas JM, Romero AA (2005) Microporous Mesoporous Mater 84:11
22. Staiti P, Freni S, Hocevar S (1999) J Power Sources 79:250
23. Okuhara T, Nishimura T, Watanabe H, Misono M (1999) J Mol Catal 74:247
24. (a) van de Water LGA, van der Waal JC, Jansen JC, Maschmeyer T (2004) J Catal 223:170; (b) El Berrichi Z, Cherif L, Orsen O, Fraissard J, Tessonnier JP, Vanhaecke E, Louis B, Ledoux MJ, Pham-Huu C (2006) Appl Catal A Gen 298:194; (c) Selvin R, Hsiu-Ling Hsu, Tze-Min Her (2008) Catal Commun 10:169; (d) Pei Chun Shih, Jung-Hui Wang, Chung-Yuan Moub (2004) Catal Today 93:365; (e) Shih PC, Wang JH, Mou CY (2004) Catal Today 365:93; (f) Yadav GD, Asthana NS, Kamble VS (2003) J Catal 217:88; (g) Parida KM, Mallick S, Pradhan GC,(2009) J Mol Catal A Chem 297:93
25. (a) Corma A, Martınez A, Martınez C,(1996) J Catal 164:422; (b) Pizzio LR, Blanco MN,(2003) Appl Catal A Gen 255:265; (c) Saemin Choi, Yong Wang, Zimin Nie, Jun Liu, Charles HF Peden (2000) Catal Today 55:117; (d) Yadav JS, Subba Reddy BV, Purnima KV, Nagaiah K, Lingaiah N (2008) J Mol Catal A Chem 285:36
26. Liu Y, Xu L, Xu B, Li Z, Jia L, Guo W (2009) J Mol Catal A Chem 297:86
27. Baeza B, Rodriguez J, Ruiz A, Ramos I (2007) Appl Catal A 333:281
28. Hu J, Wang Y, Chen L, Richards R, Yang W, Liu Z, Xu W (2006) Microporous Mesoporous Mater 93:158
29. Patel K, Shringarpure P, Patel A (2011) Trans Met Chem 36:171
30. Pope MT (1983) Heteropoly and isopoly oxometalates. Springer, Berlin
31. Mizuno N, To-oru Hirose, Tateishi M, Iwamoto M (1994) J Mol Catal 88:125
32. Knapp C, Ui T, Nagai K, Mizuno N (2001) Catal Today 71:111
33. Maurya MR, Amit Kumar (2006) J Mol Catal A Chem 250:190
34. Das DP, Parida KM (2006) Appl Catal A Gen 305:32
35. Brunauer S, Deming LS, Deming E, Teller E (1940) J Am Chem Soc 62:1723
36. Parida KM, Rath D (2004) J Mol Catal A Chem 258:381
37. Nowinska K, Waclaw A, Masierak W, Gutsze A (2004) Catal Lett 92:157
38. Narender N, Krishna Mohan KVV, Vinod Reddy R, Srinivasu P, Kulkarni SJ, Raghavan KV (2003) J Mol Catal A Chem 192:73
39. Miyaura N, Suzuki A (1995) Chem Rev 95:2457
40. Gruber M, Chouzier S, Koehler K, Djakovitch L (2004) Appl Catal A Gen 265:161
41. Jin M-J, Taher A, Kang H-J, Choi M, Ryoo R (2009) Green Chem 11:309
42. Choudary BM, Madhi S, Chowdari NS, Kantam ML, Sreedhar B (2002) J Am Chem Soc 124:14127
43. Jamwal N, Gupta M, Paul S (2008) Green Chem 10:999
44. Hagio H, Sugiura M, Kobayashi S (2006) Org Lett 8:375
45. Han W, Liu C, Jin Z (2008) Adv Synth Catal 350:501
46. Mori A, Miyakawa Y, Ohashi E, Haga T, Maegawa T, Sajiki H (2006) Org Lett 8:3279
47. Rode CV, Vaidya MJ, Jaganathan R, Chaudhari RV (2001) Chem Eng Sci 56:1299
48. Vaidya MJ, Kulkarni SM, Chaudhari RV (2003) Org Process Res Dev 7:202
49. Kruk M, Jaroniec M, Ji Man Kim, Ryoo R (1999) Langmuir 15:5279
50. Mehnert CP, Ying JY (1997) Chem Commun 2215
51. Kogan V, Aizenshtat Z, Neumann R (2002) New J Chem 26:272
52. Ali Abdalla ZE, Li B, Tufail A (2009) Colloid Surf A Physicochem Eng Asp 341:86
53. Yang H, Zhang G, Hong X, Zhu Y (2004) J Mol Catal A: Chem 210:143

Chapter 4
Vanadium-Substituted Tungstophosphoric Acid Supported on Titania: A Heterogeneous Catalyst for Selective Oxidative Cleavage of Olefins to Carbonyl Compounds at Room Temperature

N. Lingaiah, K.T. Venkateswara Rao, and P.S. Sai Prasad

Contents

1 Introduction 92
2 Experimental Section 92
  2.1 Preparation of Catalysts 92
  2.2 Preparation of VOTPA Catalyst 93
  2.3 Preparation of Titania-Supported $TPAV_1$ Catalyst 93
3 Characterisation of Catalysts 93
4 General Reaction Procedure 94
5 Results and Discussion 94
6 Conclusions 103
References 103

Abstract A series of vanadium-incorporated tungstophosphoric acid catalysts supported on titania were prepared and characterised by FT-infrared, X-ray diffraction, laser Raman spectroscopy, X-ray photoelectron spectroscopy and temperature-programmed reduction techniques. The characterisation data reveals the incorporation of vanadium into Keggin unit of TPA and its subsequent dispersion on titania. The catalytic activity of these catalysts was evaluated for oxidative cleavage of olefins to corresponding carbonyl compounds using H_2O_2 as oxidant at room temperature. Influence of vanadium presence in primary as well as secondary structure of tungstophosphoric acid on catalytic activity was also studied. Different parameters such as H_2O_2 to substrate mole ratio, solvent effect and temperature on activity and selectivity were also evaluated. The preparation of the catalyst is simple, easily recoverable and reusable many times without loss of activity and selectivity. The catalyst is also active even for oxidative cleavage of aliphatic olefins.

N. Lingaiah (✉) • K.T. Venkateswara Rao • P.S. Sai Prasad
Catalysis Laboratory, I&PC Division, Indian Institute of Chemical Technology, Hyderabad 500 607, India
e-mail: nakkalingaiah@iict.res.in

A. Patel (ed.), *Environmentally Benign Catalysts: For Clean Organic Reactions*,
DOI 10.1007/978-94-007-6710-2_4,

1 Introduction

Oxidation of organic substrates to oxygen-rich compounds is one of the paramount reactions developed in synthetic chemistry. The oxidative functionalisation and cleavage of olefins to corresponding carbonyl compounds are significant transformations in organic synthesis [1]. These compounds are having high commercial value as they have been used as intermediates for many chemical feedstocks, agrochemicals, fragrances, pharmaceuticals and polymers. The classical method for the scission of alkenes is ozonolysis followed by appropriate workup [2–5]. However, its utility is often limited as per safety concerns, and serious accidents have been reported [6, 7]. Alternatively, olefins can also be cleaved by the use of homogeneous catalysts such as $Pd(OAc)_2$ [8], AuCl in combination of TBHP oxidant [9] and $RuCl_3$ [10]. From the economical and environmental point of view, these homogeneous catalysts are not preferable. Therefore, there is a scope to develop heterogeneous catalyst for the oxidative cleavage of carbon–carbon double bonds under more eco-friendly conditions. In this regard, recently, supported Keggin type of heteropoly acid catalysts [11–13], vanadium complexes [14] and strontium ferrite spinel [15] have been reported for the oxidation of styrene. Although these processes are efficient, they have limitations: it takes longer reaction time and requires high reaction temperature, and utility of the catalyst for different olefin substrates are not reported.

From the past decade polyoxometalates especially Keggin type of heteropoly acids (HPAs) were extensively used for the acid and oxidation catalysts for several industrial applications [16–18]. HPAs possess incredibly strong Bronsted acidity and on the other hand appropriate redox properties. Both the acid and redox properties of HPAs can be tuned by varying its chemical composition at molecular levels. An essential drawback of these catalysts is their low thermal stability, low surface area and high solubility in polar solvents. Efforts are being made to overcome these difficulties by supporting the HPAs on various acidic supports or by simply exchange of protons with different metal ions. We have been working on modified Keggin-type heteropoly acids for various oxidation reactions [19–21].

Herein, we demonstrate vanadium-substituted tungstophosphoric acid supported on titania as a heterogeneous catalyst for the selective oxidative cleavage of olefins to carbonyl compounds at room temperature. The scope of the catalyst was also studied for different olefins. The effect off V content and its location in Keggin ion of TPA on the oxidative cleavage of olefins is also one of the aims of the present study.

2 Experimental Section

2.1 Preparation of Catalysts

Vanadium-substituted TPA ($H_4PW_{11}V_1O_{40}$; $TPAV_1$) was prepared according to the similar method reported in the literature [22] using $NaVO_3$, Na_3PO_4 and $Na_2WO_4{\cdot}2H_2O$. Sodium metavanadate ($NaVO_3$, 7.33 mmol) was dissolved in 50 mL of deionised water at 80 °C and mixed with disodium hydrogen phosphate (Na_2HPO_4, 7.3 mmol), which was previously dissolved in 20 mL of water. The mixture was cooled to room temperature. Concentrated H_2SO_4 (5 mL) was then added to give a red solution. Sodium tungstate dihydrate ($Na_2WO_4{\cdot}2H_2O$, 80.63 mmol) was dissolved in 50 mL of distilled water separately and added to the above red solution drop-wise with vigorous stirring followed by slow addition of concentrated H_2SO_4. $TPAV_1$ was obtained with ether extraction followed by evaporation. The yellow solid obtained was dissolved in water and concentrated until crystals appeared. Similarly, $TPAV_2$ and $TPAV_3$ catalysts were also prepared by varying the amounts of starting materials.

2.2 Preparation of VOTPA Catalyst

The catalyst with vanadium exchanged with the protons of TPA (denoted as VOTPA) was prepared by the exchange of the protons of $H_3PW_{12}O_{40}$ (TPA) with $(VO)^{+2}$ ions [23]. Calculated amount of V_2O_5 was dissolved in oxalic acid at 100 °C followed by cooling the solution to room temperature. This solution was added to the aqueous solution of TPA with constant stirring. The excess water was removed on a water bath, and the sample was further dried at 120 °C for 12 h.

2.3 Preparation of Titania-Supported $TPAV_1$ Catalyst

A series of catalysts with 10–25 wt% of $TPAV_1$ supported on TiO_2 were prepared by impregnation method. Calculated amount of $TPAV_1$ was dissolved in deionised water and added to TiO_2. The excess water was removed on a water bath followed by drying at 120 °C for 12 h and finally calcined at 300 °C for 2 h.

3 Characterisation of Catalysts

FT-IR spectra of catalysts were taken on a DIGILAB (USA) IR spectrometer by the KBr disc method. XRD patterns were measured on a RIGAKU MINI FLEX diffractometer using CuKα radiation ($\lambda = 1.54$ Å). The 2-theta angels were scanned from 2° to 60° at a rate of 2° min^{-1}.

The Raman spectra of the samples were collected with a Horiba-Jobin Yvon LabRAM-HR spectrometer equipped with a confocal microscope, 2,400/900 grooves/mm gratings and a notch filter. The visible laser excitation at 532 nm (visible/green) was supplied by a Yag doubled diode pumped laser (20 mW). The scattered photons were directed and focused onto a single-stage monochromator and measured with a UV-sensitive LN2-cooled CCD detector.

Temperature-programmed reduction (TPR) of the catalysts was carried out in a flow of 10 % H_2/Ar mixture gas at a flow rate of 30 mL/min with a temperature ramp of 10 °C/min. Before the H_2–TPR run, the catalysts were pretreated with argon gas at 250 °C for 2 h. The hydrogen consumption was monitored using a thermal conductivity detector.

XPS measurements were conducted on a KRATOS AXIS 165 with a dual-anode (Mg and Al) apparatus using Mg KR anode. The nonmonochromatised Al KR X-ray source (1486.6 eV) was operated at 12.5 kV and 16 mA. Before acquisition of the data, the sample was outgassed for about 3 h at 100 °C under a vacuum of 1.0×10^{-7} Torr to minimise surface contamination. The XPS instrument was calibrated using Au as standard. For energy calibration, the carbon 1s photoelectron line was used. The carbon 1s binding energy was taken as 285 eV. A charge neutralisation of 2 eV was used to balance the charge up of the sample. The spectra were deconvoluted using a Sun Solaris-based Vision-2 curve resolver. The location and the full width at half-maximum (fwhm) value for the species were first determined using the spectrum of the pure sample. Symmetric Gaussian shapes were used in all cases.

4 General Reaction Procedure

In a typical experiment, a 25 mL round bottom flask equipped with magnetic stirrer was charged with styrene (1 mmol), acetonitrile (3 mL), 30 % H_2O_2 (3 mmol) and 20 % TPAV1/TiO_2 (50 mg) catalyst. The resulting mixture was stirred at room temperature for 10 h. After completion of the reaction, the catalyst was separated by filtration and dried over anhydrous sodium sulphate. The products were identified by GC–MS (SHIMADZU-2010) analysis by separating them on a DB-5 column.

5 Results and Discussion

Initially, styrene was taken as model substrate for the oxidative cleavage of olefins and conducted the reaction at room temperature under atmospheric pressure using vanadium containing TPA catalyst. The results are shown in Table 4.1. The main oxidation

Table 4.1 Catalysts screening for oxidation of styrene

			Selectivity (%)[b]	
Entry	Catalyst	Conversion of styrene (%)[b]	Benzaldehyde	Others
1	TPA	–	–	–
2	TPAV	50[a]	79	21
3	$TPAV_1$	70	80	20
4	$TPAV_2$	72	77	22
5	$TPAV_3$	80	62	35

Reaction conditions: styrene (1 mmol), 30 % H_2O_2 (3 mmol), acetonitrile (3 mL), catalyst (50 mg), temp: RT, 10 h

[a]At 80 °C

[b]Conversion and selectivity based on GCMS

Scheme 4.1 Oxidative cleavage of styrene

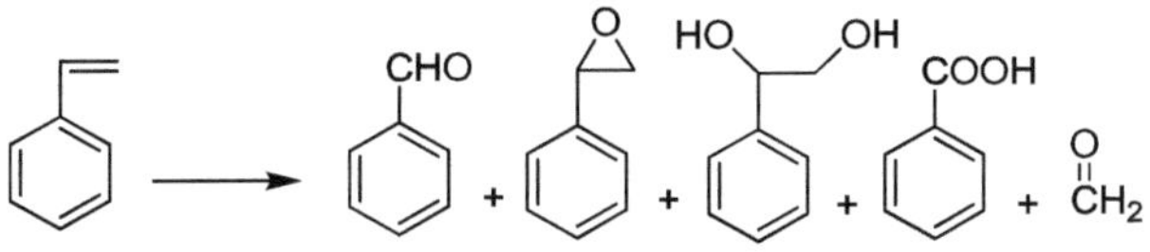

Table 4.2 Effect of $TPAV_1$ loading on TiO_2 support for oxidation of styrene

Catalyst	Conversion of styrene (%)[a]	Selectivity (%)[a]	
		Benzaldehyde	Others
10 wt% $TPAV_1/TiO_2$	65	82	17
15 wt% $TPAV_1/TiO_2$	80	80	20
20 wt% $TPAV_1/TiO_2$	97	85	14
25 wt% $TPAV_1/TiO_2$	85	72	28

Reaction conditions: styrene (1 mmol), 30 % H_2O_2 (3 mmol), acetonitrile (3 mL), catalyst (50 mg), at 30°C, 10 h
[a]Conversion and selectivity based on GCMS

product is benzaldehyde and the others are styrene epoxide, 1-phenylethane-1,2-diol, benzoic acid and formaldehyde (Scheme 4.1). Oxidation of styrene without using catalyst did not yield any product.

TPA itself is not active at room temperature. TPA showed about 50 % conversion with 79 % selectivity towards benzaldehyde when the reaction temperature increased to 80 °C. Monovanadium-substituted TPA ($TPAV_1$) catalyst gave up to 70 % conversion of styrene with 80 % selectivity to benzaldehyde at room temperature. TPA is known for its high acidic character and substitution of tungsten by vanadium in its primary structure results in generation of redox properties [23]. As the more number of tungsten atoms are replaced by vanadium atoms ($TPAV_2$, $TPAV_3$), there was an increase in conversion of styrene, but conversely decrease in selectivity to benzaldehyde was observed.

Location of vanadium in the heteropoly tungstate also has a significant effect on the oxidation ability of the catalyst [23]. Location of vanadium also played a crucial role in the oxidation of styrene. When the vanadium was present in the secondary structure of TPA (VOTPA), the activity of the catalyst is low compared (Table 4.1, entry 6) with the catalysts where the vanadium is present in the primary structure of TPA. It is observed that $TPAV_1$ was efficient catalyst for the oxidation of styrene at room temperature. However, it suffers from homogeneity.

In order to make $TPAV_1$ heterogeneous catalyst, it was supported on titania. The activity of $TPAV_1/TiO_2$ catalysts in the oxidation of styrene as function of $TPAV_1$ loading at room temperature is shown in Table 4.2. TiO_2 itself is not found to be active and its role is to disperse the active component on its surface. The conversion of styrene increased with increase in the content of $TPAV_1$ and attained maximum conversion (97) at a loading of 20 wt%. Further increasing the loading, a marginal decrease in conversion is observed.

In search of the reason for the observed variation in activity with change in the content of $TPAV_1$ on titania, these catalysts are characterised to know their surface and structural characteristics.

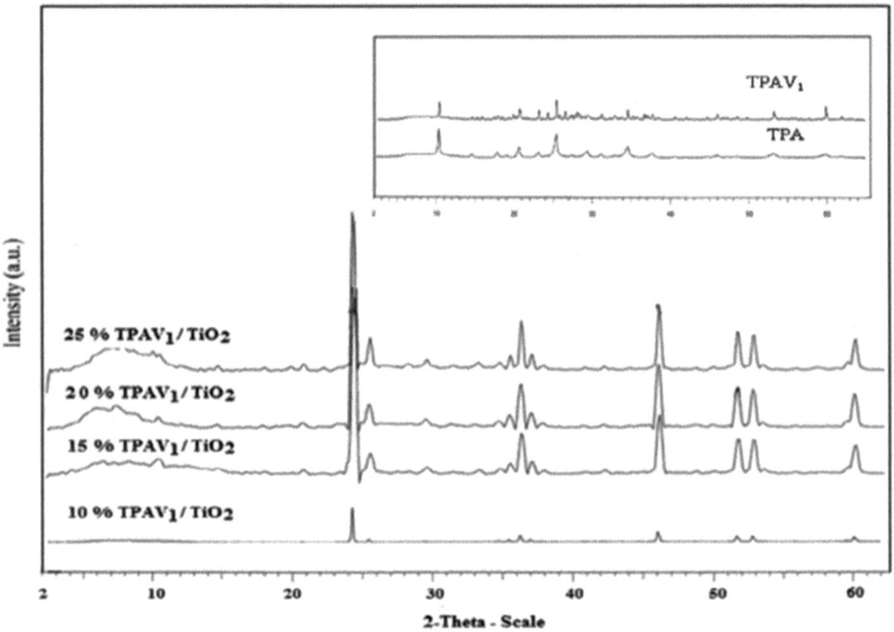

Fig. 4.1 X-ray diffraction patterns of $TPAV_1/TiO_2$ catalysts

The XRD patterns of $TPAV_1/TiO_2$ catalysts are shown in Fig. 4.1. The XRD pattern of bulk TPA and $TPAV_1$ is also shown for the sake of comparison. The characteristic Keggin peaks of pure TPA are obtained at a 2θ of 10.4°, 25.3° and 34.6° [24]. These characteristic lines are present even after incorporation of vanadium into the primary structure of TPA suggesting the presence of intact Keggin structure. The catalysts with low content of $TPAV_1$ did not reveal any diffraction peaks of crystalline $TPAV_1$. They display mainly the patterns of titania support. This might be due to the well dispersion of Keggin units on titania support. The XRD pattern related to Keggin ion is visible for the catalysts with high $TPAV_1$ content (>20 wt%).

The FT-IR patterns of the catalysts are shown in Fig. 4.2. The IR pattern of TPA and $TPAV_1$ is shown in the inset of Fig. 4.2. The TPA exhibited four bands in the region of 1,100–500 cm^{-1} with the main bands at 1,081, 986, 890 and 800 cm^{-1} which are assigned to the stretching vibrations of P–O, W=O_t, W–Oc–W and W–Oe–W, respectively, related to Keggin ion [25]. The IR spectra imply that the Keggin structure was intact after the incorporation of V in the tungsten matrix of 12-tungstophosphoric acid. With the substitution of one V atom for W in the primary structure of the oxoanion ($TPAV_1$ catalyst), the P–O and W=O bands shifted towards lower wave number due to a reduced structural symmetry [23]. This splitting suggests the incorporation of vanadium into Keggin matrix. Therefore, XRD and FT-IR results are in agreement with each other.

Figure 4.3 shows the Raman spectra of the catalysts. The pure TPA exhibited characteristic Raman bands at 1,006, 991 and 905 cm^{-1} related to W=O_t symmetric

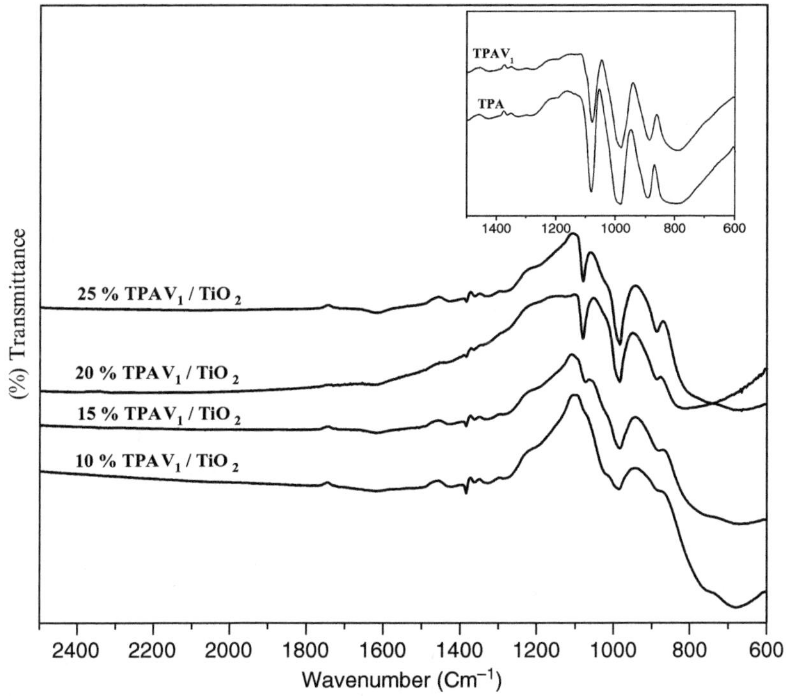

Fig. 4.2 FT-infrared patterns of $TPAV_1/TiO_2$ catalysts

Fig. 4.3 Raman spectra of bulk (*a*) $TPAV_1$ and (*b*) 20 %$TPAV_1/TiO_2$

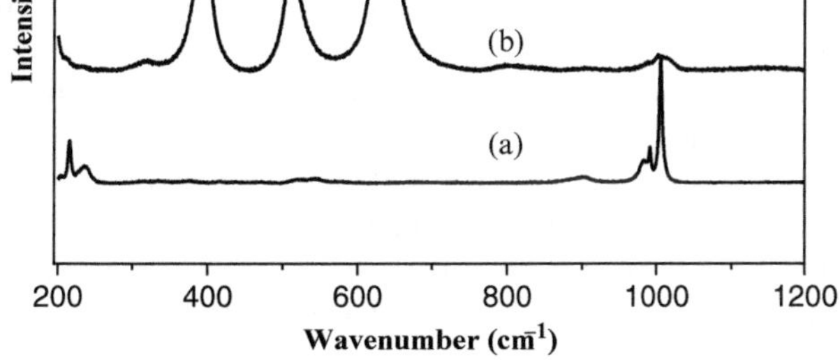

and asymmetric vibrations of Keggin ion [26]. Similar bands were observed after substitution of vanadium in TPA. This shows vanadium incorporation into the Keggin matrix of TPA. The supported catalyst mainly exhibits Raman peaks at 394, 511 and 635 cm^{-1} related to the characteristic peaks of TiO_2 [27]. The titania peaks

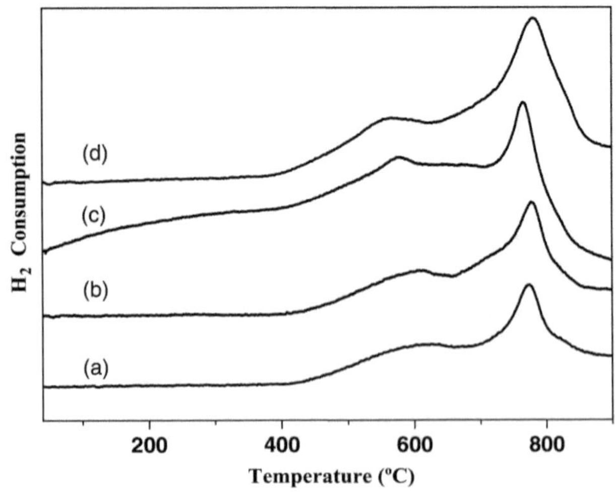

Fig. 4.4 H2–TPR of (*a*) 10 wt% $TPAV_1/TiO_2$, (*b*) 15 wt% $TPAV_1/TiO_2$, (*c*) 20 wt% $TPAV_1/TiO_2$, (*d*) 25 wt% $TPAV_1/TiO_2$

are dominating in the supported catalyst and a small peak at 1,007 cm^{-1} corresponding to the Keggin ion of $TPAV_1$ was observed. This indicates that $TPAV_1$ is well dispersed with retention of Keggin structure on titania support.

The H_2–TPR technique provides important information about the reducibility of individual species, making it a useful tool to characterise the heteropoly acid catalysts. The TPR pattern of 10–25 wt% of $TPAV_1/TiO_2$ catalysts is depicted in Fig. 4.4. Two types of reduction peaks were observed in all catalysts. Generally heteropoly acids start decomposition above 500 °C and the reduction peaks endorsed to the reduction of metal oxides generated during H_2–TPR experiment. The first reduction peak observed at around 400–700 °C centred at 603 °C corresponds to reduction of segregated vanadium oxide species [21, 28]. L. Chen et al. observed that the TPR profile of V_2O_5–WO_3 supported on titania the reduction peak at around 520 °C mentioned the reduction of V (V)–V (III) [2]. With the increase of loading from 10 to 25 wt%, the hydrogen consumption of vanadium increased and reduction peak slightly shifted to lower temperature. The high temperature reduction peak originated due to the reduction of product WO_3 generated by the decomposition of $TPAV_1$. As previously mentioned [29] it is speculated that the high temperature reduction peak corresponds to the reduction of W (VI) to W (0).

XPS is used to investigate the binding energies of the surface elements provides the information about the chemical state of individual elements. Figure 4.5 shows the XPS measurement of the highly active 20 wt% $TPAV_1/TiO_2$ catalyst. The binding energy of P 2p in core level at 133.9 eV. The value of the position of P 2p photo peak confirms that phosphorous is phosphate [30]. The BEs for the Ti 2p core level are observed at 458.2 and 464 eV corresponding to Ti $2p_{3/2}$ and Ti $2p_{1/2}$, suggesting the oxidation state of titanium is Ti^{+4} [31]. Two well-resolved peaks of the O 1s core level perceived at 530.1 and 531.5 eV. The binding energy of the O 1s peak at 530.1 eV is assigned to TiO_2. The second peak related to O 1s of W–O–W [32]. The measured binding energies of W 4f showed four photo electron peaks at 34.2, 35.5,

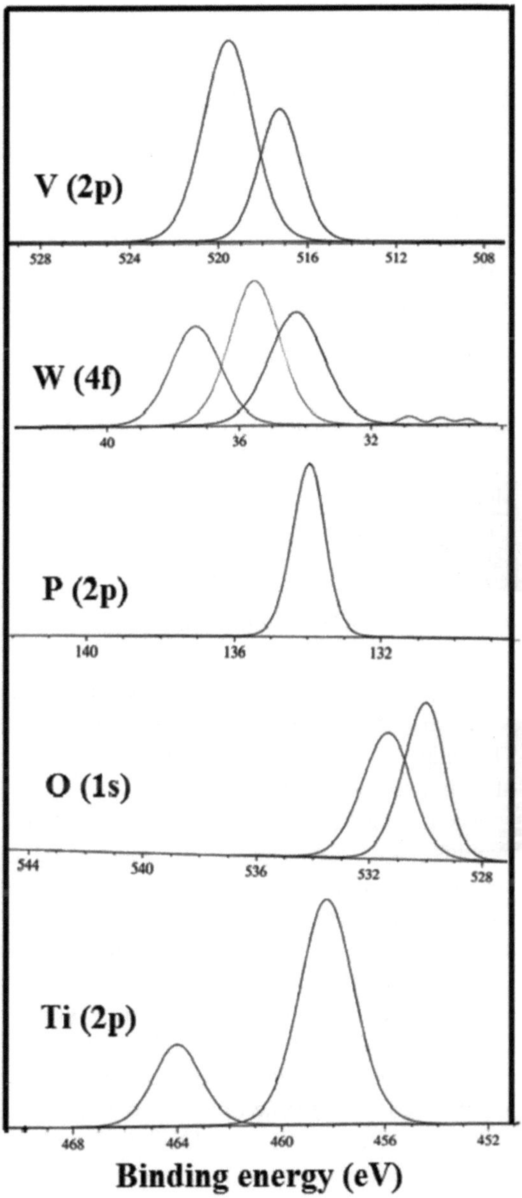

Fig. 4.5 XPS spectra of 20 % $TPAV_1/TiO_2$

36.6 and 37.3 eV. The binding energy values at 35.5 and 36.6 eV assigned to W $4f_{7/2}$ and W $4f_{5/2}$, respectively. These are related to characteristic band of Keggin ion due to spin–orbit splitting. These values are in agreement with literature data [33]. It has been reported that binding energy values at 34.2 and 37.3 related to the presence of water on the surface [32]. The BE of the V $2p_{3/2}$ core level at 517.2 eV is characteristic of V^{+5} ions [34].

Table 4.3 Effect of solvent on oxidative cleavage of styrene

		Selectivity (%)[a]	
Solvent	Conversion of styrene (%)[a]	Benzaldehyde	Others
Methanol	75	25	75
DCE	27	98	2
DCM	35	93	7
Acetonitrile	97	85	15
Toluene	–	–	–
DMF	–	–	–

Reaction conditions: styrene (1 mmol), 30 % H_2O_2 (3 mmol), solvent (3 mL), catalyst (50 mg), at 30°C, 10 h
[a]Conversion and selectivity based on GCMS

Table 4.4 Effect of mole ratio of styrene to H_2O_2

		Selectivity (%)[a]	
Mole ratio	Conversion of styrene (%)[a]	Benzaldehyde	Others
1:1	54	95	5
1:2	73	86	14
1:3	97	85	15

[a]Conversion and selectivity based on GCMS

The characterisation results support the intact Keggin ion of $TPAV_1/TiO_2$. $TPAV_1$ is well dispersed on titania and the catalyst with 20 wt% showed optimum dispersion. The catalyst with 20 % $TPAV_1/TiO_2$ showed high activity mainly due to the presence of well dispersed $TPAV_1$.

The most active 20 wt% TPAV1/TiO_2 catalyst was further studied to evaluate the reaction parameters. The reaction was carried using different solvents and the results are shown in Table 4.3. The reaction was found to proceed more selectively in polar solvents and ineffective in nonpolar solvents. In the absence of solvent, the reaction did not proceed. Acetonitrile was found to be solvent of choice for this reaction.

The effect of styrene to H_2O_2 mol ratio on activity was studied and the results are shown in Table 4.4. The reaction was carried out by varying mole ratio of styrene to H_2O_2 from 1:1 to 1:3. The increase in concentration of H_2O_2 conversion of styrene and selectivity to benzaldehyde was increased. Almost quantitative yields are obtained with styrene to H_2O_2 mol ratio of 1:3. About 97 % conversion of styrene and 85 % selectivity to benzaldehyde were obtained at this molar ratio.

The effect of reaction temperature on oxidative cleavage of styrene was also studied. The best results were obtained when the reaction was carried at room temperature. With the increase in reaction temperature, the conversion of styrene did not vary considerably, and at the same time, the selectivity to benzaldehyde was decreased. This is due to the further oxidation of benzaldehyde to benzoic acid, a stable oxidation product.

The scope of the reaction for various styrene derivatives is explored with the optimised reaction conditions, and the results are presented in Table 4.5. The reaction proceeds smoothly with electron donating and withdrawing groups on aromatic ring

Table 4.5 Oxidative cleavage of various styrene derivatives and aliphatic olefins

Entry	Substrate	Conversion (%)[a]	Product	Selectivity (%)[a]
1	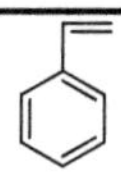	97	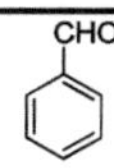	85
2	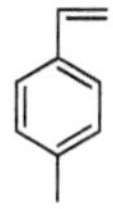	99	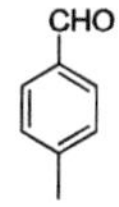	87
3	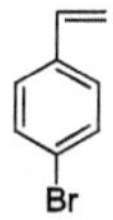	95	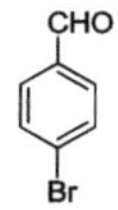	82
4	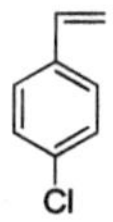	96	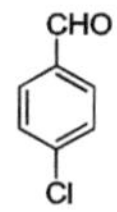	88
5	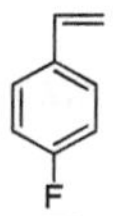	85	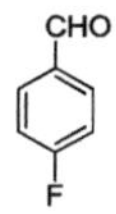	82
6	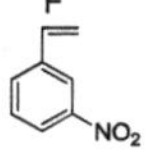	90	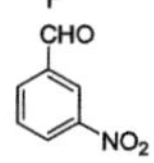	79
7	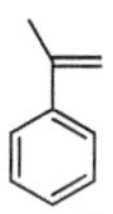	92	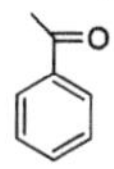	81
8	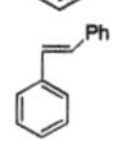	88	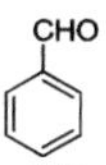	89
9	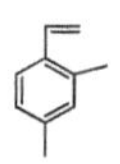	96	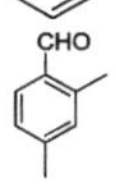	85
10	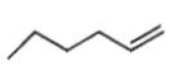	98	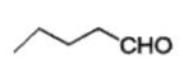	88
11	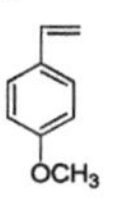	30	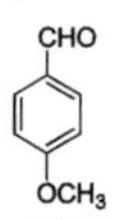	98
12	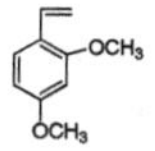	20	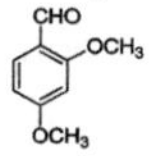	95
13	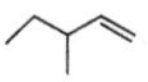	23	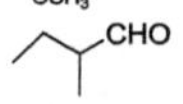	89

Reaction conditions: styrene (1 mmol), 30 % H_2O_2 (3 mmol), acetonitrile (3 mL), catalyst (50 mg), at 30°C, 10 h

[a]Conversion and selectivity based on GCMS

(entries 1–9). With α-methyl styrene (entry 10), 98 % conversion and 97 % selectivity towards acetophenone were obtained. The reaction with cinnamaldehyde (entry 12) resulted about 50 % conversion with 96 % selectivity to benzaldehyde. When stilbene (entry 13) was used, the activity of the catalyst is low with only 30 %

Table 4.6 Recycling results of 20 % $TPAV_1/TiO_2$ catalyst

Number of cycles	Conversion (%)	Selectivity (%)	Catalyst recovery (%)
1	95	85	96
2	92	86	92
3	92	85	90
4	85	90	85

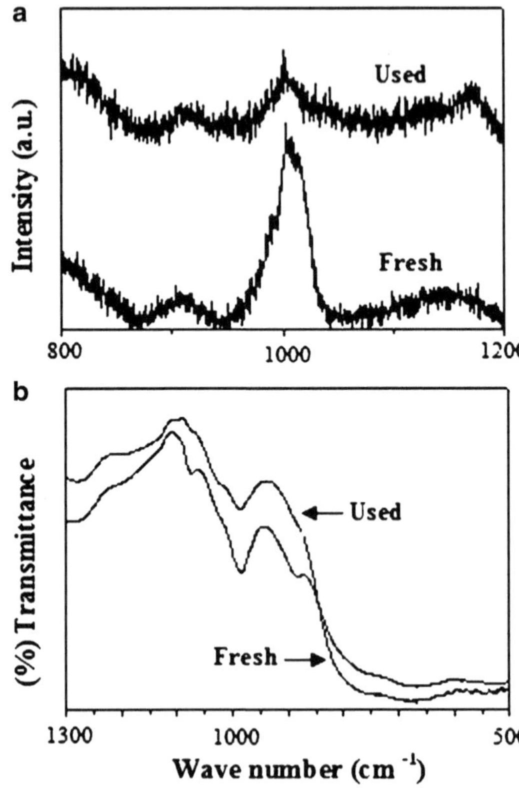

Fig. 4.6 (**a**) Raman spectra and (**b**) FT-IR pattern of fresh and used catalysts

conversion and 96 % selectivity to benzaldehyde. The low activity for this olefin might be due to its steric hindrance. The present catalyst is also active for oxidative cleavage of aliphatic olefins. Aliphatic olefin 1-hexene gave 20 % conversion and 95 % selectivity to pentanal, whereas 3-methyl-1-pentene afforded 20 % conversion with 90 % selectivity to 2-methylbunal.

The catalyst was studied for its reusability and results obtained by recycling the catalyst are summarised in Table 4.6. Even after the third cycle, the catalyst showed almost consistent activity and selectivity. The FT-IR and Raman spectra of used catalyst are shown in Fig. 4.6. The characterisation results suggest the presence of intact Keggin ion structure of $TPAV_1$ on titania for the used catalyst. The recycling results reiterate the heterogeneous nature of the catalysts for the oxidative cleavage of olefins. A separate experiment was carried to prove the heterogeneous nature of the catalyst. A styrene conversion of 50 % with about 87 % selectivity to benzaldehyde was obtained for a 4 h course of the reaction in the presence of 20 wt% $TPAV_1/TiO_2$. After quick removal of the catalyst by filtration, the reaction was further carried out for 4 h under the same conditions. There was no appreciable increase in conversion (50 %) and selectivity (86.5 %). This result reveals that the oxidation of styrene took place over the catalyst.

6 Conclusions

In conclusion a simple and highly efficient catalyst for the selective oxidative cleavage of olefins to carbonyl compounds at room temperature was demonstrated. The activity of the catalyst depends on vanadium present in the primary structure of Keggin ion rather than secondary structure. The activity also depends on the dispersed amount of $TPAV_1$ on support. This catalyst is effective for oxidation of various styrene derivatives. The present catalyst is inexpensive, noncorrosive, environmentally friendly and easy to recover and recyclable without loss of activity and selectivity.

Acknowledgement One of the authors, KTVR, thanks the Council of Scientific and Industrial Research (CSIR), India, for the financial support in the form of Senior Research Fellowship.

References

1. Kuhn FE, Fischer RW, Herrmann WA, Weskamp T (2004) In: Beller M, Bolm C (eds) Transition metals for organic synthesis. Wiley-VCH, Weinheim
2. Hudlicky M (1990) Oxidations in organic chemistry, American Chemical Society Monograph 186. American Chemical Society, Washington, DC
3. Larock RC (1999) Comprehensive organic transformations, 2nd edn. Wiley–VCH, New York
4. Bailey PS (1982) Ozonation in organic chemistry. Academic, New York
5. Criegee R (1975) Angew Chem Int Ed Engl 14:745–752
6. Koike K, Inoue G, Fukuda T (1999) J Chem Eng Jpn 32:295–299
7. Ogle RA, Schumacher JL (1998) Process Saf Prog 17:127–133
8. Wang A, Jiang H (2010) J Org Chem 75:2321–2326
9. Xing D, Guan B, Cai G, Fang Z, Yang L, Shi Z (2006) Org Lett 8:693–696
10. Yang D, Zang C (2001) J Org Chem 66:4814–4818
11. Sharma P, Patel A (2009) J Mol Catal A Chem 299:37–43
12. Pathan S, Patel A (2011) Dalton Trans 40:348–355
13. Hu J, Li K, Li W, Ma F, Guo Y (2009) Appl Catal A Gen 364:211–220
14. Maurya MR, Bisht M, Avecilla F (2011) J Mol Catal A Chem 344:18–27
15. Paradeshi SK, Pawar RY (2011) J Mol Catal A Chem 334:35–43
16. Kozhevnikov IV (1995) Catal Rev Sci Eng 37:311–352
17. Lim SS, Kim YH, Park GI, Lee WY, Song IK, Youn HK (1999) Catal Lett 60:199–204
18. Mori H, Mizuno N, Misono M (1990) J Catal 131:133–142

19. Nagaraju P, Pasha N, Sai Prasad PS, Lingaiah N (2007) Green Chem 9:1126–1129
20. Venkateswara Rao KT, Rao PSN, Nagaraju P, Sai Prasad PS, Lingaiah N (2009) J Mol Catal A Chem 303:84–89
21. Lingaiah N, Mohan Reddy K, Nagaraju P, Sai Prasad PS, Wachs IE (2008) J Phys Chem C 112:8294–8300
22. Tsigdinos GA, Hallada CJ (1968) Inorg Chem 7:437–441
23. Lingaiah N, Molinari JE, Wachs IE (2009) J Am Chem Soc 131:15544–15554
24. Dias JA, Osegovic JP, Drago RS (1998) J Catal 183:83–90
25. Thouvenot R, Fournier M, Franck R, Rocchiccioli - Deltcheff C (1984) Inorg Chem 23:598–605
26. Rocchiccioli-deltcheff C, Fournier M, Franck R, Thouvenot R (1983) Inorg Chem 22:207–216
27. Mallick K, Witcomb MJ, Scurrell MS (2004) Appl Catal A Gen 259:163–168
28. Kompio PGWA, Brückner A, Hipler F, Auer G, Löffler E, Grünert W (2012) J Catal 286:237–247. doi: 10.1016/j.jcat.2011.11.008
29. Chena L, Li J, Gea M (2011) Chem Eng J 170:531–537
30. Damyanova S, Fierro JLG, Sobrados I, Sanz J (1990) Langmuir 15:469–476
31. Damyanova S, Fierro J (1998) Chem Mater 10:871–879
32. Jalil PA, Faiz M, Tabet N, Hamdan NM, Hussain Z (2003) J Catal 217:292–297
33. Xu L, Li W, Hu J, Li K, Yang X, Ma F, Guo Y, Yu X, Guo Y (2009) J Mater Chem 19:8571–8579
34. Delichere P, Bere EK, Abon M (1998) Appl Catal A 172:295–309

Chapter 5
Supported Heteropoly Acids and Multicomponent Polyoxometalates as Eco-Friendly Solid Catalysts for Bulk and Fine Chemicals Synthesis

G.V. Shanbhag, Ankur Bordoloi, Suman Sahoo, B.M. Devassy, and S.B. Halligudi

Contents

1 Introduction 106
2 Part 1: Zirconia-Supported Heteropoly Acids as Strong Solid Acid Catalysts 108
2.1 Experimental 108
2.2 Results and Discussion 108
3 Part 2: Multicomponent Polyoxometalates Immobilized on Solid Supports as Eco-Friendly Catalysts in Oxidation Reactions 119
3.1 Results and Discussion 120
4 Conclusions 136
References 137

Abstract Supported heteropoly acids are an important class of eco-friendly solid catalysts which offer strong acidity or redox properties. Part 1 of this chapter gives the detailed description of Keggin heteropoly acids such as silicotungstic acid (STA) and phosphotungstic acid (PTA) supported on zirconia as thermally stable, reusable solid catalysts which can be used for variety of organic transformations such as alkylation, acylation, and allylation. The catalysts are characterized by various techniques such as X-ray diffraction, N_2 sorption measurements, DTG–DTA, UV–Vis spectroscopy, FTIR pyridine adsorption, NH_3-TPD, FT-Raman, and ^{31}P MAS NMR. These supported HPA catalysts found to be highly active and selective with a long catalytic

G.V. Shanbhag
Catalysis Division, National Chemical Laboratory, Pashan Road, Pune 411008, India

Poornaprajna Institute of Scientific Research, Bidalur, Ta Devanahalli, Bangalore 562110, India

A. Bordoloi • S. Sahoo • B.M. Devassy
Catalysis Division, National Chemical Laboratory, Pashan Road, Pune 411008, India

S.B. Halligudi (✉)
Centre of Excellence GIDC, Vapi 396 195, Gujarat, India
e-mail: sb_halligudi123@yahoo.co.in

A. Patel (ed.), *Environmentally Benign Catalysts: For Clean Organic Reactions*, DOI 10.1007/978-94-007-6710-2_5, © Panch Tattva Publishers, Pune, India 2013

lifetime in the discussed reactions. Part 2 of this chapter describes the preparation of inorganic–organic hybrid materials by immobilization of molybdo-vanadophosphoric acids onto mesoporous silicas such as MCM-41, MCM-48, and SBA-15. The study has been further extended to mesoporous carbon and ethane-bridged SBA-15. All catalyst materials were characterized by elemental analysis, FTIR, N_2 physisorption measurements, XRD, UV/Vis, XPS, CP-MAS-NMR, SEM, and TEM for their structural integrity and physicochemical properties. These materials were applied for various selective and controlled oxidation processes to develop environmentally benign protocols for synthesis of fine chemicals and tried to study their mechanisms. Further, a simple cation exchanged form of $H_5[PMo_{10}V_2O_{40}]\cdot 32.5H_2O$ supported on ionic liquid-modified SBA-15 (V2ILSBA) and its application in catalyzed aerobic oxidation of primary and secondary alcohols to corresponding aldehydes and ketones with no trace of over oxidation has been discussed.

1 Introduction

Heteropoly acids (HPAs) are an exclusive class of materials, active both as redox and acid catalysts [1, 2]. These polyoxometalates are made up of large heteropoly anions having metal–oxygen octahedra as the basic structural unit. Based on the primary structure, HPAs are classified into Keggin, Wells–Dawson, and Anderson–Evans, etc. Among them, Keggin-type HPAs are most important in catalysis due to their high thermal stability, high acid strength, redox properties, and easiness of preparation. They are strong Brönsted acid catalysts and strength of acidity is higher than those of conventional solid acids like zeolites and mixed oxides.

HPAs catalyze a wide range of reactions in homogeneous liquid phase and have stronger acidity than mineral acids. Further, HPA catalysis avoids side reactions such as sulfonation, chlorination, and nitration, which occur with mineral acids. Also, as stable, relatively nontoxic crystalline materials, HPAs are preferable with respect to safety and ease of handling [3].

The major common problem, however, with homogeneous catalysis is the difficulty in catalyst recovery and recycling which minimizes their applications. Since HPAs are expensive than mineral acids, the recycling of HPA catalysts is the key issue to their application [3]. HPAs can be used either directly as a bulk material or in supported form as heterogeneous catalysts which can solve the above mentioned problems. The catalyst in supported form is preferable because of its high surface area compared to the bulk material (5–8 $m^2\ g^{-1}$) and better accessibility of reactants to the active sites. Acidic or neutral solids, which interact weakly with HPAs such as silica, active carbon, and acidic ion-exchange resin, have been reported to be suitable as HPA supports [4]. However, heteropoly acids on these conventional supports are not highly stable in polar reaction media, and part of the reaction occurs due to homogeneous catalysis. Serious problems associated with this type of materials are their susceptibility to deactivation during organic reactions due to the formation of carbonaceous deposit (coke) on the catalyst surface. The thermal stability of HPAs is not high enough to carry out conventional regeneration by burning coke at 500–550 °C, as routinely used in the case of zeolites and aluminosilicates [2]. Thus, the preparation of an active and stable HPA in supported form is essential in order to utilize fully the potential of these materials as catalysts.

In recent years, zirconia is attracting much attention as both a catalyst and catalyst support because of its high stability, amphoteric character, and redox properties [5, 6]. Extensive studies have been carried out on zirconia modified by 4 anions like SO^{2-} and WO_4^{2-}, which act as efficient solid acid catalysts [7–10]. But relatively few works are reported on zirconia modified by heteropolytungstates as catalyst in bulk and fine chemical synthesis [11, 12]. The interaction of heteropoly acid with zirconia support is an important aspect of this catalyst system. Further the activity can be greatly enhanced by dispersing zirconia-supported HPAs on mesoporous silica support such as SBA-15.

Polyoxometalates (POMs) supported directly on mesoporous materials by immobilization and encapsulation methods are the approaches to design and fabricate catalyst systems that give nearly 100 % selectivity to the desired product without sacrificing activity and while using less energy [13]. In constrained environments, the active POMs lose some of the degrees of freedom that they had in the bulk state, adopt a particular geometry, hook onto the functional groups available on the support surfaces, change their coordination sphere geometry, and relax or restrict their sphere of influence, depending on whether or not they reside inside the channels of the mesoporous supports. Thus, they exhibit improved reactivity so as to promote the reaction in sterically controlled pathways [14].

Recent reports have described the immobilization of POMs and transition metal-substituted POMs on various supports, including silica, carbon, mesoporous silicas like MCM-41, and SBA-15, through an organic linker, and explored their use in various academically and industrially important organic transformations [15]. Molybdenum-based heteropoly acids are better catalysts for oxidation reactions than their tungsten counterparts. The activity of these catalysts can be improved by partial substitution of vanadium for molybdenum in the Keggin structure, thus changing the stoichiometric formula from $H_3PMo_{12}O_{40}$ to $H_4PMo_{11}VO_{40}$. The higher number of charge balancing protons associated with the V-substituted material is due to the difference in formal charge between a molybdenum ion (+6) and a vanadium ion (+5). Heteropoly acids having a second vanadium atom ($H_5PMo_{10}V_2O_{40}$) are even more active in oxidation chemistry [16].

Kholdeeva et al. has reported the recent studies in the immobilization of polyoxometalate using different techniques, such as embedding into a silica matrix by sol–gel method, irreversible adsorption on active carbon, electrostatic attachment to NH_2-modified mesoporous silica, and incorporation within nanocages of the metal- organic framework MIL-101 [17]. Their catalytic performances in liquid-phase selective oxidations are compared, with special attention to catalyst stability and recyclability.

This chapter is divided into two parts. In Part 1 of this chapter, we discussed the supported heteropoly acids and their application in acid-catalyzed reactions. In Part 2, immobilization of multicomponent polyoxometalate onto solid supports using different strategies and study of some model oxidation reactions using the supported catalysts is discussed.

2 Part 1: Zirconia-Supported Heteropoly Acids as Strong Solid Acid Catalysts

2.1 *Experimental*

2.1.1 Catalyst Preparation

Zirconium oxyhydroxide was first prepared by the hydrolysis of 0.5 M zirconyl oxychloride aqueous solution by the dropwise addition of aqueous ammonia (10 M) to a final pH of 10. The precipitate was then filtered and washed with ammoniacal water (pH = 8) until free from chloride ions by the silver nitrate test. Zirconium oxyhydroxide thus obtained was dried at 120 °C for 12 h, powdered well, and dried in an oven. The catalysts were prepared by suspending a known amount of phosphotungstic acid (PTA). Each time, 4 ml of methanol per gram of solid support was used, and the mixture was stirred in a rotary evaporator for 8–10 h. After stirring, the excess methanol was removed at ca. 50 °C under vacuum. The resulting solid materials were dried at 120 °C for 24 h and ground well. A series of catalysts with different HPA loading were prepared by changing the HPA concentration in methanol. The dried samples were then calcined at specific temperature in air [18].

2.1.2 Catalyst Characterization

The catalysts were characterized by X-ray diffraction, N_2 sorption measurements, DTG–DTA, UV–Vis spectroscopy, FTIR pyridine adsorption, NH_3-TPD, FT-Raman spectroscopy, and ^{31}P MAS NMR spectroscopy measurements [19].

2.1.3 Catalytic Activity Measurements

Typically all the liquid-phase reactions are carried out either in glass batch reactor or 50-ml Parr autoclave. The continuous reactions are carried out in down-flow fixed-bed quartz reactor.

2.2 *Results and Discussion*

2.2.1 Physicochemical Characterization

X-Ray Diffraction

The bulk structure of pure ZrO_2 and of supported STA catalysts was analyzed by powder X-ray diffraction (Fig. 5.1). The support showed amorphous behavior up to 350 °C and crystallized to a mixture of monoclinic and tetragonal phases, and the

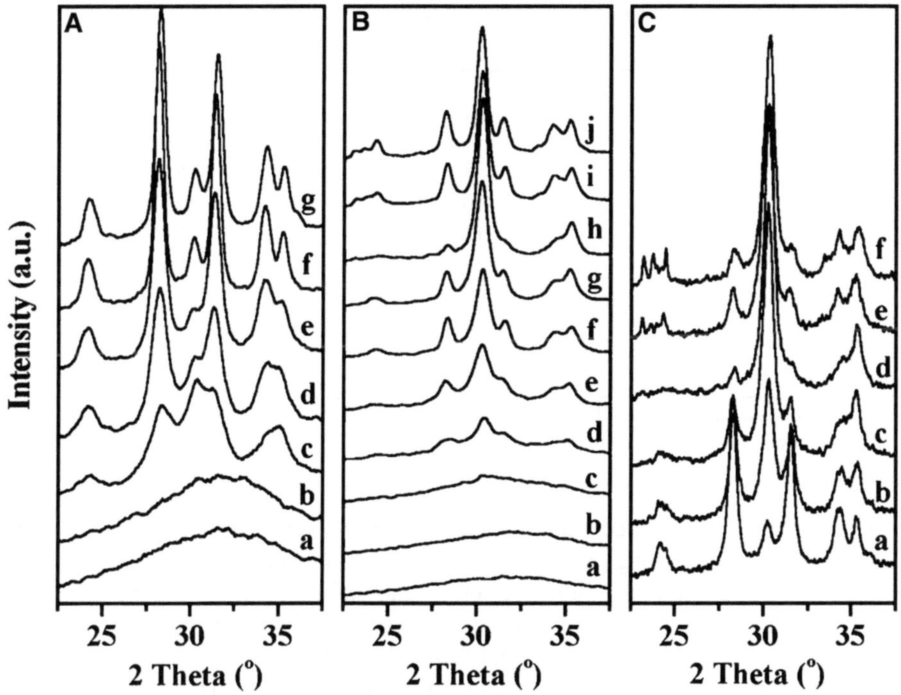

Fig. 5.1 X-ray diffractograms of (**A**) ZrO_2 calcined at (*a*) 120, (*b*) 250, (*c*) 350, (*d*) 450, (*e*) 550, (*f*) 650, and (*g*) 750 °C; (**B**) 15 SZ calcined at different temperatures: (*a*) 120, (*b*) 250, (*c*) 350, (*d*) 450, (*e*) 550, (*f*) 650, (*g*) 700, (*h*) 750, (*i*) 800, and (*j*) 850 °C; (**C**) catalysts with different STA loading calcined at 750 °C (*a*) 0, (*b*) 5, (*c*) 10, (*d*) 15, (*e*) 20, and (*f*) 25 %

intensity of monoclinic phase increased with the increase in calcination temperature. The XRD pattern of the catalysts with different STA loading calcined at 750 °C showed the presence of STA which strongly influences the crystallization of zirconium oxyhydroxide into zirconia. Pure zirconia calcined at 750 °C was mainly monoclinic with only a small amount of tetragonal phase. For catalysts with low STA loading calcined at 750 °C, the XRD pattern could be described as the sum of the monoclinic and tetragonal phases of zirconia; this latter phase became dominant for catalyst with 15 % STA. As shown in Fig. 5.1, 15 % catalyst was amorphous up to 450 °C and as the calcination temperature increased, zirconia crystallizes progressively to tetragonal phase and at 750 °C, the catalyst exists mainly in tetragonal phase. Above 750 °C, formation of monoclinic phase of zirconia was observed [20]. Similar observations were made for ZrO_2-supported PTA.

N_2 Sorption Studies

Pure zirconium oxyhydroxide dried at 120 °C had a surface area of 332 $m^2\ g^{-1}$, and after calcination at 750 °C, the surface area was decreased to 16 $m^2\ g^{-1}$. Addition of heteropoly acid (e.g., silicotungstic acid) to the support resulted in an increase of the

Table 5.1 Surface area, surface density, and acidity of various catalysts

Catalyst	Surface area ($m^2 g^{-1}$)	Surface density ($W nm^{-2}$)	[a]Acidity ($NH_3 nm^{-2}$)
Z-750	16	0	[b]n. e
5 SZ-750	43	3.1	2.63
10 SZ-750	50	5.4	2.77
15 SZ-750	55	7.4	2.81
20 SZ-750	54	10	2.71
25 SZ-750	52	13.1	2.65
15 SZ-600	125	3.2	n. e.
15 SZ-650	108	3.7	1.93
15 SZ-700	80	5.0	2.43
15 SZ-800	46	8.7	2.61
15 SZ-850	36	11.4	n. e.

[a]Acidity values obtained from NH_3-TPD
[b]Not evaluated

surface area, which reached maximum at 15 % HPA loading (Table 5.1). This could be explained as the strong interaction of HPA with the support reduces the surface diffusion of zirconia and inhibits sintering and stabilizes the tetragonal phase of zirconia, which leads to an increase in surface area. Above 15 % STA loading, surface area does not change appreciably, which could be due to the formation of crystalline WO_3, and this probably narrows or plugs the pores of the samples.

The nominal WO_3 loading corresponding to different STA loading and surface area of the catalysts with different STA loading and calcination temperature was determined to calculate the nominal tungsten (W) surface density. The tungsten surface densities were obtained by the equation W surface density = {[WO_3 loading (wt%)/100] × 6.023 × 1,023}/[231.8 (formula weight of WO_3) × BET surface area ($m^2 g^{-1}$) × 10^{18}] and are presented in Table 5.1. The values show that an increase of STA loading results in an increase of W surface density. The specific surface area of SZ catalyst also depends on the calcination temperature. The W surface density increased with the calcination temperature because of the concomitant decrease in the ZrO_2 surface area (Table 5.1).

TPD of Ammonia

Ammonia adsorption–desorption technique is used for the determination of the strength of acid sites present on the catalyst surface together with total acidity. The NH_3-TPD profiles of the catalysts with different STA loading and of 15 % STA on ZrO_2 (denoted as 15 SZ) catalyst calcined at different temperatures are shown in Fig. 5.2, and the amount of NH_3 desorbed per nm^2 are presented in Table 5.1. All samples which showed a broad TPD profile revealed that the surface acid strength has been widely distributed. It is seen from Table 5.1 that the acidity increased up to 15 % loading and thereafter decreased with further loading. For 15 SZ catalysts calcined at different temperatures, the amount of desorbed ammonia increased with

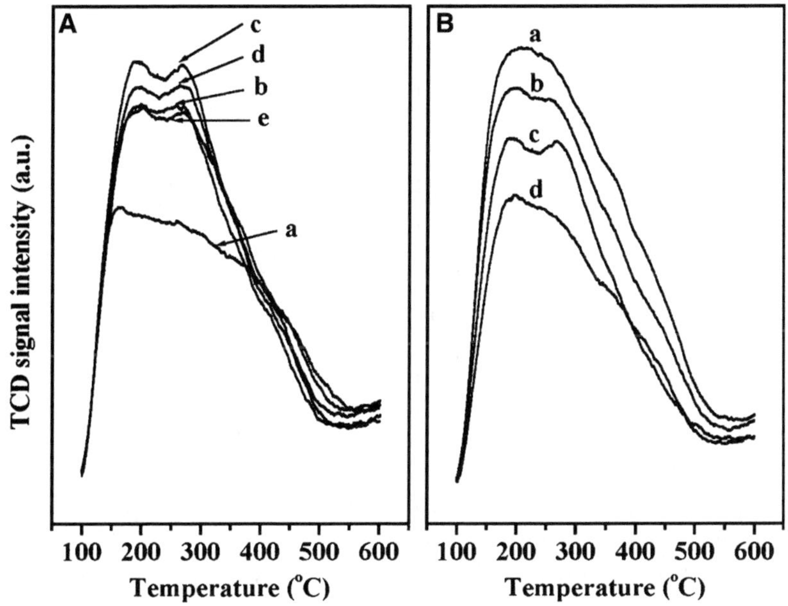

Fig. 5.2 NH_3-TPD profile of (**A**) catalyst with different STA loading calcined at 750 °C (*a*) 5, (*b*) 10, (*c*) 15, (*d*) 20, and (*e*) 25 % and (**B**) 15 SZ calcined at different temperatures (*a*) 650, (*b*) 700, (*c*) 750, and (*d*) 800 °C

calcination temperature and reached maximum at 750 °C [20]. The total acidity of 15 % STA/ZrO_2 was higher than 15 % PTA/ZrO_2 [19].

FTIR Pyridine Adsorption

The use of IR spectroscopy to detect the adsorbed pyridine enables to distinguish different acid sites. The FTIR pyridine adsorption spectra of catalysts with different STA loading calcined at 750 °C and 15 SZ catalyst calcined at different temperatures are shown in Fig. 5.3. The spectra showed sharp pyridine absorption bands at 1,604, 1,485, 1,444, 1,636, and 1,534 cm^{-1}. Pyridine molecules bonded to Lewis acid sites absorbed at 1,604 and 1,444 cm^{-1}, while those responsible for Brönsted acid sites (pyridinium ion) absorbed at 1,534 and 1,636 cm^{-1} [21]. The band at 1,485 cm^{-1} is a combined band originating from pyridine bonded to both Brönsted and Lewis acid sites. The catalyst showed mainly Lewis acidity at lower STA loading. Lewis acidity decreased with an increase in loading, whereas Brönsted acidity increased and reached a maximum at 15 % STA loading. An increase in STA loading above 15 % reduced Brönsted acidity, but Lewis acidity remained the same as that of 15 % catalyst. The decrease in acidity above 15 % could be due to the formation of crystalline WO_3, which prevents the accessibility of pyridine to the active sites. As in the case of catalysts with different loadings, the nature of acidity also depends

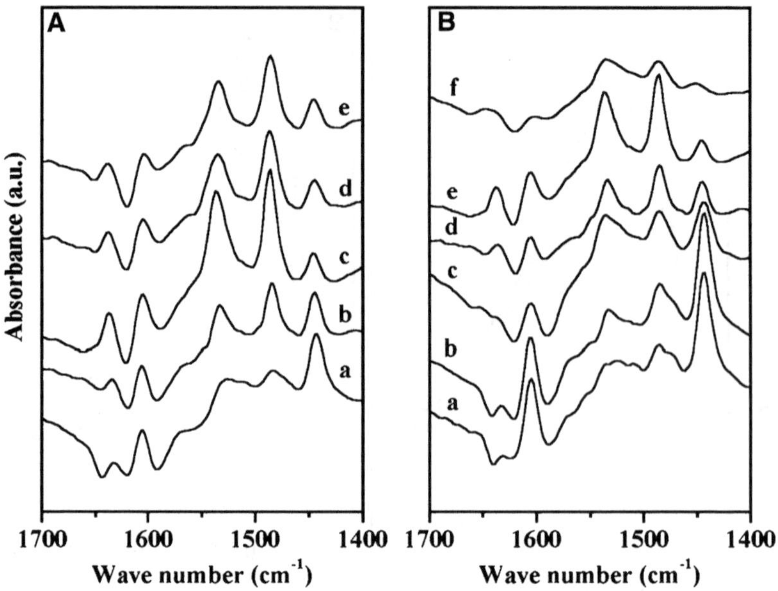

Fig. 5.3 FTIR pyridine adsorption spectra of (**A**) catalyst with different STA loading calcined at 750 °C (*a*) 5, (*b*) 10, (*c*) 15, (*d*) 20, and (*e*) 25 % and (**B**) 15 SZ calcined at different temperatures (*a*) 450, (*b*) 550, (*c*) 650, (*d*) 700, (*e*) 750, and (*f*) 800 °C after in situ activation at 300 °C

on calcinations temperature. At low calcination temperature (450 °C), the catalyst 15 SZ showed mainly Lewis acidity with very low Brönsted acidity that increases with calcination temperature up to 750 °C. Thus, the Brönsted acidity of the catalyst increased with both loading and calcination temperature up to a monolayer of STA on ZrO_2 [19]. It is observed that Brönsted acidity is higher in 15 % STA/ZrO_2 compared to that of 15 % PTA/ZrO_2.

Raman Spectroscopy

To understand the interaction of heteropoly acid with ZrO_2 support, FT-Raman spectra of pure STA and 15 SZ catalyst calcined at different temperatures were taken as shown in Fig. 5.4. The strong Raman bands of the support below 700 cm^{-1} interfere with the diagnostic Raman bands of tungstate species, while the range above 700 cm^{-1} is free from characteristic bands of ZrO_2 and, hence, relevant to the determination of tungsten species structure. Pure STA shows a sharp band at 998 cm^{-1} with a distinct shoulder at 974 cm^{-1}. The bands at 998 and 974 cm^{-1} were attributed to $\nu_{(W=O)}$ symmetric and asymmetric stretching modes. In addition, another broad band observed at 893 cm^{-1} was assigned to the $\nu_{(W\text{–}O\text{–}W)}$ asymmetric stretching mode [22, 23]. Pure phosphotungstic acid (PTA) showed similar type of Raman spectrum as that of pure STA. The 15 SZ-120 catalyst showed a broad band with two components at 973 cm^{-1} and at 946 cm^{-1}. These bands appeared to be redshifted

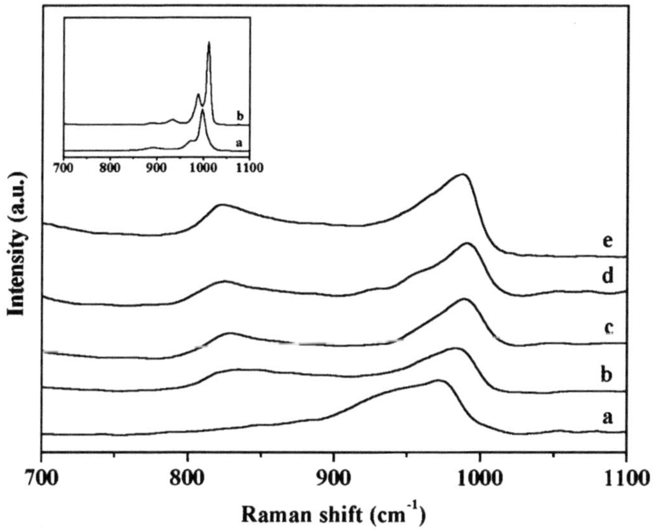

Fig. 5.4 Raman spectra of the catalysts (*a*) 15 SZ-120, (*b*) 15 SZ-650, (*c*) 15 SZ-700, (*d*) 15SZ-750, and (*e*) 15 PZ-750. Inset: (*a*) STA and (*b*) PTA

in comparison with pure STA, and it could be due to the partial weakening of W=O bonds, possibly due to the interaction of STA with the support [24]. The position of $\nu_{(W=O)}$ band was shifted from 973 to 992 cm^{-1} with an increase in calcination temperature from 120 to 750 °C, indicating an increase in W=O bond order. The increase in $\nu_{(W=O)}$ wave number with calcination temperature could be due to an increase in the agglomeration of silicotungstate species [25]. This was also evidenced from the increase in tungsten surface density with calcination temperature (Table 5.1). At a calcination temperature of 650 °C and above, Raman spectra showed a new band centered near 825 cm^{-1}, and Scheithauer et al. assigned this band to $\nu_{(W-O-W)}$ stretching mode [26]. Recently, Loridant et al. proved that this band originates from $\nu_{(W-O-Zr)}$ vibration and not from $\nu_{(W-O-W)}$ vibration [25]. In addition to the bands at 825 and 992 cm^{-1}, the catalyst 15 SZ-750 showed an additional band at 910 cm^{-1} due to $\nu_{(W-O-W)}$ stretching mode. Raman spectra of 15 SZ catalyst did not show bands at 720 and 807 cm^{-1} that are characteristic of WO_3 up to calcination temperature of 750 °C which showed that STA did not decompose into oxides. The overall results indicate that Keggin structure of STA is intact and the support ZrO_2 stabilizes the heteropoly acid. The 15 PZ-750 catalyst (15 % phosphotungstic acid on zirconia calcined at 750 °C) showed bands at 988 and 822 cm^{-1} due to $\nu_{(W=O)}$ and $\nu_{(W-O-Zr)}$ vibrations, respectively.

^{31}P MAS NMR Spectroscopy

This is one of the most useful characterization techniques to study the state of phosphorous in heteropoly acids. The chemical shift depends upon the phosphorous

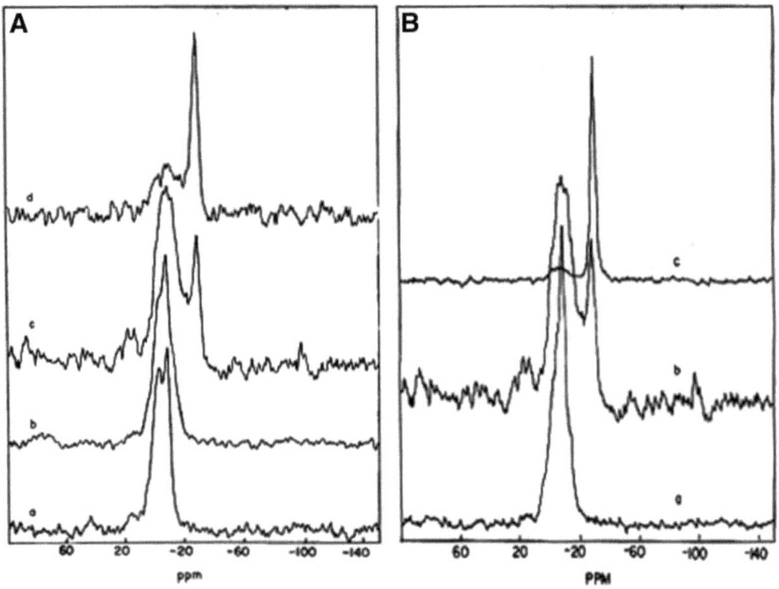

Fig. 5.5 ^{31}P MAS NMR spectra of (**A**) catalysts with different PTA loading calcined at 750 °C (*a*) 5, (*b*) 10, (*c*) 15, and (*d*) 20 % and (**B**) 15 PZ calcined at different temperatures (*a*) 650, (*b*) 750, and (*d*) 850° C

environment, and it depends on factors such as hydration number, addenda metal ion, and support [27]. The ^{31}P MAS NMR spectra of the catalysts with 5–20 % phosphotungstic acid (PTA) and 15 % catalyst calcined from 650 to 850 °C showed that the state of phosphorous in the catalyst depended on PTA loading and calcination temperature (Fig. 5.5). For low loadings and at calcination temperature of 750 °C, broad signal above –20 ppm was observed and was attributed to P–OH group associated with phosphotungstate, which is in interaction with zirconia [27a, e]. While at higher loadings and at higher calcination temperatures, new signal appeared below – 20 ppm, attributed to phosphorous oxide (P–O–P) resulting from the decomposition of the polyoxometalate [12]. This phosphorous oxide represents 20 and 45 % of the total phosphorous for 15 and 20 PTA/ZrO_2-750 °C (denoted as 15 PZ-750 and so on) catalysts and 80 % for 15 PZ-850 catalyst, respectively. The ^{31}P MAS NMR spectra of the catalysts with different PTA loading and calcination temperature showed that PTA starts decomposing at a loading of ca. 15 % at 750 °C calcination. The results from different characterization techniques suggest that a geometric monolayer of PTA on zirconia was attained for 15 PZ-750 catalyst. This could be a decisive proof of existence of Keggin anion in the zirconia-supported PTA catalyst calcined at 750 °C.

Thus, the results from different characterizations could be summarized as follows. The X-ray diffractograms of the catalysts with different STA loading and calcination temperature along with Raman spectra of 15 SZ catalyst showed the presence of zirconia-anchored mono-oxotungstate as the major tungsten species up to STA loading of ca. 15 % at 750 °C. The results from other characterization techniques such as NH_3-TPD and FTIR pyridine adsorption indicated that 15 SZ-750 catalyst had the highest total and Brönsted acidity. ^{31}P MAS NMR spectra of PZ catalysts showed the existence of phosphotungstate in zirconia support.

2.2.2 Catalytic Activity Studies

Synthesis of Linear Alkyl Benzenes (LAB)

Alkylation of benzene with C_{10-14} linear alkenes is used for the synthesis of linear alkyl benzenes (LAB), which are the primary raw materials for the production of LAB sulfonates, a surfactant detergent intermediate [28]. The reaction of benzene with 1-octene was carried out in liquid phase in a glass batch reactor. The main reactions occurred with these catalysts were alkene double-bond-shift isomerization and benzene alkylation. Monoalkylbenzenes (MOB) was the main reaction product, whereas dialkylbenzenes (DOB) and alkene dimers (DIM) formed in low amount. The liquid-phase alkylation reactions were carried out in a 50-ml glass batch reactor (slurry reactor) with an anhydrous $CaCl_2$ guard tube.

Zirconia-supported catalyst with different PTA loadings was screened initially for this reaction. Five percent loading gave 1.3 % octene conversion, and it increased to 53 % at 15 % PTA loading but decreased at higher loadings [18]. This suggests that 15 % loading is required for the monolayer coverage on ZrO_2 support. To study the effect of calcination temperature on the conversion of octene and product selectivity, 15 PZ catalyst calcined between 650 and 850 °C were used. The catalyst calcined at 650 °C gave 8.7 % octene conversion, which increased to 53.4 % at 750 °C calcination (Table 5.2). With an increase in calcination temperature, selectivity to dialkylation increased up to 750 °C and then decreased. The highest activity of 15 PZ-750 catalyst was due to its higher Brönsted and total acidity. Under optimized reaction conditions, namely, catalyst weight = 2 wt% of total reactants, temperature = 84 °C, benzene/1-octene (molar ratio) = 10, reaction time = 1 h, and alkylation of benzene with 1-octene gave 98.1 % conversion of octene with 93.7 % MOB selectivity (isomer distribution 57.2 % 2-PO, 26 % 3-PO, and 16.8 % 4-PO) and 6.3 % DOB selectivity.

The used 15 PZ-750 was filtered, washed with solvent, dried, and calcined at 600 °C for the regeneration. Then the catalyst was recycled twice with minimal loss in activity. After 2nd regeneration, the catalyst gave 96 % initial conversion.

Allylation of Anisole with Allyl Alcohol

Allylation of aromatic compounds is industrially important as the allylated aromatics find application in polymer synthesis and in perfume industry. *p*-Allylanisole (or estragole), the main product of allylation of anisole, is used in perfumes and flavoring agent in food and liquors. The reaction was carried out in a 50-ml Parr stirring autoclave under N_2 pressure.

Table 5.2 Conversion of octene and product selectivity over various catalysts

Sample	Octene conversion (mol%)	TOF (10^{-3} mol mol^{-1} W s^{-1})	MOB selectivity (%)	DOB selectivity (%)
5 PZ-750	1.3	3.7	100	0
10 PZ-750	14.9	20.8	100	0
15 PZ-750	53.4	50.0	95.5	4.5
20 PZ-750	47.6	33.6	97.9	2.1
25 PZ-750	41.1	23.1	98.7	1.3
15 PZ-650	8.7	8.1	100	0
15 PZ-700	34.6	32.5	97.8	2.2
15 PZ-800	24.0	22.5	99.0	1.0
15 PZ-850	5.4	5.0	100	0

Conditions: total weight = 25 g, catalyst weight = 0.125 g, temperature = 84 °C, benzene/1-octene (molar ratio) = 10, and time = 1 h

Table 5.3 Comparison of different catalysts

Catalyst	Si/Al	Surface area (m^2 g^{-1})	Acidity (mmol g^{-1})	Anisole conversion (wt%)	Selectivity (%, *ortho* + *para*)
15 % PTA/ZrO_2	–	53. 2	0.46	25.9	86.3
15 % PTA/SiO_2	–	16.0	Nil	0.5	50
H-beta	15	540	0.94	34.0	67.6
H–Y	3	530	2.25	22.0	93.7
H-mordenite	10	490	0.72	24.7	80.8
H-ZSM-5	50	364	0.82	2.3	92.0
PTA[a]	–	–	–	45.7	17.6

Conditions: catalyst wt = 1.25 g, wt of the reaction mixture = 25 g, temperature = 180 °C, allyl alcohol to anisole molar ratio = 1, and time = 2 h

[a]Homogeneous

Activities of zirconia and silica which supported heteropoly acids were compared with zeolite catalysts in allylation of anisole under similar reaction conditions [29]. It is seen from the results presented in Table 5.3 that PTA (homogeneous) gave the highest anisole conversion (45.7 %) among the catalysts tested for their catalytic activities in allylation of anisole followed by H-beta, 15 % PTA/ZrO_2, H-mordenite, H–Y, and the rest, while H–Y gave the highest *ortho*- and *para*-product selectivity (93.7 %) followed by H-ZSM-5, 15 % PTA/ZrO_2, H-mordenite, and the rest. Interestingly, 15 % PTA/SiO_2 catalyst showed least activity in allylation of anisole, which is due to the absence of acidity required for acid-catalyzed reactions. The absence of acidity in 15 % PTA/SiO_2 calcined at 750 °C confirms that PTA is not stable and decomposes on SiO_2. Similarly, among the zeolite catalysts, H-ZSM-5 having Si/Al = 50 having smaller pore size gave poor catalytic activity (2.3 %). In general, it appears that the performance of the catalysts in allylation of anisole depends upon the compound effect of acidity and surface area among the zeolite catalysts, while, with PTA-supported catalysts, the catalytic activity depends

Table 5.4 Benzoylation of veratrole with benzoic anhydride

Catalyst	Veratrole conversion (mol. %)	TOF (mol mol^{-1}_{H+} h^{-1})
H–Y[a]	75	14
15 SZ-750	40	27

Conditions: veratrole = 1.38 g (10 mmol), benzoic anhydride = 2.26 g (10 mmol), chlorobenzene = 50 ml, catalyst weight = 0.5 g, temperature = 130 °C, and time = 1 h
[a]Si/Al = 15

strongly on the stability (Keggin structure intact) of PTA over the support and their acidities. In addition to this, it is found that zirconia-supported heteropoly acid (15 % PTA/ZrO_2) calcined at 750 °C showed comparable activity (25.9 %), with that of H- mordenite (24.7 %).

Veratrole Acylation with Benzoic Anhydride

The benzoylation of veratrole with benzoic anhydride (BA) over supported HPA catalysts lead to the formation of 3,4-dimethoxy benzophenone as the benzoylated product. In this reaction, activity of STA/ZrO_2 was compared with that of PTA/ZrO_2, STA/SiO_2, and Zeolite HY [19]. Fifteen percent STA loading on ZrO_2 showed highest activity among 5–30 % loadings. This was attributed to the monolayer formation of STA on ZrO_2 surface at 15 % loading. The activity of STA/ZrO_2 catalyst was compared with that of strong Brönsted acidic zeolite HY (Si/Al = 15). Turnover frequencies (TOF, mol $mol^{-1}(H^+)$ h^{-1}) indicated that STA/ZrO_2 catalyst was more active and the TOF was nearly two times higher than that of H–Y zeolite (Table 5.4). The two most common Keggin heteropoly acids are silicotungstic acid and phosphotungstic acid. It is important to understand the difference between the two, when loaded on ZrO_2 support. It is interesting to note that the heterogeneous catalysts STA/ZrO2 and PTA/ZrO_2 showed different catalytic behavior. For STA/ZrO_2, conversion of BA increased constantly with time, and it attained 76 % after 480 min, whereas PTA/ZrO_2 clearly deactivated, reaching a plateau between 240 and 480 min before reaching maximum conversion. This profile of PTA/ZrO_2 catalyst is similar to that reported for zeolite and HPA-catalyzed acylation reactions, where the catalyst deactivation is attributed mainly to the strong adsorption of the acylation product on the catalyst, which blocks the accessibility of the reactants to the active sites. Acidity measurements of the catalysts by FTIR pyridine adsorption and NH_3-TPD showed that 15 % STA/ZrO_2 catalyst possess the highest total acidity as well as Brönsted acidity as compared with 15 % PTA/ZrO_2. However, FTIR pyridine adsorption studies showed that the Lewis acidity was higher for PTA/ZrO_2. Anderson et al. [30] have stated that catalyst with higher Lewis acidity is more susceptible to deactivation by product inhibition. Hence, PTA/ZrO_2 with higher Lewis acidity showed catalyst deactivation. Hence, each Keggin heteropoly acids supported on ZrO_2 can show different catalytic performance as they vary in properties such as type of acidity and total acidity.

Alkylation of Phenol and *p*-Cresol with *tert*-Butanol

The alkylation of phenols with different alcohols is industrially important as it is used for the production of a variety of products. Among these, alkylation of phenol or *p*-cresol with *tert*-butanol to *tert*-butylated phenols is important. For instance, 2-*tert*-butyl phenol (2-TBP) is a starting material for the synthesis of antioxidants and agrochemicals, whereas 4-*tert*-butyl phenol (4-TBP) is used to make fragrances and phosphate esters, 2,4-di-*tert*-butyl phenol (2,4-DTBP) is used in the synthesis of substituted triaryl phophites. 2-*tert*-butyl-*p*-cresol (TBC) and 2,6-di-*tert*-butyl-p-cresol (DTBC), commercially known as butylated hydroxytoluene (BHT), are widely used in the manufacture of phenolic resins, as antioxidants and polymerization inhibitors. Commercially *tert*-butyl phenols are produced by the reaction of phenol (or cresol) with isobutene in the presence of homogeneous catalysts. These reactions are carried out in a continuous down-flow quartz reactor.

Heteropoly acids supported on zirconia are reported to be highly active and stable heterogeneous catalysts for alkylation of phenols. At optimized reaction conditions, namely, 140 °C, LHSV of 4 h^{-1}, *tert*-butanol/phenol molar ratio of 2, 15 % STA/ZrO_2 (calcined at 750 °C) showed phenol conversion of 95 % with selectivity to 4-TBP 59 %, 2,4-DTBP 36 %, and 2-TBP 4 % [31]. Thus, the activity of this catalyst is found to be similar to that of large pore zeolites and much higher than that of other microporous and mesoporous materials. There was no change in the conversion of phenol and selectivity to different products after a 50-h reaction that showed a high stability of catalyst toward deactivation Fig. 5.6. Phosphomolybdic acid [PMA] supported on ZrO_2 is also applied for this reaction. But the catalyst deactivated due to the leaching of active sites [32]. For *tert*-butylation of *p*-cresol with *tert*-butanol, 15 % PTA/ZrO_2 (calcined at 750 °C) was also studied under continuous down-flow conditions [33]. The catalyst gave good *p*-cresol conversion (~60 %) and high selectivity (~90 %) for TBC. Moreover, the catalyst showed reasonably good stability toward deactivation in a time on stream of 100 h, retaining 90 % of its original conversion at the end of the run.

HPA supported on zirconia are applied in various other commercially important organic transformations such as acylation of diphenyl oxide with benzoyl chloride, alkylation of resorcinol with butanol, n-heptane isomerization, synthesis of biodiesel by transesterification of vegetable oil with methanol, and acetalization of benzaldehyde and 2-naphthyl ketone [34].

Comparison of HPA/ZrO_2 with Tungstated Zirconia

Tungstated zirconia is a well-known solid acid catalyst with strong acidity which belongs to the class of supported isopoly tungstate catalyst. It is appropriate to compare the activity with its counterpart HPA on zirconia catalysts. The reactions chosen for that is benzylation of phenol with benzyl alcohol and acylation of 2-methoxy-naphthalene with acetic anhydride. The major products are ortho- and para-benzyl phenol and 1-acetyl-2-methoxynaphthalene (1-AcMN), respectively.

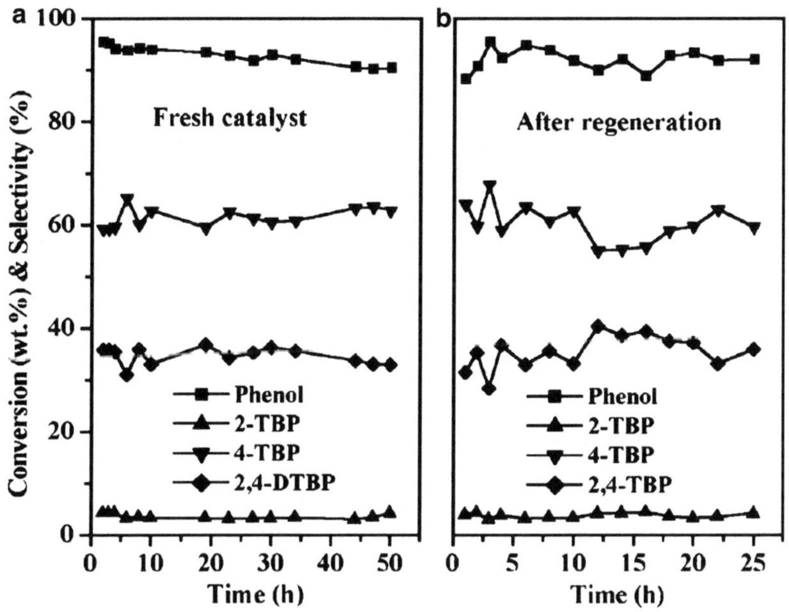

Fig. 5.6 Time on stream on the conversion of phenol and product selectivity (Conditions: temperature = 140 °C, LHSV = 4 h^{-1}, and *tert*-butanol/phenol molar ratio = 2)

The reactions were carried out in a 50-ml glass batch reactor. Fifteen percent PTA/ ZrO2 (750 °C) showed 80 % benzyl alcohol conversion, whereas it showed 50 % 2-methoxy-naphthalene conversion with 100 % selectivity for (1-AcMN). The activity of PTA catalyst is about 2–3 times higher than tungstated zirconia catalyst which is mainly attributed to higher acidity of PTA catalyst [35].

3 Part 2: Multicomponent Polyoxometalates Immobilized on Solid Supports as Eco-Friendly Catalysts in Oxidation Reactions

Chemical immobilization of polyoxometalates onto mesoporous solid support combines the advantages of molecular complexes and reusable solids; furthermore, new activities and selectivities may arise due to confinement effects. Immobilization of heteropoly acid catalyst on mesoporous materials such as mesostructured cellular foam silica and mesoporous carbon and their use in some model oxidation reactions has been reported by Kim et al. [36]. Iron-substituted polyoxometalate supported on cationic silica was found to be active for oxidation of sulfides and aldehydes [37]. $H_5PV_2Mo_{10}O_{40}$ supported on a mesoporous molecular sieve, both by adsorption onto MCM-41 and by electrostatic binding to MCM-41 modified with amino groups, was active in the aerobic oxidation of alkenes and alkanes in the presence of isobutyraldehyde as sacrificial reagent [38].

Immobilization of $H_5PV_2Mo_{10}O_{40}$ was also carried out by solvent anchored supported liquid-phase strategy in which $H_5PV_2Mo_{10}O_{40}$ was supported onto polypropylene glycol-modified silica synthesized by sol–gel process [39]. Since polyoxometalates are anionic, preparation of silica particles with quaternary ammonium moieties on the surface led to a useful catalytic assembly with $\{[WZnMn_2(H_2O)_2][(ZnW_9O_{34})_2]\}$ as the active species. Importantly, using the sol–gel synthesis for the preparation of silica, the surface hydrophobicity could be controlled by choice of the organosilicate precursors [40].

In the study of gas-phase oxidation of methacrolein to methacrylic acid by Kanno et al., SiO_2-supported $H_4PMo_{11}VO_{40}$ showed low activity and selectivity of methacrylic acid than $H_4PMo_{11}VO_{40}$ supported on NH_3-modified SiO_2 synthesized by impregnation method (~90 %) [41]. The activity of $H_4PMo_{11}VO_{40}/NH_3$-modified SiO_2 was more than five times higher than that of the corresponding unsupported catalysts.

Amine-functionalized SBA-15 immobilized with mono-vanadium-substituted molybdovanadophosphoric acid [42] was assessed in the oxidation of selected substrates, norbornene, cyclooctene, cyclohexene, and styrene with aqueous hydrogen peroxide, and the results were compared with that of the neat molybdovanadophosphoric acid catalyst (homogeneous). Although lower conversion was obtained, the selectivity of the desired products was higher with the immobilized catalyst. Ressler et al. also used SBA-15 as a support in the immobilization of heteropolyoxomolybdates ($H_4PMo_{11}VO_{40}$) by incipient wetness and investigated the selective oxidation propene [43]. In this material, the Keggin ions were found to be thermally unstable than the bulk in the reaction conditions.

Here, we are discussing our work of the immobilization of molybdovanadophosphoric acid onto solid supports using different strategies and study of some model oxidation reactions using the supported catalysts.

3.1 Results and Discussion

3.1.1 Immobilization onto Amine-Modified Mesoporous Silica

We have immobilized molybdovanadophosphoric acid (V1PA, V2PA, and V3PA, where V1, V2, and V3 are number of vanadium atom) onto acid-treated aminopropyl-modified mesoporous silica (Scheme 5.1) [44, 45]. The mesoporous silica support was made cationic prior to the treatment with the polyoxometalates for better anchoring and less leaching in the catalytic run. We have assessed these catalyst systems in some model oxidation reactions. The successful immobilization was confirmed by different characterization techniques showing the retention of the structure of molybdovanadophosphoric acid after immobilization.

Mass NMR Spectroscopy

The ^{31}P NMR spectra of (a) V1PA–NH_3^+–SBA-15, (b) V2PA–NH_3^+–SBA-15, and (c) V3PA–NH_3^+–SBA-15 are depicted in Fig. 5.7. There is a slight shift in ^{31}P NMR peaks of anchored samples compared to the corresponding neat polyoxometalates

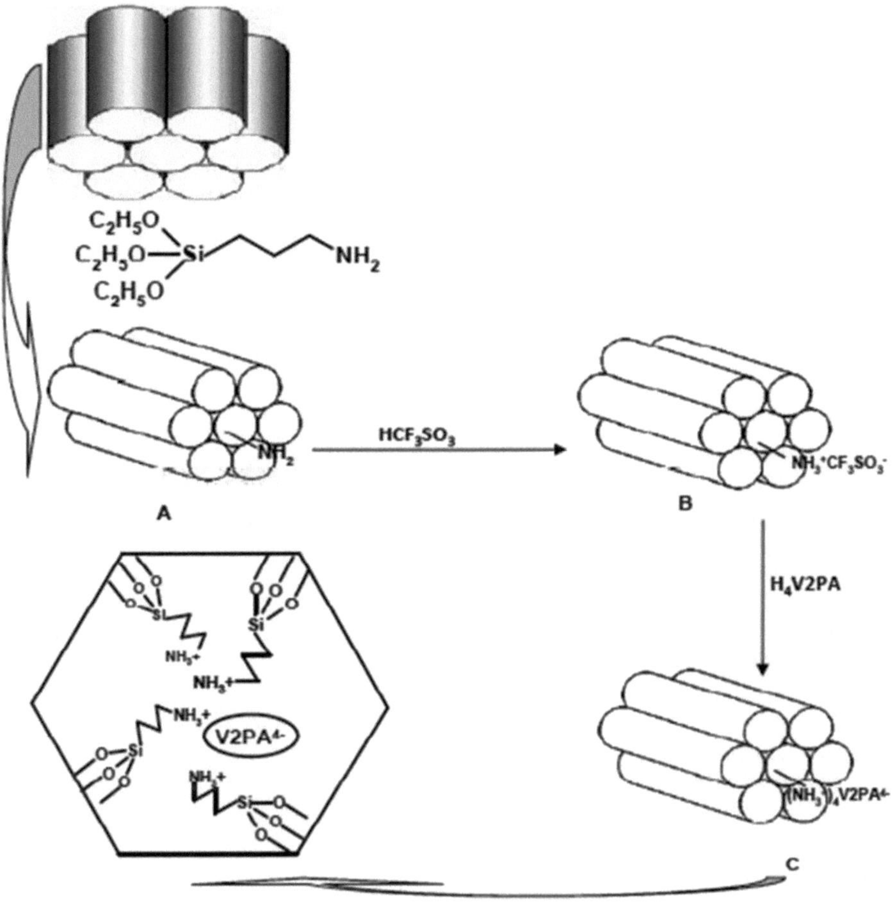

Scheme 5.1 Immobilization process of V2PA on SBA-15

as reported in literature [38]. ^{31}P NMR peaks corresponding to V1 (+4.05 ppm), V2 (+3.09 ppm), and V3 (+1.68 ppm) peak positions as per literature shifted to +2, +1.5, and +1 ppm, (Fig. 5.7a, b, c), respectively, which could be either due to the interaction of polyoxometalates with mesoporous silica support or because of the differences in the degree of hydration of polyoxometalates upon immobilization. These NMR results would support that structure of polyoxometalate is retained upon immobilization onto mesoporous supports [46].

^{51}V NMR spectra of neat V2PA and its anchored form V2PA–NH_3^+–SBA-15 are depicted in Fig. 5.8. Vanadium spectra show the presence of numerous spinning side bands centered around –400 ppm attributed to various possible stereoisomers present. In fact, from ^{51}V NMR spectra, it is not possible for us to make a clear distinction between different isomers; however, we could only say that vanadium is in an octahedrally distorted form in anchored form as reported in literature [47]. Figure 5.9a, b shows ^{29}Si MAS NMR spectra of modified SBA-15 samples with

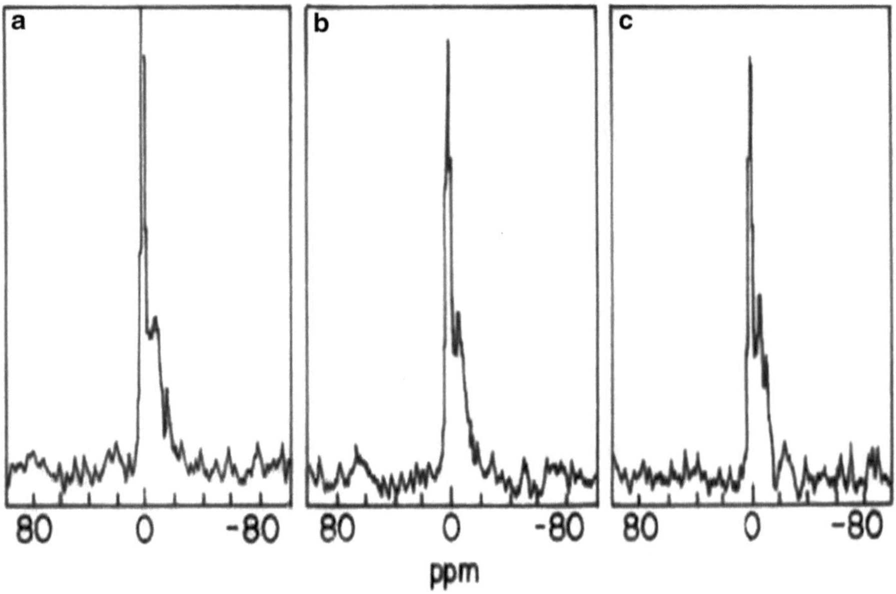

Fig. 5.7 ^{31}P MAS NMR of (**a**) V1PA-NH_3^+–SBA-15, (**b**) V2PA-NH_3^+–SBA-15, and (**c**) V3PA-NH_3^+–SBA-15

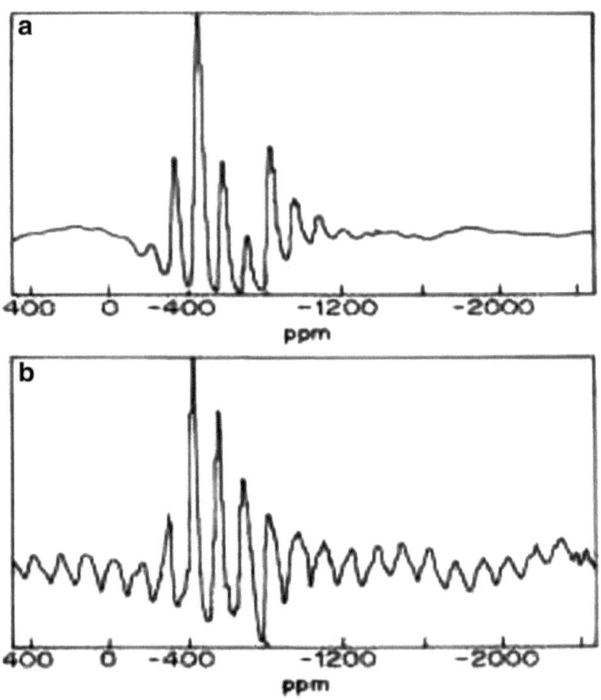

Fig. 5.8 ^{51}VMAS NMR profile of (**a**) V2PA and (**b**) V2PA-NH_3+–SBA-15

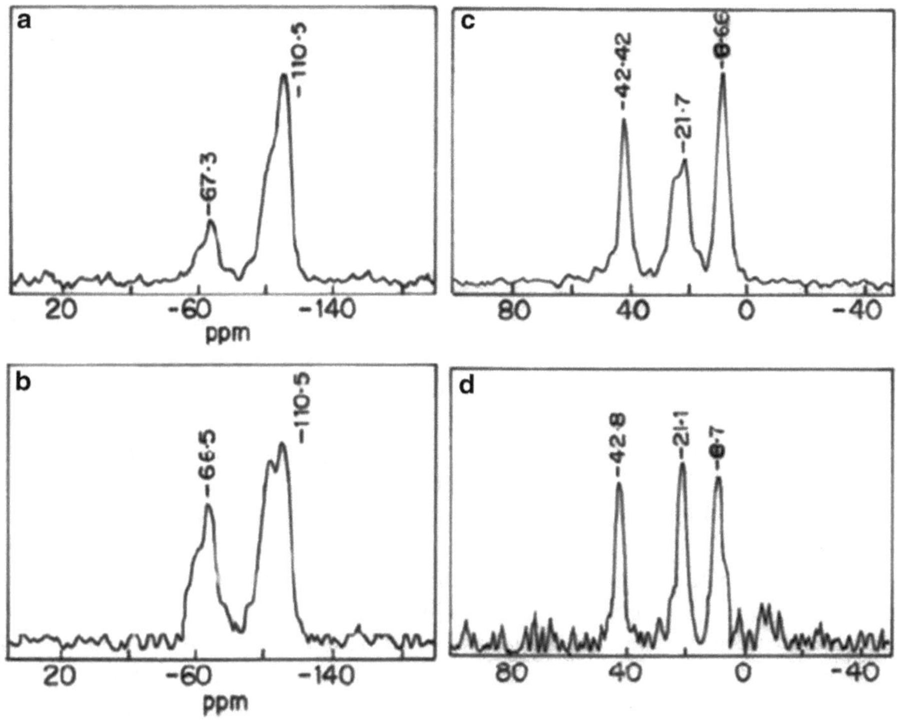

Fig. 5.9 ^{29}Si MAS NMR profile of (**a**) NH_3^+–SBA-15, (**b**) V2PA-NH_3^+–SBA-15, ^{13}C CP-MAS NMR profile of (**c**) NH_3^+–SBA-15 and (**d**) V2PA-NH_3^+–SBA-15

aminopropyl groups. ^{29}Si MAS NMR spectra of parent SBA-15 give a broad peak and dominated by an intense peak at −110 ppm assigned to $Si(OSi)_4$ and two shoulder peaks at −100 and −90 ppm. Shoulder peaks appear due to $Si(OSi)_3OH$ (Q3) and $Si(OSi)_2(OH)_2$ (Q2) structural units present in SBA-15. Upon incorporation of aminopropyl groups, in addition to above three peaks, two more peaks at –56 and –67 ppm appeared, and their intensities greatly enhanced by ^{1}H cross polarization. No peak appeared at –45 ppm corresponding to chemical shift of silicon in liquid (3-aminopropyl) trialkoxysilane, indicating the absence of free silane molecules physically adsorbed on SBA-15 surface. A peak at –67 ppm indicates the formation of new siloxane linkages (Si–O–Si) of aminopropylsilane to surface silicon atoms of SBA-15 via three siloxane bonds, $(–O–)_3Si\text{-}CH_2CH_2CH_2NH_2$ (T_3), and peak at –56 ppm via two siloxane bonds, $(–O–)_2Si\text{-}CH_2CH_2CH_2NH_2$ (T2). 13C MAS NMR spectra shown in Fig. 5.9c, d provided useful information on nature of incorporated aminopropyl groups on internal surface of SBA-15. Three well-resolved peaks at –8.7, 22, and 42 ppm observed have been assigned to C1, C2, and C3 carbons of incorporated aminopropyl groups, $(–O–)_3SiCH_2(1)CH_2(2)CH_2(3)NH_2$, respectively. Structure of aminopropyl groups remains intact during the incorporation process. Broad peak at 25 ppm indicates existence of some protonated aminopropyl groups in the presence of extra surface hydroxyl groups or moisture. Absence of peaks due to residual ethoxy carbons (18 and 60 ppm) in spectra confirms that hydrolysis and/or condensation of (3-aminopropyl) triethoxysilane molecules inside internal surface of SBA-15 is virtually complete.

Catalytic Study

Anthracene Oxidation

Intense interest has been focused on the anthraquinone (AQ) derivatives due to their potential utility in cancer chemotherapy as antioxidants and antitumor agents. 9,10-Anthraquinone is the most important quinone derivative of anthracene (AN) [48]; its most widespread use is as a hydrogen carrier in the industrial production of H_2O_2 [49]. It is commercially produced in several ways, including oxidation of AN with chromic acid [50], condensation of benzene with phthalic anhydride, and through Diels–Alder reaction [51]. Vapor-phase oxidation of AN by air has been carried out over a supported iron vanadate–potassium catalyst at 390 °C [53]. Over the last two decades, a number of homogeneous catalyst systems have been used for the oxidation of AN. Because of the drawbacks of homogeneous catalyst systems, polymeric metal chelates, immobilized enzymes, and supported vanadium POM ($H_5PV_2Mo_{10}O_{40}\cdot32\cdot5H_2O$) have been studied for the oxidation of AN. Liquid-phase catalytic oxidation of AN was conducted at atmospheric air using newly synthesized immobilized catalysts and was compared with their homogeneous analogues [45]. Results on catalytic activities of different catalysts like neat and immobilized in oxidation of AN by TBHP to give mainly AQ as product and reaction conditions used for oxidation are presented in Table 5.5. Under reaction conditions (conditions: catalyst, 0.1 g, 56 mmol (1 g) of AN, 3-ml Benzene, AN/Catalyst mole ratio, 430 and AN/TBHP molar ratio 5, 80 °C, and 12 h) studied, homogeneous catalysts (V1PA, V2PA, and V3PA) gave AQ (selectivity in range 66–72 %) and remaining anthrone and oxanthrone as oxidation products of AN. As seen from the results (Table 5.5), conversion of AN was nearly the same (~58 %) for V2PA and V3PA, while it was 50 % for V1PA. Lower activity of V1PA could be due to the lower reduction potential or the presence of variable vanadium isomers [52]. Hence, immobilized catalysts were prepared using more active V2PA and mesoporous silica supports like MCM-41, MCM-48, and SBA-15, and their performance in AN oxidation was studied. It could be seen from the catalytic activity data (Table 5.5) that the immobilized catalysts were quite active and AN conversions were comparable to that of the neat catalysts; however, these differ in product selectivities. Selectivity for AQ with neat catalysts was in the range 66–72 % (entries 1–3, Table 5.5), and interestingly it was 100 % with immobilized catalysts, which is the most important function of these catalysts. Among immobilized catalysts, though $SiO_2V_2(30)$ amorphous silica gave good AN conversion (59 %), nevertheless, had poor selectivity for AQ (69 %). Immobilized catalysts retained the activities (TOFs) of their homogeneous analogues (neat) and selectively formed AQ (100 %) in AN

Table 5.5 Catalytic activity data on the oxidation of anthracene

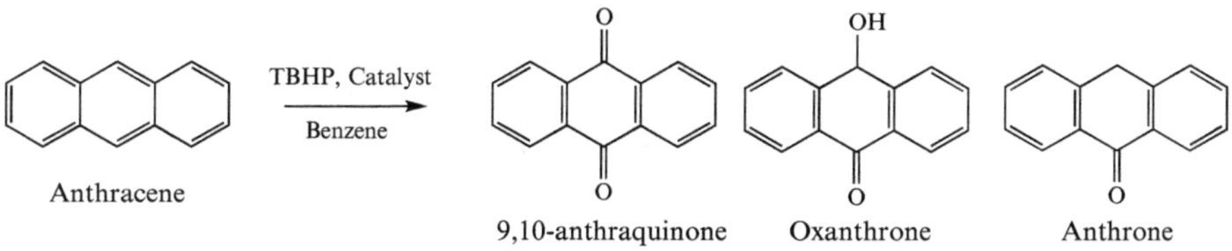

No.	Catalyst	Convn. of AN (mol %)	TOF	Selectivity of products (mol %) AQ	Anthrone	Oxanthrone
1	V1PA (neat)	50	17	66	22	12
2	V2PA (neat)	58	20	72	15	13
3	V3PA (neat)	57	20	70	16	14
4	SiO_2V2(30)	59	11	69	16	15
5	MCM41V2(30)	50	17	100	0	0
6	MCM48V2(30)	49	17	100	–	–
7	SBAV2(10)	30	10	100	–	–
8.	SBAV2(20)	46	16	100	–	–
9.	SBAV2(30)	60	21	100	–	–
10.	SBAV2(40)	55	19	100	–	–

Legend: SiO_2V2(30) = 30 wt% V2PA loaded on SiO_2-NH_3^+, MCM-41V2(30) = 30 wt% V2PA loaded on NH_3^+-MCM-41, MCM48V2(30) = 30 wt% V2PA loaded on NH_3^+-MCM-48, SBAV2(10 to 30) = +10–30 wt% V2PA loaded on NH_3^+-SBA-15, V1PA = $H_4[PMo_{11}VO_{40}]\cdot32.5H2O$, V2PA = $H_5[PMo_{10}V_2O_{40}]\cdot32.5H_2O$, and V3PA = $H_6[PMo_9V_3O_{40}]\cdot34H_2O$. *AN* Anthracene and *AQ* 9,10-Anthraquinone. *Reaction conditions*: 0.013 mmol catalyst (0.1 g), 56 mmol of substrate (1 g), 280 mmol TBHP (2.5 g), 3-ml Benzene, 80 °C, and 12 h. Substrate/catalyst mole ratio 430 and TBHP/substrate mole ratio 5

oxidation. The catalyst was separated by filtration, washed and dried in oven (100 °C) for 1 h, and then reused in the second cycle with fresh reaction mixture under the same reaction condition. The first cycle gave 60 % AN conversion with 100 % selectivity to AQ. It is seen from the results that the AN conversions were nearly the same in four cycles with AQ selectivities essentially constant (100 %), indicating that the NH_3^+-SBA-15 (SBAV2 (30)) catalyst could be used in repeated cycles without loss in its activity. There was a marginal decrease in conversion after each cycle, which could be due to handling losses and perhaps not due to the leaching of active species into the reaction medium, which was further confirmed with hot filtration test.

Oxyfunctionalization of Adamantane: Kinetics and Mechanistic Study

Activation of carbon–hydrogen bond of an alkane is considerably more difficult due to its kinetic stability. The energy required to overcome the kinetic stability of alkanes leads to deep oxidation rather than selective oxidation [53]. Adamantane has been used as a model compound by numerous researchers to investigate C–H bond activation because the substituted adamantane derivatives, especially mono- or

Table 5.6 Catalytic data on the oxidation of adamantane

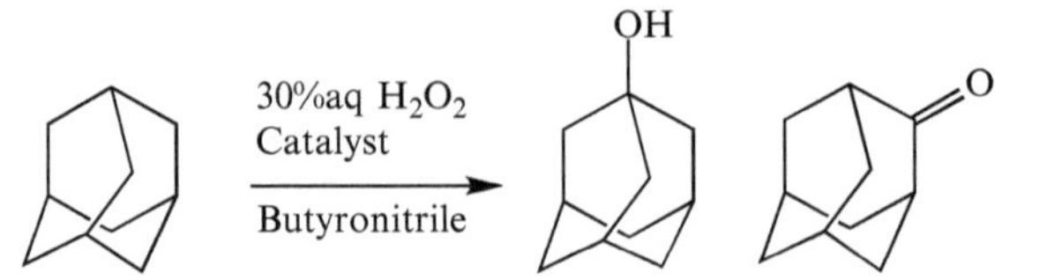

Adamantane 1-Adamantol Adamantanone

No.	Catalyst	Conv. of adamantane (mol%)	Selectivity of products (mol %)	
			Adamantanol	Adamantanone
1	SBA-15WOx(30)	No conversion	–	–
2	SBA-15MoOx(30)	No conversion	–	
3	SBA-15VOx(30)	46	82	18
4	SBA-15PTA(30)	No conversion	–	–
5	SBA-15STA(30)	No conversion	–	–
6	SBA-15MPA(30)	No conversion	–	–
7	SBA-15V1(30)	49	81	19
8	SBA-15V2(30)	65	80	20
9	SBA-15V3(30)	61	76	24

Legend: SBA-15V2(30) = 30 wt% $H_5[PMo_{10}V_2O_{40}]\cdot 32.5H_2O$ loaded on NH_3^+-SBA-15, V1 = $H_4[PMo_{11}VO_{40}]\cdot 32.5H_2O$, and V3 = $H_6[PMo_9V_3O_{40}]\cdot 34H_2O$; STA = silicotungstic acid; PTA = tungstophosphoric acid; and MPA = molybdophosphoric acid. WOx = sodium tungstate, MoOx = sodium molybdate and VOx = sodium metavanadate. *Reaction conditions*: 0.02 M of adamantane, 0.02 M of oxidant (30 % aq. H_2O_2); catalyst, 2.6×10^{-5} M 10 cm^3, butyronitrile (10 ml, solvent), 358 K, and 12 h

disubstituted ones, are used as important precursors for photoresistance and medicines [54]. It is difficult to obtain tertiary mono- and di- oxygenated products of adamantane. Conventionally, 1-adamantanol and 1, 3-adamantanediol are obtained by the bromination of adamantane with molecular bromine, followed by the hydrolysis of corresponding brominated derivatives, while 2-adamantanol is obtained by the rearrangement of tertiary hydroxylated adamantanes using concentrated H_2SO_4. Due to the number of side products and harmful reagents, conventional techniques are unacceptable for their preparation. Therefore, there is still a challenge to develop an ecological and economical process for adamantane oxidation. Moreover, we have used our immobilized catalyst systems (in the previous section) in the kinetic and mechanistic study of adamantane oxidation [55].

The major product of adamantane oxidation was tertiary C–H bond-oxygenated 1-adamantanol. It is seen from the results that the vanadium-containing catalysts exhibited higher catalytic activities compared to the others, which shows that vanadium centers are essential for the catalysts to get higher yields of oxygenated products. As we can see from the results (Table 5.6, entries 1,2,4,5, and 6), it is clear that the catalysts consisting of isopoly and heteropoly acids having no vanadium content in it are inactive in the oxidation of adamantane by 30 % aq H_2O_2, while, vanadium-containing SBA-15VOx(30) and vanadium-substituted {SBA-15V1(30)}, {SBA-15V2 (30)}, {SBA-15V3 (30)}, respectively (Table 5.2, entry 3,7,8, and 9), have shown

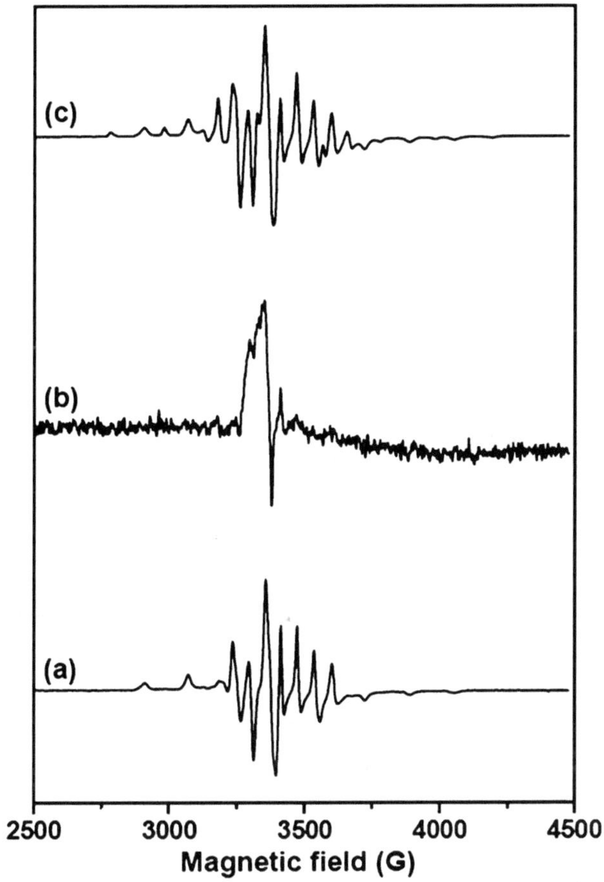

Fig. 5.10 EPR spectra of (*a*) SBA-15 V2 (30) in butyronitrile, (*b*) SBA-15 V2 (30) + 30 % aq. H_2O_2 in acetonitrile, and (*c*) the reaction mixture at 0.5 h

good activities in the oxidation of adamantane (conversion of adamantane in the range 46–65 %). Among these, {SBA-15V2 (30)} has shown the highest activity with adamantane conversion (65 %). Catalytic studies show that the main product of adamantane oxidation reaction catalyzed by vanadium-substituted phosphomolybdic acid anchored to amine-functionalized SBA-15 gave 1-adamantanol as the main product and adamantanone as the minor.

To understand the reaction mechanism, we have made an attempt to look into the reaction intermediates by conducting EPR experiments. The spectrum shown in Fig. 5.10a is of pure catalyst at room temperature, and it exhibits isotropic ^{51}V hyperfine lines due to the presence of trace amounts of V(4+) species in the catalyst. The EPR spectrum contains eight hyperfine line patterns due to the interaction of paramagnetic electron of V(4+) with its nucleus ($I = 7/2$). The Hamiltonian parameters of this species based on the computer simulation are $giso = 1.92$ and $Aiso = 158$ G. A small difference between the eight different lines is noticed due to the second-order hyperfine effect. The mixture of catalyst and 30 % aq H_2O_2 was heated at 358 K and the EPR spectra recorded. The disappearance of V(4+) species (Fig. 5.10b)

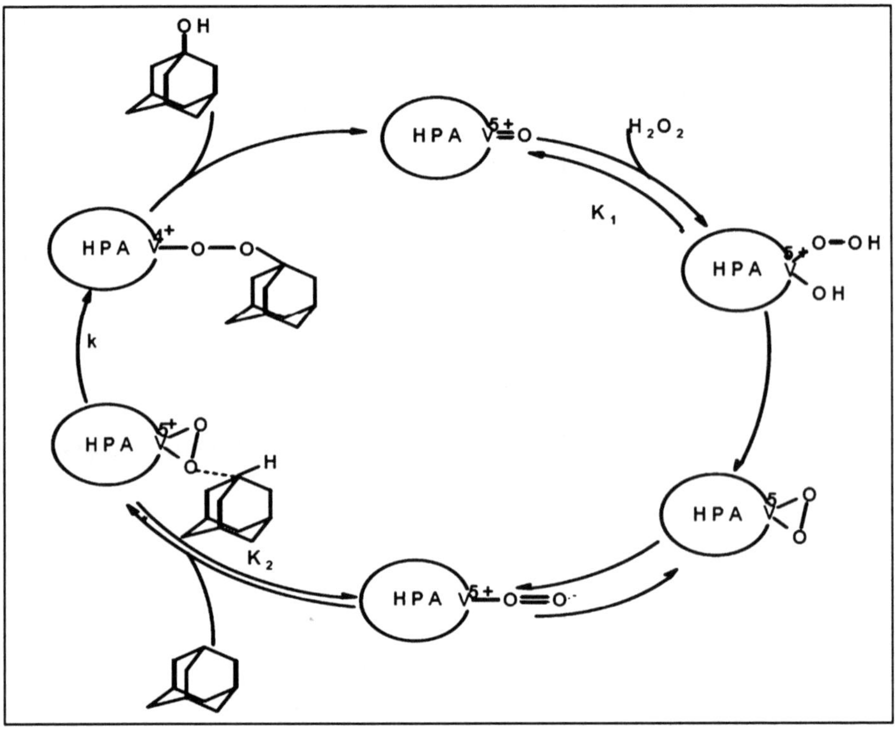

Scheme 5.2 Mechanism for the oxidation of adamantane catalyzed by SBA-15 V2 (30) in the presence of 30 % aq. H_2O_2

indicates that V(4+) center is oxidized to V(5+). The species generate superoxo radical which corresponds to V(5+) (Scheme 5.2) on interaction with H_2O_2, which may partly be in equilibrium with V(5+)-peroxo species (Scheme 5.2). The V(5+)-superoxo species might have formed via unstable vanadium (V) hydroxy–hydroperoxy species [46b]. Based on the EPR studies, a plausible reaction mechanism has been proposed in Scheme 5.2. The spectrum indicates that the mechanism of oxidation of adamantane involves the formation of stable vanadium–superoxo species, as shown in Scheme 5.2, which is partially in equilibrium with vanadium peroxo species.

As it is seen in Fig. 5.10c, a slightly different spectra of V(4+) is observed, which could be attributed to metalloperoxy adamantane intermediate species formed by the interaction of vanadium–superoxo species with adamantane molecule. The minor product adamantanone might have formed by secondary oxidation, where a portion of adamantanol might have converted to adamantanone. The reaction was promoted mainly by the vanadium species, and the rest of the Keggin ion enhances the activity. The catalyst generates adamantyl radicals by the abstraction of one electron from adamantane. The oxidation is believed to proceed through the reactive intermediates, V(5+)-peroxo and V(4+)-superoxo species. We were able to provide, for the first time, EPR spectroscopic evidence for the participation of V(4+)-superoxo species in selective oxidations over vanadium-substituted molybdovanadophosphoric acid immobilized on amine-functionalized SBA-15.

3.1.2 Inorganic–Organic Hybrid Materials Based on Functionalized Organosilica and Mesoporous Carbon

Because of their extremely high surface area and uniform pore-size distributions, ordered carbon materials promise the opportunity of exploring them in heterogeneous catalysis [56]. $H_5PV_2Mo_{10}O_{40}$ ($PMo_{10}V_2$) catalyst was chemically immobilized as a charge-matching component on the nitrogen-containing mesostructured cellular foam carbon (N-MCF-C), which was synthesized by a templating method using mesostructured cellular foam silica (MCF-S) and polypyrrole as a templating agent and a carbon precursor [15u]. Characterization results showed that the $H_5PV_2Mo_{10}O_{40}$ ($PMo_{10}V_2$) catalyst was finely dispersed on the N-MCF-C support via strong chemical interaction. In the vapor-phase oxidation of benzyl alcohol, the immobilized catalyst showed higher conversion and a higher oxidation activity (formation of benzaldehyde) than the unsupported counterpart and a suppressed acid catalytic activity. Similar reactivity trend was also observed in the vapor-phase conversion of 2-propanol using $H_5PV_2Mo_{10}O_{40}$ ($PMo_{10}V_2$) catalyst chemically immobilized onto modified nitrogen-containing spherical carbon (N-SC) [57] and nitrogen-rich macroporous carbon (N–MC) [15t]. $[PV_2Mo_{10}O_{40}]^{5-}$ and $[PV_2Mo_{10}O_{40}]^{9-}$ impregnated on carbon have been used to catalyze oxidation of alcohols, amines, and phenols [15k, 58]. Similarly, $[PV_2Mo_{10}O_{40}]^{5-}$ on several supports, such as carbon or textile fibers, were active for oxidation of various odorous volatile organics such as acetaldehdye, 1-propanethiol, and thiolane [59].

Our previous discussion shows that vanadium-containing polyoxometalates immobilized onto SBA-15 are active catalysts in the liquid-phase oxidation reactions [45, 55]. To broaden the scope of immobilization, we have immobilized vanadium-containing polyoxometalates onto mesoporous silica, organosilica, and carbon supports and studied their influence on catalytic activities and 2MNQ selectivity in the oxidation of 2MN [66]. Hydrophilic–hydrophobic interactions in these heterogeneous catalyst systems have been specially addressed.

Organo SBA-15 (OSBA-15) was synthesized using 1,2-bis-(trimethoxylsilyl) ethane (BTME) as the organic source in the standard procedure for the synthesis of SBA-15 [67]. Then it was modified with amine and treated with acid to form cation for efficient immobilization of molybdovanadophosphoric acid [68]. Mesoporous carbon (MC) was synthesized using silica as template [69] (Scheme 5.3) and was oxidized and modified by thio group [70], then modified by bis(3-aminopropyl) amine (trien) [71], and subsequently immobilized with molybdovanadophosphoric acid. The diagnostic techniques such as small-angle XRD, N_2 physisorption studies, SEM, and HRTEM together provided an evidence for the structural integrity of the newly synthesized catalyst systems.

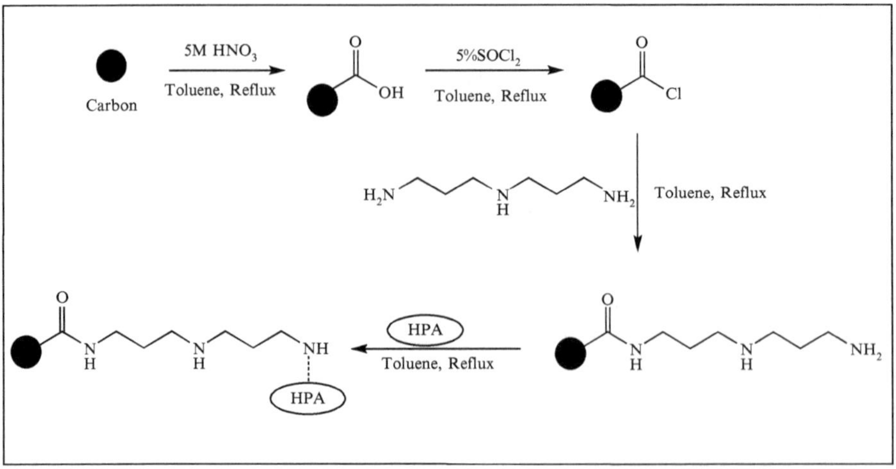

Scheme 5.3 Functionalization of mesoporous carbon

Catalytic Oxidation of 2-Methylnaphthalene by 30 % aq. H_2O_2

2-MNQ (menadione, vitamin K3) is one of the important compounds produced by the oxidation of arenes and used as a supplement for vitamins K1 and K2 in veterinary medicine [60]. Many methods have been reported for the controlled oxidation of 2MN using a series of oxidizing agents in the presence of various catalyst systems [61]. Stoichiometric oxidation of 2MN using sulfuric acid and chromic acid, about 18 kg of chromium, is produced per kilo gram product (30–60 % yield), causing environmentally hazardous chromium containing waste water [62]. The use of acetic acid, hydrogen peroxide, and methyl trioxorhenium, as well as, Pd-polystyrene sulfonic acid resin, metalloporphyrin along with potassium monopersulfate and zeolites catalyst systems, has been reported for the oxidation of 2MN [63, 64]. The best result till now for the selective oxidation of 2MN was achieved using glacial acetic acid as solvent, sulfuric acid as catalyst, and acetic anhydride as the dehydration reagent to get about 80 % yield for 2MNQ [65]. The disadvantages of conventional protocols for the oxidation of 2MN are the use of acidic solvents and mineral acid catalysts, which causes environmental pollution on a larger scale. Thus, to improve the product selectivities of 2MN and its derivatives under neutral environment by developing heterogeneous catalyst systems using 30 % aq. hydrogen peroxide is still a significant and demanding goal.

The results on the catalytic activities of different catalysts like neat and immobilized in the oxidation of 2MN by 30 % aq. H_2O_2 to give mainly 2-methy-1,4-naphthoquinone as the product and the reaction conditions used for the oxidation are presented in Table 5.7. To compare the activities of both homogeneous (neat) and immobilized catalysts in 2MN oxidation with 30 % aq. H_2O_2 oxidant, the amount of different catalysts were used in such a way that they all had the same concentrations of vanadium content. As discussed in our previous section, we have observed that the 30 %

Table 5.7 Physicochemical properties and catalytic data of the oxidation of 2MN

No.	Materials	V (mmol g^{-1})	Mo (mmol g^{-1})	Surfacearea (m^2 g^{-1})	BJH pore diameter (nm)	Conv. (mol%)	Selectivity (mol%)
1	V2PA	2.51	13.01	–	–	40	59
2	SBA	–	–	970	9.1	0	0
3	V2SBA	2.48	12.89	422	6.5	33	99
4	OSBA	–	–	894	5.8	0	0
5	V2OSBA	2.49	12.92	209	5.7	38	99
6	MC	–	–	1,432	9.3	0	0
7	V2MC	2.45	12.84	1,133	4.3	24	99

Legends: V2PA = $H_5[PMo_{10}V_2O_{40}]\cdot 32.5H_2O$, SBA = SBA-15, V2SBA = 30 wt% V2PA loaded on NH_3^+-SBA-15, OSBA = Organo SBA-15, V2OSBA = 30 wt% V2PA loaded on NH_3^+-OSBA-15, MC = Mesoporous carbon, and V2MC = 30 wt% V2PA loaded on MC. *Reaction condition*: 0.1 g (0.7 mmol) of 2MN, 0.24 g (7 mmol) of oxidant (30 % aq. H_2O_2); catalyst 0.1 g (0.013 mmol), acetonitrile 10 ml (solvent), 353 K, and 10 h

polyoxometalates loading onto the mesoporous support was the optimum loading to perform as an effective catalyst system, and hence the same loading was used in the present studies. Under the reaction conditions studied, homogeneous catalyst V2PA showed a conversion of 2MN 40 and 59 % 2-methyl-1,4-naphthoquinone selectivity (Table 5.7). The remaining products are 2-methyl-1-naphthone and 2-naphthoic acid. Hence, the immobilized catalysts were prepared using the V2PA and mesoporous silica supports like mesoporous silica (SBA-15), mesoporous organosilica (OSBA-15), and mesoporous carbon (MC), and their performance in 2MN oxidation was studied. It could be seen from the catalytic activity data (Table 5.7) that the immobilized catalysts are quite active and 2MN conversions are comparable to that of the neat catalysts. However, these differ in product selectivities. The selectivity for 2-methy-1,4-naphthoquinone with neat catalyst is in the range 59 % (entries, 1–3, Table 5.7), and interestingly, it is 99 % with immobilized catalysts, which is one of the important functions and achievements of our functionalized mesoporous catalyst system. Among the immobilized catalysts, V2SBA-15 gave 2MN conversion (33 %), with good selectivity for 2-methyl-1,4-naphthoquinone (99 %). V2OSBA exhibited better activity (38 % 2MN conversion) than V2SBA and similar selectivity (99 %). This improved catalytic activity of the material V2OSBA could be attributed to the hydrophobicity of the material. V2MC showed lower activity, 24 % 2MN conversion, and 99 % ketone selectivity in the oxidation of 2MN, which could be due to the blocking of channels by polyoxometalates loaded onto the mesoporous carbon material as the pore size of MC is comparatively lower than that of SBA-15 and OSBA-15 (Table 5.7). The immobilized catalyst, V2OSBA retained the activity of the homogeneous analogue (neat) and gave a clean product 2-methyl-1,4-naphthoquinone (99 %) in 2MN oxidation.

The new materials are truly heterogeneous and are efficient for synthesis of 2-methy-1,4-naphthoquinone (menadione, vitamin K3 precursor). Compared to the conventional preparation of Vitamin K3, this method could be economically viable and environmentally benign as the use of mineral acid and chromium salts is avoided.

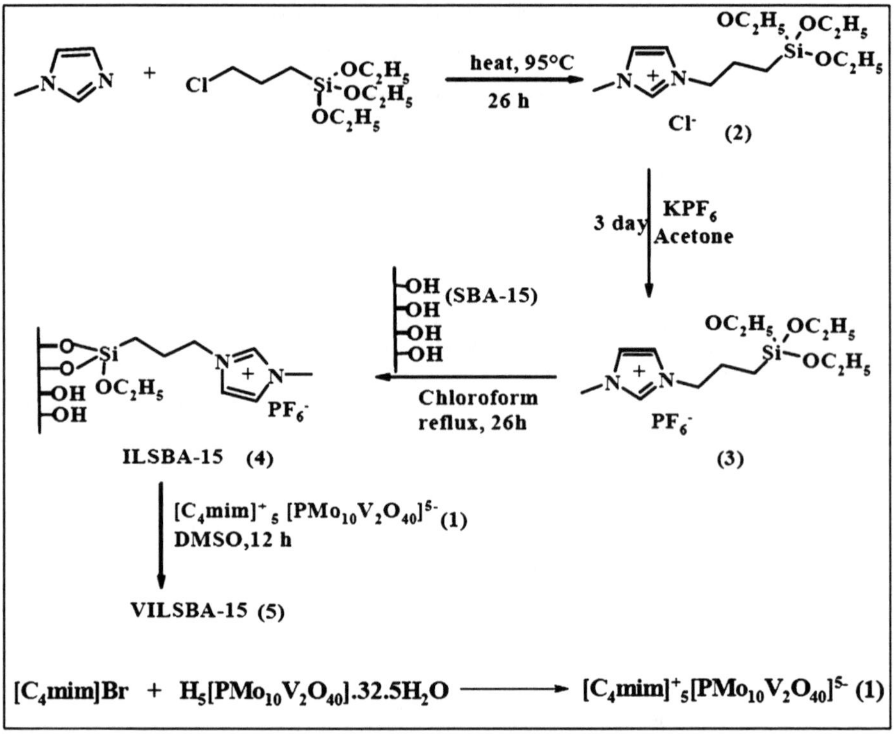

Scheme 5.4 Synthesis of V2ILSBA

3.1.3 $H_5[PMo_{10}V_2O_{40}]\cdot 32.5H_2O$ Immobilized on Ionic Liquid-Modified Mesoporous Silica

The use of ionic liquids, composed entirely of ions with a melting point below 100 °C, has become one of the most prolific areas of ionic liquids research to date, due to their unique properties, including low volatility, high polarity, good stability over a wide temperature range, and selective dissolving capacity by a proper selection of cation and anion [72]. Moreover, since ionic liquids are expensive, it is desirable to minimize the amount of ionic liquid used in usual biphasic reaction system. Dissolving organometallic complexes in supported films of ionic liquids has recently been introduced as a strategy to immobilize molecular catalysts [73, 74], which is known as supported ionic liquid-phase (SILP) strategy. This allows fixing molecular catalysts in a widely tailorable environment. Shi et al. have reported recently the immobilization of peroxotungstates on multilayer ionic liquid brushes-modified silica and as an efficient and reusable catalyst for selective oxidation of sulfides with H_2O_2 [74]. We have tried to developed an effective method for the aerobic oxidation of primary and secondary alcohols to the aldehydes and ketones under mild conditions with good to excellent conversions, using cation-exchanged $H_5[PMo_{10}V_2O_{40}]\cdot 32.5H_2O$ (V) immobilized on 1-butyl-3-methylimidazolium hexafluorophosphate ionic liquid-modified SBA-15 (V2ILSBA) catalyst (Scheme 5.4) [75]. Based on our earlier studies along with some other research groups, we have chosen SBA-15 as mesoporous support for immobilization of ionic liquids and $H_5[PMo_{10}V_2O_{40}]\cdot32.5H_2O$ as active catalyst species.

^{31}P Mass NMR Spectroscopy

The structure of heteropoly anion after immobilization has been evaluated by using NMR spectroscopy. Due to the nature of the compound, ionic liquid-modified $H_5[PMo_{10}V_2O_{40}]\cdot32.5H_2O$ (relatively viscous), we were not able to have a good rotation, so the ^{31}P one shows numerous spinning sidebands around 4.2 ppm (Fig. 5.11a). The ^{31}P NMR spectrum of ionic liquid-modified SBA-15 shows that only the resonance of PF_6^- corresponds to the formula BMIM+. PF_6^- as this anion gives a signal at ca. −150 ppm (Fig. 5.11b). The phosphorus spectrum of ionic liquid-modified $H_5[PMo_{10}V_2O_{40}]\cdot32.5H_2O$ catalyst supported on ionic liquid-modified SBA-15 (V2ILSBA) shows the two signals at 4.2 and 150 ppm of the heteropolyanion and PF_6^- (the multiplet corresponds to the P–F couplings).

Catalytic Oxidation of Alcohol

The oxidation of 1-(Napthylen-2-yl) ethanol, as a model substrate, was first investigated using V2ILSBA under atmospheric oxygen (Table 5.8) in acetonitrile solvent, α,α-azobisisobutyronitrile (AIBN), or *tert*-butylhydroperoxide (TBHP) as radical initiator. We were pleased to find that the alcohol was completely oxidized to methylnapthylketone in 99 % yields. To evaluate the scope of this protocol, the oxidation of other alcohols was further studied. As indicated in Table 5.8 (entries 2–6), secondary alcohols such as diphenylmethanol, cyclohexanol, phenylethanol, 2-hexanol, and 2-phenylpropanol oxidized to the corresponding ketones in high yields. A similar reactivity was observed in the oxidation of substrates having electron-withdrawing and electron-donating groups in the aromatic ring, i.e., 4-methoxy-, 4-methyl-, 4-chloro-, 4-bromo-, and 4-nitro-phenylethanol to corresponding ketones in good yields (entries 7–11).

The oxidation proceeds well with some complicated alcohols also such as benzoin, menthol, [1,7,7]trimethylbicyclo[1,2,2]heptan-2-ol, and 3,5,5-trimethylcyclohex-2-enol, in high yields (entries 12–15). Benzyl alcohol was oxidized to benzaldehyde within 12 h with excellent yield (entry 16). The oxidation profile of a 1:1 mixture of benzyl alcohol and phenyl ethanol (Fig. 5.12) was found that the rate of oxidation of the secondary alcohol (phenyl ethanol) is faster compared to primary alcohol (benzyl alcohol).

The selective oxidation of primary and secondary alcohols to aldehydes and ketones, respectively, without over oxidation of the aldehydes to the carboxylic acids, found for heteropolyoxometallates is electron transfer redox-type oxidations utilizing molybdenum-vanadium mixed-addenda Keggin anions [45]. As per literature, such reactions are thought to proceed by the generalized mechanism, shown in Eq. 1 (Scheme 5.5).

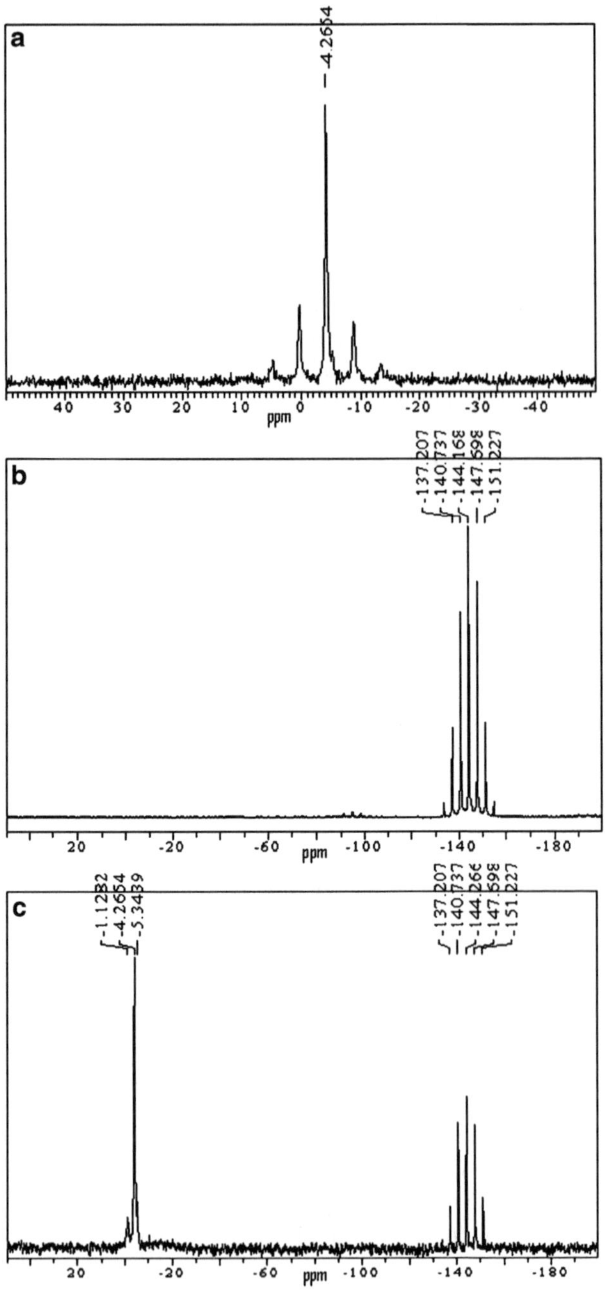

Fig. 5.11 31P CP MAS NMR of (**a**) [C4mim]+ salt of a [PMo10V2O40]5− (**b**) ILSBA and (**c**) V2ILSBA

Table 5.8 Aerobic oxidation of alcohols to aldehydes and ketones[a]

Entry	Substrate	Product	*t* (h)	Yield[b] (%)
1	1-(Napthylen-2-yl) ethanol	2-Methylnapthylketone	7	99
2	Diphenylmethanol	Benzophenone	7	99
3	Cyclohexanol	Cyclohexanone	6	99
4	Phenylethanol	Acetophenone	6	99
5	2-Hexanol	2-Hexanone	5	98
6	2-Phenylpropanol	Propiophenone	7	93
7	4-Methoxy phenylethanol	4-Methoxy acetophenone	6	98
8	4-Methyl phenylethanol	4-Methyl acetophenone	6	96
9	4-Chloro phenylethanol	4-Chloroacetophenone	6	98
10	4-Bromo phenylethanol	4-Bromo acetophenone	6	98
11	4-Nitro phenylethanol	4-Nitro acetophenone	6	94
12	Benzoin	Benzil	7	95
13	Menthol	Menthone	6	96
14	[1, 7, 7]Trimethylbicyclo [2,2,1]heptan-2-ol	Camphor	8	95
15	3,5,5-Trimethylcyclohex-2-enol	3,5,5-Trimethylcyclohex-2-enone	7	94
16	Benzyl alcohol	Benzaldehyde	12	98
17	1, 3-Butanediol	4-Hydroxybutane-2-one	8	83
18	Geraniol	Geranial	10	97
19	Cinnamyl alcohol	Cinnamaldehyde	13	98
20	Pyridine-2-methanol	2-Pyridinecarboxaldehdye	11	96

[a]Substrate (1 mmol) and catalyst (V2ILSBA) (100 mg, 0.02 mol% of V2) were stirred at *ca.* 100 °C in acetonitrile (20 ml) under air pressure (3.4 atm) in Parr autoclave (pressure reactor) and α,α azobisisobutyronitrile (AIBN) or *tert*-butylhydroperoxide (TBHP) as radical initiator being used

[b]GC yield

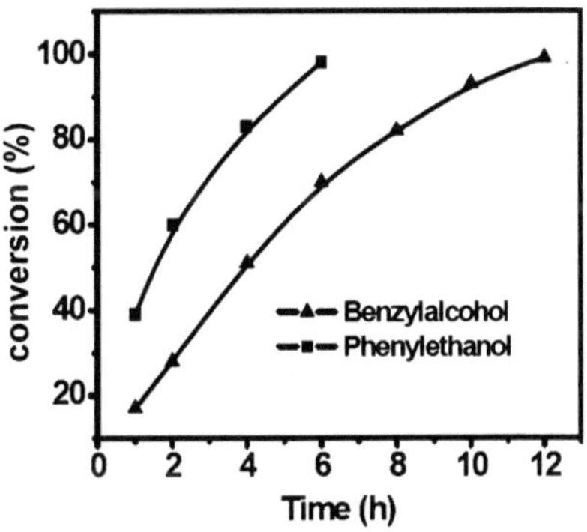

Fig. 5.12 Progress of the oxidation of benzyl alcohol and phenylethanol

Scheme 5.5 Possible mechanism of oxidation of alcohols

$$ArCH_2OH + PMo_{10}V^{V}_{2}O_{40}^{5-} \xrightarrow{-2H^+; -2e^-} ArCHO + H_2PMo_{10}V^{IV}_{2}O_{40}^{5-}$$

$$H_2PMo_{10}V^{IV}_{2}O_{40}^{5-} + 1/2O_2 \longrightarrow PMo_{10}V^{V}_{2}O_{40}^{5-} + H_2O \text{ — (1)}$$

4 Conclusions

Part 1 of the chapter gives a detailed description of zirconia-supported Keggin heteropoly acids as eco-friendly solid catalysts for various kinds of acid-catalyzed reactions. The amount of HPA is based on the monolayer formation on ZrO_2. Fifteen percent of HPA can be supported on ZrO_2 and could be calcined up to 750 °C for high catalytic performance. Keggin structure of heteropoly anion is stable up to 750 °C. High thermal stability of heteropoly acid on zirconia is due to the formation of W–O–Zr bond. STA/ZrO_2 catalyst possesses the higher acidity as compared with 15 % PTA/ZrO_2. These catalysts can be effectively used for variety of acid-catalyzed reactions such as alkylation and acylation reactions of aromatics with alcohols, alkenes, and anhydrides. The catalysts showed high activity, reusability with longer lifetime in most of the reactions. PTA/ZrO_2 showed better performance in benzylation of phenol with benzyl alcohol and acylation of 2-methoxy-naphthalene compared to tungstated zirconia which could be attributed to higher acidity of PTA catalyst.

In Part 2, immobilization of multicomponent polyoxometalates onto solid supports using different strategies, and study of some model oxidation reactions using the supported catalysts is discussed. The structural integrity of mesoporous supports as well as heteropoly acids has been well preserved after immobilization of heteropoly acids to amine functionalized mesoporous network. It has been proven with the help of various characterization techniques and effectively utilized in the selective oxidation of anthracene. Furthermore, C–H activation was carried out in adamantane oxidation by IOHM. Primarily it was observed that the reaction was mainly promoted by vanadium species and the rest of the Keggin ion enhances the activity. Evidence of presence of V(4+)-superoxo species in the reaction mechanism has been demonstrated by EPR spectroscopy.

Mesoporous carbon and organofunctionalized mesoporous silica have also been explored to synthesize a new sets of IOHM. Compared to the conventional preparation of vitamin K3 by oxidation of 2-methylnaphthalene, using these particular IOHM catalyst system could be economical and environmentally benign.

An efficient strategy (supported ionic liquid phase) has been developed for the immobilization of molybdovandophosphoric acid and was utilized in the oxidation of broad range alcohols. High degree of chemoselectivity was obtained using V2ILSBA catalyst. Oxidation rate of the secondary alcohols was found to be faster than that of the primary ones.

Acknowledgments G.V. Shanbhag, A. Bordoloi, S. Sahoo, and B.M. Devassy gratefully acknowledge CSIR (INDIA) and NCL, Pune, for their support. S.B. Halligudi thanks governing council of Centre of Excellence, Vapi.

References

1. Hill CL (1998) Chem Rev 98:1
2. Okuhara T, Mizuno N, Misono M (1996) Adv Catal 41:113
3. Kozhevnikov IV (1998) Chem Rev 98:171
4. Wu Y, Ye X, Yang X, Wang X, Chu W, Hu Y (1996) Ind Eng Chem Res 35:2546
5. Tanabe K, Yamaguchi T (1994) Catal Today 20:185
6. Chuah GK (1999) Catal Today 49:131
7. Yadav GD, Nair JJ (1999) Microporous Mesoporous Mater 33:1
8. Barton DG, Soled SL, Iglesia E (1998) Topics Catal 6:87
9. Kuba S, Lukinskas P, Grasselli RK, Gates BC, Knözinger H (2003) J Catal 216:353
10. Gregorio FD, Keller V (2004) J Catal 225:45
11. Lopez-Salinas E, Hermandez-Cortez JG, Cortes-Jacome MA, Navarrete J, Llanos ME, Vazquez A, Armendariz H, Lopez T (1998) Appl Catal A 175:43
12. Lopez-Salinas E, Hermandez-Cortez JG, Schifter I, Torres-Garcia E, Navarrete J, Gutierrez-Carrillo A, Lopez T, Lottici PP, Bersani D (2000) Appl Catal A 193:215
13. Song H, Rioux RM, Hoefelmeyer JD, Komor R, Niesz K, Grass M, Yang P, Somorjai GA (2006) J Am Chem Soc 128:3027
14. Ganesan R, Viswanathan B (2000) Bull Catal Soc India 10:1–10
15. (a) Mazeaud A, Dromzee Y, Thouvenot R (2000) Inorg Chem 39:4735–4740; (b) Kukovecz Â, KoÂnya Z, Kiricsi I (2001) J Mol Struct 565:121–124; (c) Johnson BJS, Stein A (2001) Inorg Chem 40:801–808; (d) Kaleta W, Nowinska K (2001) Chem Commun 535–536; (e) Maldotti A, Molinari A, Varani G, Lenarda M, Storaro L, Bigi F, Maggi R, Mazzacani A, Sartori G (2002) J Catal 209:210–216; (f) Li HL, Perkas N, Li QL, Gofer Y, Koltypin Y, Gedanken A (2003) Langmuir 19:10409–10413; (g) Wang J, Zhu HU (2004) Catal Lett 93:209–212; (h) Errington RJ, Petkar SS, Horrocks BR, Houlton A, Lie LH, Patole SN (2005) Angew Chem Int Ed 44:1254–1257; (i) Kato N, Tanabe A, Negishi S, Goto K, Nomiya K (2005) Chem Lett 34:238–239; (j) Izumi Y, Urabe K (1981) Chem Lett 663–663; (k) Neumann R, Levin M (1991) J Org Chem 56:5707–5710; (l) Fujibayashi S, Nakayama K, Nishiyama Y, Ishii Y (1994) Chem Lett 1345–1345; (m) Karimi Z, Mahjoub AR, Harati SM (2011) Inorg Chimi Acta 376:1–9; (n) Hajian R, Tangestaninejad S, Moghadam M, Mirkhani V, Mohammadpoor B, Iraj K, Reza A (2011) J Coord Chem 64:4134–4144; (o) Aoki S, Kurashina T, Kasahara Y, Nishijima T, Nomiya K (2011) Dalton Trans 40:1243–1253; (p) Ferreira P, Fonseca IM, Ramos AM, Vital J, Castanheiro JE (2010) App Catal B 98:94–99; (q) Panchenko VN, Borbath I, Timofeeva MN, Goboeloes S (2010) J Mol Catal A 319:119–125; (r) Qi W, Wang Y, Li W, Wu L (2010) Chem Eur J 16:1068–1078; (s) Zhang R, Yang C (2008) J Mater Chem 18:1691–2703; (t) Lee J, Kim H, La KW, Park DR, Jung JC, Lee SH, Song IK (2008) Catal Lett 123:90–95; (u) Kim H, Jung JC, Park DR, Lee H, Lee J, Lee SH, Baeck SH, Lee KY, Yi J, Song IK (2008) Catal Today 132:58–62
16. (a) Okuhara T, Mizuno N, Misono M (1996) Adv Catal 41:113–252; (b) Mizuno N, Misono M (1998) Chem Rev 98:199–218; (c) Kozhevnikov IV (1998) Chem Rev 98:171–198
17. Kholdeeva OA, Maksimchuk NV, Maksimov GM (2010) Catal Today 157:107–113
18. Devassy BM, Lefebvre F, Halligudi SB (2005) J Catal 231:1–10
19. Devassy BM, Halligudi SB (2005) J Catal 236:313–323
20. Devassy BM, Halligudi SB (2006) J Mol Catal A Chem 253:8–15
21. Busca G (1998) Catal Today 41:191–206
22. Rocchiccioli-Deltcheff C, Fournier M, Franck R, Thouvenot R (1983) Inorg Chem 22:207–216
23. Teague CM, Li X, Biggin ME, Lee L, Kim J, Gewirth AA (2004) J Phys Chem B 108:1974–1985
24. Echizen MT, Nagata K, Yoshinaga Y, Okuhara T (2003) J Mol Catal A 201:145–153
25. Loridant S, Feche C, Essayem N, Figueras F (2005) J Phys Chem B 109:5631–5637
26. Scheithauer M, Grasselli RK, Knözinger H (1998) Langmuir 14:3019–3029
27. (a) Misono M (2001) Chem Commun 1141; (b) Dillon CJ, Holles JH, Davis RJ, Labinger JA, Davis ME (2003) J Catal 218:54; (c) Ghanbari-Siahkali A, Philippou A, Dwyer J, Anderson

MW (2000) Appl Catal A 192:57; (d) Molnar A, Beregszaszi T, Fudala A, Lentz P, Nagy JB, Konya Z, Kiricsi I (2001) J Catal 202:379; (e) Uchida S, Inumaru K, Misono M (2000) J Phys Chem B 104:8108
28. Kocal JA, Vora BV, Imai T (2001) Appl Catal A Gen 221:295–301
29. Shanbhag GV, Devassy BM, Halligudi SB (2004) J Mol Catal A Chem 218:67–72
30. Bachiller-Baeza B, Anderson JA (2004) J Catal 228:225–233
31. Devassy BM, Shanbhag GV, Mirajkar SP, Böhringer W, Fletcher J, Halligudi SB (2005) J Mol Catal A Chem 233:141–146
32. Devassy BM, Shanbhag GV, Halligudi SB (2006) J Mol Catal A Chem 247:162–170
33. Devassy BM, Shanbhag GV, Lefebvre F, Halligudi SB (2004) J Mol Catal A Chem 210:125–130
34. (a) Sawant DP, Devassy BM, Halligudi SB (2004) J Mol Catal A Chem 217:211–217; (b) Devassy BM, Halligudi SB, Elangovan SP, Ernst S, Hartmann M, Lefebvre F (2004) J Mol Catal A Chem 221:113–119; (c) Devassy BM, Lefebvre F, Böhringer W, Fletcher J, Halligudi SB (2005) J Mol Catal A Chem 236:162–167; (d) Sunita G, Devassy BM, Vinu A, Sawant DP, Balasubramanian VV, Halligudi SB (2008) Catal Commun 9:696–702; (e) Justus J, Vinu A, Devassy BM, Balasubramanian VV, Böhringer W, Fletcher J, Halligudi SB (2008) Catal Commun 9:1671–1675
35. (a) Devassy BM, Halligudi SB, Hegde SG, Halgeri AB, Lefebvre F (2002) Chem Commun 1074–1075; (b) Devassy BM, Shanbhag GV, Lefebvre F, Böhringer W, Fletcher J, Halligudi SB (2005) J Mol Catal A Chem 230:113–119
36. Kim H, Jung JC, Song IK (2007) Catal Surv Asia 11:114–122
37. (a) Okun NM, Anderson TM, Hill CL (2003) J Am Chem Soc 125:3194–3195; (b) Okun NM, Anderson TM, Hill CL (2003) J Mol Catal A 197:283–290
38. Khenkin AM, Neumann R, Sorokin AB, Tuel A (1999) Catal Lett 63:189–192
39. (a) Neumann R, Cohen M (1997) Angew Chem Int Ed 36:1738–1740; (b) Cohen M, Neumann R (1999) J Mol Catal A 146:291–298
40. Neumann R, Miller H (1995) J Chem Soc Chem Commun 2277–2278
41. Kanno M, Yu-ki M, Yasukawa T, Hasegawa T, Ninomiya W, Ooyachi K, Imai H, Tatsumi T, Kamiya Y (2011) Catal Commun 13:59–62
42. Kala Raj NK, Deshpande SS, Ingle RH, Raja T, Manikandan P (2004) Catal Lett 98:217–223
43. Ressler T, Dorn U, Walter A, Schwarz S, Hahn AHP (2010) J Catal 275:1–10
44. Mizuno N, Yamaguchi K, Kamata K (2011) Catal Sur Asia 15:68–79
45. Bordoloi A, Lefebvre F, Halligudi SB (2007) J Catal 247:166–175
46. (a) Siahkali AG, Philippou A, Dwyer J, Anderson MW (2000) Appl Catal A 192:57–69; (b) Kanda Y, Lee KY, Nakata S, Asaoka S, Misona M (1988) Chem Lett 1:139–139; (c) Mastikhin VM, Kulikov SM, Nosov AV, Kozhevnikov IV, Mudrakovsky IL, Timofeeva MN (1990) J Mol Catal 60:65–70; (d) Huang W, Todaro L, Yap GPA, Beer R, Francesconi LC, Polenova T (2004) J Am Chem Soc 126:11564–11573
47. (a) Kasai J, Nakagawa Y, Uchida S, Yamaguchi K, Mizuno N (2006) Chem Eur J 12:4176–4184; (b) Casarini D, Centi G, Jiru P, Lena V, Tvaruzkova Z (1993) J Catal 143:325–344
48. (a) Giles GI, Sharma RP (2005) J Peptide Sci 11:17–423; (b) Priebe W (ed) (1993) Anthracycline antibiotics, new analogues, methods of delivery, mechanisms of action. In: American Chemical Society Symposium Series 574; (c) Driscoll JS, Hazard GF, Wood HB, Goldin A (1974) Cancer Chemother Rep 2:1–362; (d) Cairns D, Michalitsi E, Jenkins TC, Mackay SP (2002) Bioorg Med Chem 10:803–807; (e) Lown JW (1993) Pharmacol Ther 60:185–214; (f) Cheng C, Zee-Cheng RKY (1983) Prog Med Chem 20:83–118; (g) Huang HS, Chiou JF, Fong Y, Hou CC, Lu YC, Wang JY, Shih JW, Pan YR, Lin JJ (2003) J Med Chem 46:3300–3307; (h) Ge P, Russell RA (1997) Tetrahedron 53:7469–17476; (g) Lee CP, Singh KH (1982) J Nat Prod 45:206–210
49. Monneret C (2001) Eur J Med Chem 36:483–493
50. Ullmann F, Gerhartz W, Yamamoto YS, Campbell FT, Pfefferkorn R (1985) Ullmann's encyclopedia of industrial c hemistry, vol 2. VCH, Weinheim, pp 347–348
51. Butterworth BE, Mathre OB, Ballinger K (2001) Mutagenesis 16:169–177

52. Kozhevnikov I (1995) Catalyst for fine chemical synthesis: catalysis by polyoxometalates. Wiley, New York, p 138
53. (a) Hill CL (1998) In: Baumstark AL (ed) Advances in oxygenated processes, vol 2. JAI Press, London, pp 1–30; (b) Hudlucky M (1990) In: Oxidations in organic chemistry, ACS monograph series. American Chemical Society, Washington, DC; (c) Shilov AE, Shul'pin GB (2000) In: Activation and catalytic reactions of saturated hydrocarbons in the presence of metal complexes. Kluwer Academic Publishers, Dordrecht; (d) Shilov AE, Shul'pin GB (1997) Chem Rev 97:2879–2932; (e) Mizuno N, Kamata K, Yamaguchi K (2010) Topics Catal 53:876–893
54. (a) Barton DHR (1996) Chem Soc Rev 25:237–239; (b) Perkins MJ (1996) Chem Soc Rev 25:229–236
55. Bordoloi A, Lefebvre F, Halligudi SB (2007) Appl Catal A Gen 333:143–152
56. (a) Bang Y, Park DR, Lee YJ, Jung JC, Song IK (2011) J Chem Engg 28:79–83; (b) Ferreira P, Fonseca IM, Ramos AM, Vital J, Castanheiro JE (2011) Catal Commun 12:573–576
57. Hong UG, Park DR, Park S, Seo JG, Bang Y, Hwang S, Youn MH, Song IK (2009) Catal Lett 132:377–382
58. Fujibayashi S, Nakayama K, Hamamoto M, Sakaguchi S, Nishiyama Y, Ishii Y (1996) J Mol Catal A 110:105–117
59. Xu L, Boring E, Hill CL (2000) J Catal 195:394–405
60. Bordoloi A, Mathew N, Lefebvre F, Halligudi SB (2008) Microporous Mesoporous Mater 115:345–355
61. Guo W, Park JY, Oh MO, Jeong HW, Cho WJ, Kim I, Ha CS (2003) Chem Mater 15:2295–2298
62. Zhang WH, Daly B, O'Callaghan J, Zhang L, Shi JL, Li C, Morris MLA, Holmes JD (2005) Chem Mater 17:6407–6415
63. Hu Q, Pang J, Wu Z, Lu Y (2006) Carbon 44:1298–1352
64. Jarrais B, Silva AR, Freire C (2005) Eur J Inorg Chem 2005:4582–4589
65. Khenkin AM, Vigdergauz I, Neumann R (2000) Chem Eur J 6:875–882
66. Fieser LF, Tushler M, Sampson WL (1941) J Biol Chem 137:659–692
67. (a) Periasamy M, Bhatt MV (1978) Tetrahedron Lett 4:4561–4562; (b) Kreh RP, Spotnitz RM, Lundquist JT (1989) J Org Chem 54:1526–1531; (c) Skarzewski J (1984) Tetrahedron 40:4997–5000; (d) Hiranuma H, Miller SI (1982) J Org Chem 47:5083–5088; (e) Kowalski J, Ploszynska J, Sobkowiak A (1998) J Appl Electrochem 28:1261–1264
68. Fieser LF (1940) J Biol Chem 133:391–396
69. (a) Sheldon RA (1993) Top Curr Chem 164:21–43; (b) Adam W, Herrmann WA, Lin J, Saha-Moller CR, Fischer RW, Correia JDG (1995) Angew Chem Int Ed 33:2475–2477; (c) Adam W, Herrmann WA, Lin J, Saha-Moller CR (1994) J Org Chem 59:8281–8283
70. (a) Song R, Sorokin A, Bernadou J, Meunier B (1997) J Org Chem 62:673–678; (b) Anunyiata OA, Pierella LB, Beltramone AR (1999) J Mol Catal A 149:255–261
71. Bohle A, Schubert A, Sun Y, Thiel WR (2006) Adv Synth Catal 348:1011–1015
72. (a) Welton T (1999) Chem Rev 99:2071–2084; (b) Zhao D, Wu M, Kou Y, Min E (2002) Catal Today 74:157–189
73. (a) Riisager A, Fehrmann R, Flicker S, Van Hal R, Haumann M, Wasserscheid P (2005) Angew Chem Int Ed 44:815–819; (b) Gruttadauria M, Riela S, Aprile C, Meo PL, D'Anna F, Noto R (2006) Adv Synth Catal 348:82–92; (c) Mehnert CP (2004) Chem Eur J ,1:50–56; (d) Sahoo S, Kumar P, Lefebvre F, Halligudi SB (2009) Appl Catal A Gen 354:17–25
74. Shi X, Han X, Maa W, Wei J, Li J, Zhang Q, Chen Z (2011) J Mol Catal 341:57–62
75. Bordoloi A, Sahoo S, Lefebvre F, Halligudi SB (2008) J Catal 259:232–239

Chapter 6
Glycerol Etherification with Light Alcohols Promoted by Supported $H_3PW_{12}O_{40}$

Rodrigo Lopes de Souza, Wilma Araujo Gonzales, and Nadine Essayem

Contents

1 Introduction ... 142
2 Experimental Part ... 143
2.1 Catalyst Preparation ... 143
2.2 Microcalorimetry of NH_3 Adsorption ... 143
2.3 Catalytic Tests ... 144
3 Results and Discussion ... 144
3.1 Catalyst Characterisations ... 144
3.2 Catalytic Activities ... 148
4 Conclusion ... 151
References ... 152

Abstract Glycerol etherification with *tert*-butyl alcohol or ethanol was investigated using $H_3PW_{12}O_{40}$ supported on various supports (SiO_2, active carbon, $ZrOH_X$ and $NbOH_X$) and the sulfonated resins Amberlyst 35, as reference catalyst. Supported catalysts with optimised $H_3PW_{12}O_{40}$ dispersion were prepared via an incipient wetness impregnation of the vacuum-treated support with an aqueous $H_3PW_{12}O$ solution under reduced pressure. The support coverage with the Keggin anions represents 0.4

R.L. de Souza
Institut de Recherche sur la Catalyse et l'Environnement de Lyon (IRCELYON)-CNRS-Lyon1, Lyon, France

Instituto Militar de Engenharia (IME), Rio de Janeiro, Brazil

W.A. Gonzales
Instituto Militar de Engenharia (IME), Rio de Janeiro, Brazil

N. Essayem (✉)
Institut de Recherche sur la Catalyse et l'Environnement de Lyon (IRCELYON)-CNRS-Lyon1, Lyon, France
e-mail: nadine.essayem@ircelyon.univ-lyon1.fr

A. Patel (ed.), *Environmentally Benign Catalysts: For Clean Organic Reactions*,
DOI 10.1007/978-94-007-6710-2_6, © Panch Tattva Publishers, Pune, India 2013

monolayer. Microcalorimetry of NH_3 adsorption reveals significant influence of the nature of the support on the acid strength of the supported catalysts which can be correlated with the strength of the $H_3PW_{12}O_{40}$ interaction with the support, studied by DTA and XRD. It was shown that the glycerol etherification with ethanol is a more acidic-demanding reaction than glycerol etherification with *tert*-butyl alcohol. Indeed, while all the supported catalysts catalyse the glycerol etherification with *tert*-butyl alcohol at 120 °C such as Amberlyst 35, only $H_3PW_{12}O_{40}$ supported on carbon or silica promotes the glycerol etherification with ethanol at 200 °C according to their higher strength. Finally, $H_3PW_{12}O_{40}$ supported on carbon shows a water tolerance comparable to that of the sulfonated resins.

1 Introduction

The Renewable Directive Energy (RED) fixed the percentage of biofuels to 10 % in 2020 in the EU. Hence, biofuels are expected to grow at 7–8 % per year. This concerns biodiesel as well as bioethanol, bio-oils or ethers. The production of biodiesel by transesterification of vegetable oils with alcohols leads to the co-production of glycerol which represents 10 wt% of the raw material. The economic equilibrium of the transesterification process depends on the glycerol valorisation. Glycerol ethers are glycerol derivatives which find applications as fuel additives. Mixture of glycerol ethers may be used as blending in biodiesel [1]. More generally, glycerol ethers were described as possible base for fuel composition [2, 3]. The addition of glycerol ethers to bioethanol is known to lower the vapour pressure, and their addition to the biodiesel is described to lower its viscosity [1]. Moreover, glycerol ethers may advantageously substitute conventional oxygenates additives as MTBE which presents adverse environmental impacts [3]. The positive influence of glycerol ethers on the reduction of particle emission was described as well. Note that it was also recognised that hydroxyl groups of partially etherified glycerol ethers would help to incorporate small amounts of water in the fuels which would contribute to reduce the NO_X emission [3].

tert-Butyl-glycerol ethers may be produced by reaction of glycerol with isobutene in the presence of acidic resins [1, 4] or zeolites [5]. Their preparation by etherification with *tert*-butyl alcohol was also reported [6, 7]. Both reactions were shown to be catalysed by acidic resins. The use of an olefin instead of an alcohol presents the main advantage to avoid the co-production of water which may weaken the activity of the acid catalyst. It is clear that ethyl glycerol ethers present a higher interest as fuel additives since ethanol can be used in the transesterification process. However, this is a more challenging target addressed by only a few teams [8]. Here, we report a comparative study of glycerol etherification with *tert*-butyl alcohol and ethanol in order to highlight the catalyst requirement in terms of acidity strength.

To reach this objective, solid acid catalysts were synthesised via dispersion of $H_3PW_{12}O_{40}$ on different supports using an optimised method, and Amberlyst 35 was used as a reference catalyst. The catalysts were characterised in detail, and their acidity was studied by NH_3 adsorption monitored by microcalorimetry. The catalytic performances were investigated both in the glycerol etherification with *tert-butyl* alcohol and ethanol.

2 Experimental Part

2.1 Catalyst Preparation

$H_3PW_{12}O_{40}{\cdot}xH_2O$ was supplied from Aldrich. The silica support was supplied from Grace Davidson. This is a mesoporous silica with pore diameters ranging from 5 to 10 nm and a BET surface area of 320 $m^2\ g^{-1}$. Niobium oxide hydrate (HY-340) from Companhia Brasileira de Metalurgia e Mineração (CBMM) has a BET surface area of 130 $m^2\ g^{-1}$. The activated carbon, supplied from Sigma-Aldrich, has a BET surface area equal to 1600 $m^2\ g^{-1}$. $ZrOH_X$ was prepared following the method reported previously [9]. $ZrOH_X$ has a BET surface area of 211 $m^2\ g^{-1}$. The HPA loadings were adjusted in order to cover 40 % of the surface of the support by the Keggin anions, considering that a Keggin unit occupies a surface of $1.44{\cdot}10^{-18}\ m^2$. The preparation of the supported catalysts was optimised and reported previously [10]. The method consists first in a vacuum treatment of a support at 100 °C for 1 h in a rotary evaporator, and then the incipient wetness impregnation was performed by contacting an aqueous solution of $H_3PW_{12}O_{40}$ with the vacuum-treated support at ambient temperature by sorption, drop by drop. Then, the wet solid is kept under agitation in the rotary evaporator equipment at ambient temperature for 1 h. Finally, the solid is dried by lyophilisation. The chemical compositions are reported in Table 6.1.

2.2 Microcalorimetry of NH_3 Adsorption

The acidic properties were measured by NH_3 adsorption at 80 °C, using a Tian-Calvet calorimeter coupled to volumetric equipment. The catalyst samples (~0.1 g) were first evacuated at 200 °C for 1 h under secondary vacuum, then placed into the

Table 6.1 Composition of the supported catalysts

	Target values				Experimental values
Samples	*S* BET ($m^2\ g^{-1}$)	$S_{HPA}/S_{Support}$	$H_3PW_{12}O_{40}$ (wt%)	*W* (wt%)	*W* (wt%)
HPA/SiO_2	320	0.4	31	21	19.0
HPA/C	1600	0.4	70	48	47.2
$HPA/NbOH_X$	130	0.4	16	11	10.8
$HPA/ZrOH_X$	211	0.4	24	16	16.0

calorimeter up to the stabilisation of the experiment temperature, 80 °C. Then, the solid was contacted with small doses of gas up to equilibrium, and the differential enthalpy of adsorption was recorded together with the amount of sorbed ammonia.

2.3 Catalytic Tests

Glycerol, *tert*-butyl alcohol and ethanol were of analytical grade purity. The etherification reactions were carried out in a stainless steel autoclave reactor (70 cm^3) with a magnetic stirring. The reactions were carried out with an excess of alcohol as solvent. The reaction medium was pressurised at 17 bars with Ar atmosphere. In a typical run, the following quantities were loaded in the autoclave: 5 g (0.0543 mol) of glycerol, 0.375 g of catalyst (7.5 wt%/glycerol) and 0.22 mol of alcohol (molar ratio alcohol/glycerol = 4). Prior to the reaction test, the solid acid catalysts were carefully dehydrated. The sulfonated resin was dried for 12 h at 110 °C. To get the supported 12-tungstophosphoric acid in their anhydride form, the supported catalysts were placed in a glass cell connected to a vacuum equipment. The temperature program was as follows: 2 h at 200 °C using a heating rate of 1.6 °C min^{-1}. The reaction mixtures were analysed by gas chromatography, using a polar chromatographic capillary column CP-WAX 52CB (30 m × 0.25 mm × 0.25 μm) and the following temperature program: from 50 to 220 °C at a heating rate of 8 °C·min^{-1}, then 10 min at 220 °C.

3 Results and Discussion

3.1 Catalyst Characterisations

3.1.1 Differential Thermal Analysis (DTA)

The various supports, the bulk and supported $H_3PW_{12}O_{40}$ were characterised by TGA-DTA. Only the DTA curves are presented for clarity reasons. The DTA curves of the supports are shown in Fig. 6.1a. The endothermic phenomenon observed at $T < 100$ °C is due to the loss of hydration water, while the ones between 100 and 200 °C can be ascribed to a partial dehydroxylation of $ZrOH_X$ and $NbOH_X$. The exothermic peaks observed, respectively, at 420 and 560 °C for $ZrOH_X$ and $NbOH_X$ are ascribed to the formation of the oxides: ZrO_2 and Nb_2O_5. Figure 6.1b shows the DTA curves of the bulk and supported $H_3PW_{12}O_{40}$. The DTA curve of 12-tunsgtophosphoric acid is the expected one showing two endothermic phenomena at temperature lower than 300 °C due to the release of crystallisation water molecules, and the exothermic fine peak is ascribed to the crystallisation of the simple oxides WO_3 and P_2O_5 occurring when the Keggin structure was destroyed.

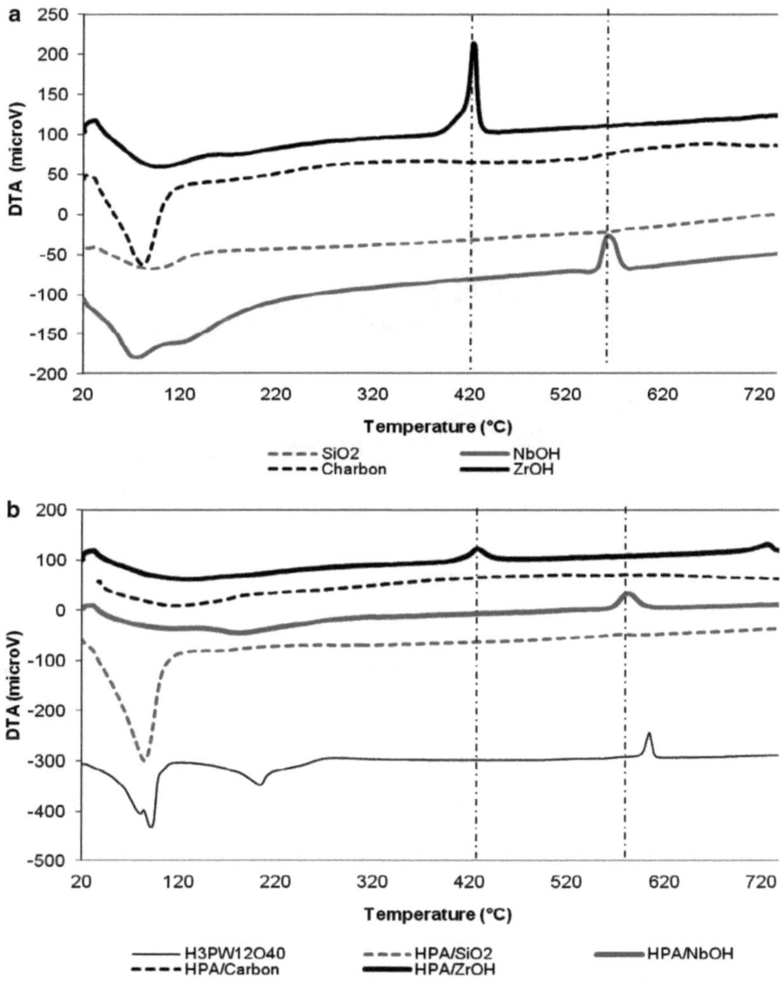

Fig. 6.1 (**a**) DTA curves of the supports: SiO_2, carbon, $NbOH_X$ and $ZrOH_X$. (**b**) DTA curves of bulk and supported $H_3PW_{12}O_{40}$

The intense endothermic peak observed at $T < 100$ °C on HPA/SiO_2 seems to be due to loss of crystallisation water associated by the HPA part of the solid. On the opposite, the DTA curve of HPA/C does not show a clear endothermic phenomenon at low temperature which excludes a significant loss of water of hydration/crystallisation. This may account for a relative hydrophobicity of HPA/C. On the DTA curves of HPA/SiO_2 and HPA/C, it is not possible to distinguish any exothermic peaks around 600 °C, which might have been attributed to the crystallisation of WO_3 and P_2O_5 as for the bulk acid. This distinct behaviour indicates the probable high dispersion of the HPA on the silica and carbon support. This will be confirmed by the XRD patterns reported in the following. The DTA curves recorded on HPA/

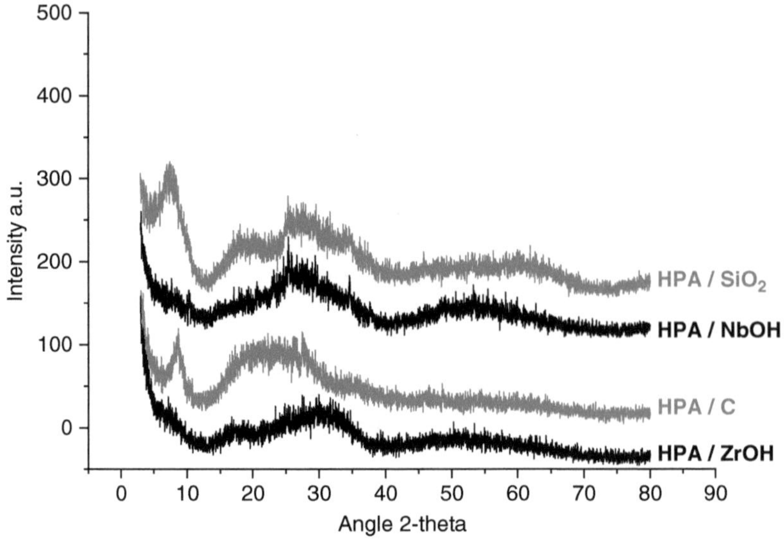

Fig. 6.2 XRD patterns of the supported $H_3PW_{12}O_{40}$

$NbOH_X$ and HPA/$ZrOH_X$ are characterised by the absence of significant endothermic peaks at low temperature, while exothermic peaks are observed at high temperature. The temperature of the exothermic peaks corresponds to the ones observed on the HPA-free supports. For HPA/$ZrOH_X$, an additional exothermic peak is seen at 720 °C. As a general comment, one can say that the DTA curves of the supported HPA are significantly different to that of the bulk acid suggesting the occurrence of modifications of the primary and secondary structures of the dispersed polyoxometalate as regards the bulk ones.

3.1.2 XRD Analysis

The X-ray diffraction patterns of the supported catalysts are shown in Fig. 6.2.

These XRD patterns are characterised of amorphous materials where the XRD peaks of the stable crystallised hydrates of the 12-tunsgtophosphoric acid cannot be distinguished ($H_3PW_{12}O_{40}$, $21H_2O$, $H_3PW_{12}O_{40}$, $13H_2O$, $H_3PW_{12}O_{40}$, $6H_2O$). The absence of 12-tungstophosphoric acid crystallites confirms the high dispersion of the Keggin clusters on the supports and/or the occurrence of modifications of the anion primary and/or secondary structures which might prevent any long-range organisation. These results highlight that our method of impregnation is efficient to disperse high amount of $H_3PW_{12}O_{40}$ even on relatively inert supports such as silica or carbon. In preliminary investigations, it was shown that for higher HPA loadings (>50 wt%), it was not possible to highly disperse the polyoxometalate on a silica support; the XRD patterns showed the presence of 12-tungstophosphoric hexahydrate [11].

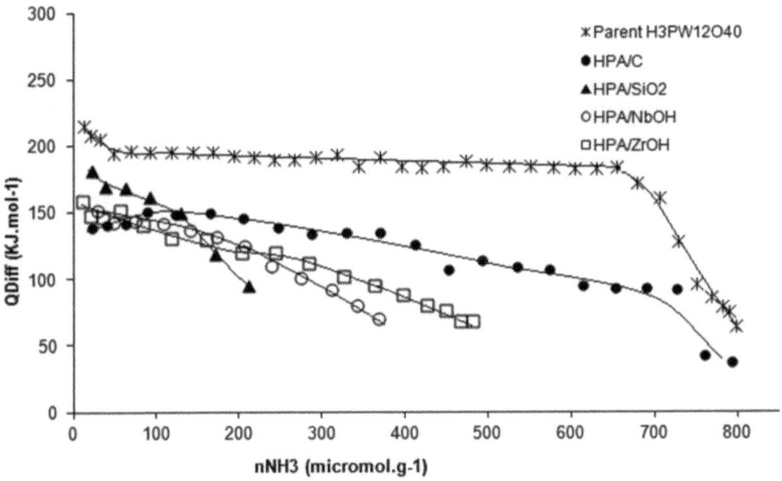

Fig. 6.3 Acid force of the different catalysts of the HPA supported

3.1.3 Microcalorimetry of NH_3 Adsorption

The acidic features of the catalysts were investigated using NH_3 adsorption monitored by microcalorimetry. Prior to the NH_3 adsorption, the bulk and the supported $H_3PW_{12}O_{40}$ were pretreated under a secondary vacuum at 200 °C for 1 h. The calorimetric curves are shown in Fig. 6.3. As a general comment, one can observe that all the supported HPA do not possess acid sites of homogeneous strength contrary to the bulk acid $H_3PW_{12}O_{40}$ or its acidic cesium salts [12]. A continuous decrease of the differential heat of adsorption with the NH_3 coverage is observed on all the supported catalysts accounting for the presence of acid sites of heterogeneous strengths. This indicates the occurrence of HPA/support interactions resulting in modifications of the acid strength. Based on the average heat of NH_3 adsorption at half NH_3 coverage, the following ranking is observed:

$$H_3PW_{12}O_{40} > HPA/SiO_2 > HPA/C > HPA/NbOH_x > HPA/ZrOH_x$$

This ranking is the expected one since the interaction of the HPA with the support will depend on the acido-basicity of the support. Thus, zirconium hydroxide was expected to interact more strongly than silica or carbon with $H_3PW_{12}O_{40}$.

The supported samples were prepared in order to keep unchanged the HPA coverage on the different supports. Then distinct HPA loadings were used (see Table 6.1). It is interesting to note that if HPA/C is the catalyst with the highest amount of acid sites in agreement with the highest HPA loading, the total amount of acid sites measured on HPA/SiO_2 is surprisingly lower than $HPA/ZrOH_X$ and $HPA/NbOH_X$. On the two latter catalysts, the supports might contribute to the total amount of acid sites.

Table 6.2 Glycerol etherification with *tert*-butyl alcohol and ethanol in the presence of Amberlyst A35

			Selectivity (%)		
Alcohol	*T* (°C)	Conversion (%)	Mono-alkyl ether	Di-alkyl ether	Tri-alkyl ether
tert-Butyl alcohol	60	32	92	8	–
	80	54	91	9	–
	120	55	79	21	–
Ethanol	60	1	–	–	–
	80	2	–	–	–
	120	2.5	–	–	–
	160	9	95	5	–

Conditions: alcohol/glycerol = 4 mol/mol, time = 4 h, m cata = 0.39 g, glycerol = 0.275 mol

3.2 Catalytic Activities

3.2.1 Glycerol Etherification in the Presence the Reference Solid Acid Catalyst: Amberlyst 35

The catalytic performances were investigated first in the presence of Amberlyst 35 which can be considered as a reference solid acid catalyst for etherification reactions. Glycerol etherification with *tert*-butyl alcohol and ethanol was compared (Table 6.2). It was clearly shown that glycerol etherification with ethanol is a much more demanding reaction than glycerol etherification with *tert*-butyl alcohol. Indeed, Amberlyst 35 promotes the glycerol etherification with *tert*-butyl alcohol at relatively low temperature, 120 °C, and the glycerol conversion achieved 55 %. When *tert*-butyl alcohol is replaced by ethanol, the glycerol is only slightly converted in ethers in equivalent conditions.

A higher reaction temperature is required to observe a significant conversion; 9 % is achieved at 160 °C. However, higher temperatures cannot be investigated in the presence of the sulfonated resin due to its limited thermal stability. The initial activation step of the reaction is most likely the protonation of the light alcohols leading to the formation of an intermediate carbocation. This helps to rationalise the differences in the reactivity of ethanol and *tert*-butyl alcohol: the *tert-butyl* carbenium ion, a tertiary carbenium ion, is by far more stable than the primary carbenium ion issued from the protonation of ethanol.

3.2.2 Catalytic Activity of Supported Heteropolyacids

Higher temperatures can be investigated in the presence of supported heteropolyacids: the catalysts were evaluated in glycerol etherification reactions with *tert*-butyl alcohol at 120 °C and with ethanol at 200 °C (Table 6.3).

For glycerol etherification with *tert*-butyl alcohol, $H_3PW_{12}O_{40}$ dispersed on the carbon support and on silica are the most active catalysts, with respectively conversions of 48.5 and 37 %. These values are not correlated with the total amount of acid

Table 6.3 Glycerol etherification with *tert*-butyl alcohol and ethanol catalysed by supported heteropolyacids

Catalysts	*tert*-Butyl alcohol Tr = 120 °C conversion (%)	Ethanol Tr = 200 °C conversion (%)	Qdiff (kJ·mol^{-1} NH3) *Average value at NH3 half coverage*
HPA/C	48.5	35	150
HPA/SiO2	37	23	160
HPA/NbOHX	19	0.5	135
HPA/ZrOH$_X$	14	–	130

Conditions: Tr = 120 °C (*tert*-butyl alcohol), Tr = 200 °C (ethanol), alcohol/glycerol = 4 mol/mol, time = 4 h, m cata = 0.39 g, glycerol = 0.275 mol

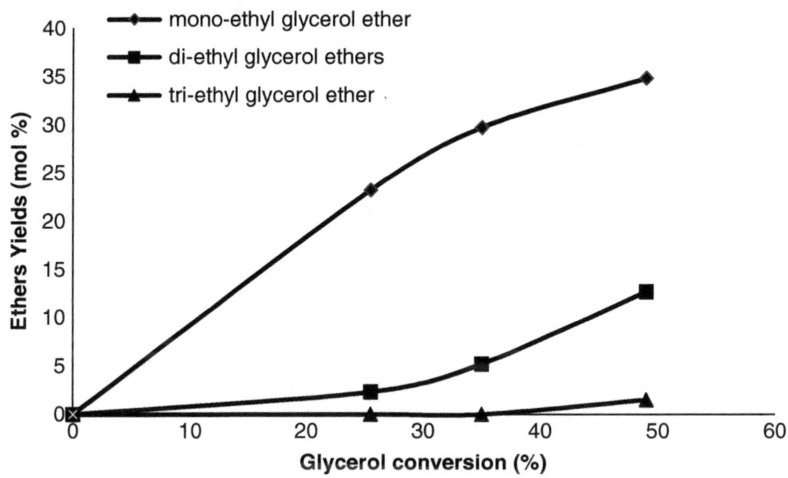

Fig. 6.4 Evolution of the ethers yields with the glycerol conversion in the presence of HPA/C

sites; the carbon-based catalyst possesses three times more acid sites than the silica-based catalyst. Their efficiency seems rather to be related to the strength of the acid sites. These remarks still hold through for the HPA dispersed on $NbOH_X$ and $ZrOH_X$; HPA/$NbOH_X$ which is more active than HPA/$ZrOH_X$ possesses also stronger acid sites than the latter catalyst. Again, it was observed that the glycerol etherification with ethanol is a more acidic-demanding reaction than the etherification with *tert*-butyl alcohol: only the strongest acids, HPA/SiO_2 and HPA/C, promote the etherification with ethanol. The supported $H_3PW_{12}O_{40}$ catalysts which present acid sites of lowest strength, HPA/$NbOH_X$ and HPA/$ZrOH_X$, do not catalyse the glycerol etherification with ethanol (Table 6.3). The two active catalysts, HPA/SiO_2 and HPA/C, possess acid sites characterised by heat of NH_3 adsorption $\geq$150 kJ·mol^{-1}.

The reaction selectivity was observed to depend strongly on the reaction extent. Figure 6.4 shows the quasi exclusive formation of mono-ethyl glycerol

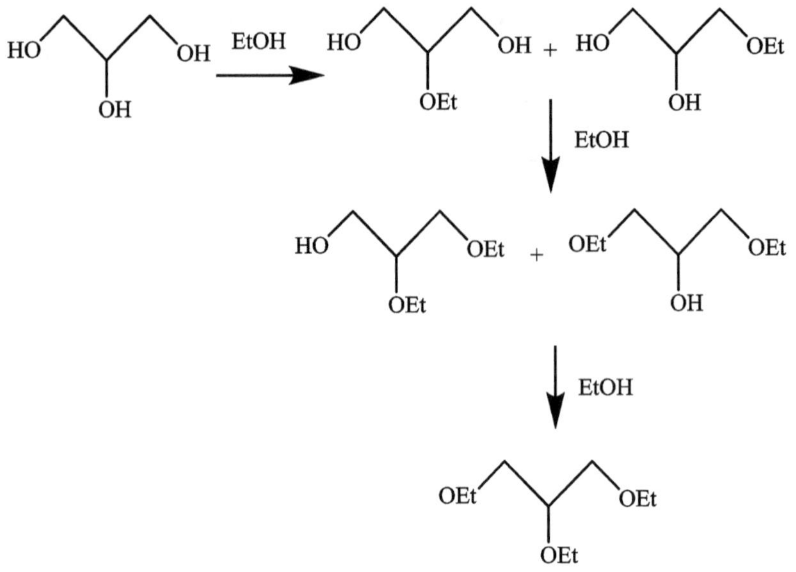

Scheme 6.1 Successive pathway of ethyl glycerol ether formation

ether up to a glycerol conversion of 20 %. Then, di-ethyl glycerol ethers are formed and reach a total yield of 10 % when the glycerol conversion achieved 50 %. Tri-ethyl glycerol ethers were only slightly detected when the glycerol conversion overreaches 50 %.

The evolution of the ethers yields with the reaction extent is in favour of a successive pathway where the primary products are the mono-ethyl-glycerol ethers, which are further converted to di-ethyl glycerol ethers and finally to tri-ethyl glycerol ether (Scheme 6.1).

3.2.3 Water Tolerance of HPA/C and Amberlyst 35

The influence of water addition in the reaction media was investigated on the most active catalysts, the carbon-based materials. This was done for two reasons: first, the reaction itself produces water which might cause a catalyst deactivation; second, the use of raw feedstocks, which may content water, are generally preferred for the process economy. The reaction of glycerol etherification with *tert*-butyl alcohol was performed in the presence of 2 and 5 % of water. It was observed that the glycerol conversion in ethers is poorly sensitive to the addition of water at least up to 5 % (Table 6.4). The close behaviour of Amberlyst 35 and HPA/C as regards the water addition may be explained by their hydrophobicity brought by their carbon content.

Table 6.4 Influence of water addition on the reaction of glycerol etherification with *tert*-butyl alcohol catalysed with Amberlyst 35 and HPA/C

			Yield (%)	
			Mono-*tert*-butyl	Di- and tri-*tert*-butyl
Catalyst	Water (wt %)	Glycerol conversion (%)	Glycerol ether	Glycerol ether
Amb 35	0	55	42	10
Amb 35	2	66	50	16
Amb 35	5	60	49	11
HPA/C	0	52	42	10
HPA/C	2	55	43	12
HPA/C	5	58	47	11

Conditions: Tr = 120 °C, *tert*-butyl alcohol/glycerol = 4 mol/mol, Catalyst: HPA/C, m cata = 0.39 g, glycerol = 0.275 mol

Table 6.5 Chemical analysis of leached W in the reaction liquid media during the glycerol-ethanol etherification reaction

Catalysts	HPA/SiO_2	HPA/$NbOH_X$	HPA/C	HPA/$ZrOH_X$
W (wt%)	0.21	0.03	0.76	0.24

Conditions of reaction: Tr = 200 °C (ethanol), ethanol/glycerol = 4 mol/mol, time = 4 h; Catalyst: m cata = 0.39 g, glycerol = 0.275 mol

3.2.4 Catalyst Resistance Towards Leaching

To check the occurrence of HPA leaching, W chemical analysis were performed on the liquid reaction media recovered at the end of the glycerol etherification with ethanol performed at 200 °C for 4 h (Table 6.5).

These data clearly show the partial dissolution of W species for HPA/SiO_2, HPA/C and HPA/$ZrOH_X$. Interestingly, HPA/$NbOH_X$ is the most resistant supported catalysts against the W leaching.

4 Conclusion

12-Tungstophosphoric acids highly dispersed on silica, carbon, $ZrOH_X$ and $NbOH_X$ supports were prepared in order to get an HPA/support coverage of 0.4. The XRD patterns do not reveal the presence of $H_3PW_{12}O_{40}$ crystallites which demonstrates the efficiency of the method of impregnation done under vacuum. The nature of the support influences the acid strength of the supported catalyst. The following ranking was established based on microcalorimetry of NH_3 adsorption: HPA/SiO_2 > HPA/C > HPA/ZrOH > HPA/NbOH. This ranking reflects the affinity of the HPA with the support.

The catalytic performances of the supported 12-tungstophosphoric acid in the reaction of glycerol etherification with *tert*-butyl alcohol were mainly governed by the catalyst acid strength. The glycerol etherification with ethanol was seen to be more acidic demanding than the etherification with *tert*-butyl alcohol: the strongest acid catalysts such as HPA/C and HPA/SiO_2 promote the reaction why HPA/$NbOH_X$ and HPA/ZrOH do not catalyse the etherification with ethanol.

Nevertheless, the strongest acid was poorly stable in the polar alcoholic medium at 200 °C due to partial HPA leaching. By contrast, HPA/$NbOH_X$ exhibits a remarkable resistance towards the W leaching in these reaction conditions which is of high interest for further applications in acidic catalysed reactions in liquid polar media.

Acknowledgements The authors wish to thank CAPES/Brazil for awarding a Ph.D grant to R. Lopes de Souza and the support of the CAPES-COFECUB project 512/05.

References

1. Noureddini H, US2000/6015440, University of Nebraska
2. Kesling H Jr, Karas LJ, Liotta F Jr, US1994/5308365, ARCO Chemical Technology
3. Bradin D, Grune G, WO2007/061903, CPS Biofuels INC
4. Delford B, Hillion G, WO2005/093015, IFP
5. Xiao L, Mao J, Zhou J, Gao X, Zhang S (2011) Appl Catal A Gen 393:88–95
6. Arredondo VM, Back DJ, WO2007/113776, Procter&Gamble
7. Frusteri F, Arena F, Bonura G, Cannilla C, Spadaro L, DiBlas O (2009) Appl Catal A Gen 367:77–83
8. Pariente S, Tanchoux N, Fajula F (2008) Green Chem 11:1256–1261
9. Hamad B, Perrard A, Figueras F, Rataboul F, Prakash S, Essayem N (2010) J Catal 269:1–4
10. Ivanov E, Zausa E, Ben Taârit Y, Essayem N (2003) Appl Catal A Gen 256(1–2):225
11. Zausa E, PhD Thesis, University Lyon1, 26 septembre 2002
12. Hamad B, Lopes de Souza RO, Sapaly G, Carneiro Rocha MG, Pries de Oliveira PG, Gonzalez WA, Andrade Sales E, Essayem N (2008) Catal Commun 10:92–97

Chapter 7
Alkoxylation of Terpenes over Tungstophosphoric Acid Immobilised on Silica Support

M. Caiado, D.S. Pito, and J.E. Castanheiro

Contents

1 Introduction 154
2 Experimental Section 155
2.1 Catalyst Preparation and Characterisation 155
2.2 Catalytic Experiments 156
3 Results and Discussion 157
3.1 Alkoxylation of α-Pinene 158
3.2 Alkoxylation of β-Pinene 159
3.3 Alkoxylation of Limonene 161
4 Conclusion 163
References 163

Abstract The alkoxylation of α-pinene, β-pinene and limonene with Cl–C4 alcohols (methanol, ethanol, 1-propanol and 1-butanol) to α-terpinyl alkyl ether was carried out in the presence of silica-occluded tungstophosphoric acid (PW/S) in liquid phase. Different linear and branched alcohols (C1–C4 alcohols) are compared in relation to their activity for the heterogeneously catalysed alkoxylation of alpha-pinene.

The catalytic activity decreased with the increase of chain length of the linear alcohols, in the alkoxylation of α-pinene, β-pinene and limonene, which can be explained due to the presence of sterical hindrance and diffusion limitations inside the porous system of PW/S.

The reactivity of the terpenes increases as follows: limonene < α-pinene < β-pinene. This behaviour can be explained due to limonene to be higher in stability than α-pinene and β-pinene. The pinenes have the angle strain of cyclobutane ring that makes them more reactive than limonene.

M. Caiado • D.S. Pito • J.E. Castanheiro (✉)
Departamento de Química, Centro de Química de Évora,
Universidade de Évora, 7000-671 Évora, Portugal
e-mail: jefc@uevora.pt

A. Patel (ed.), *Environmentally Benign Catalysts: For Clean Organic Reactions*, DOI 10.1007/978-94-007-6710-2_7, © Panch Tattva Publishers, Pune, India 2013

A decrease of the PW/S selectivity to the α-terpinyl alkyl ether with the increase of chain length of the alcohols was observed.

The catalyst (PW/S) can be easily recovered and reused.

1 Introduction

Monoterpenes are an important renewable feedstock, which are used in the pharmaceutical, flavour and fragrance industries [1, 2]. Acid-catalysed alkoxylation of terpenes are important synthesis routes to valuable terpenic ethers with many applications in perfumery and pharmaceutical industry [1–3]. α-Pinene, β-pinene and limonene are converted to α-terpinyl alkyl ethers that are used in the formulation of soap, cosmetic perfumes and medicines [4–6] (Scheme 7.1). In these reactions, mineral acids, e.g. sulphuric acid, are often used as catalysts, in over stoichiometric amounts, which lead to large amounts of waste. This problem can be overcome by the use of solid acid catalysts. These catalysts are preferable, offering easy separation from the reactants and products by filtration, are less harmful to the environment and have no corrosion or disposal of effluent problems. The heterogeneous catalyst can also be reused.

The α-pinene and limonene alkoxylation to 1-methyl-4-[alpha-alkoxy-isopropyl]-L-cyclohexene in the presence of beta zeolite in liquid phase has been studied in both a batch reactor and a continuous fixed bed reactor [7]. In both reactor types, methanol and limonene as feedstock react to 1-methyl-4-[alphamethoxy-isopropyl]-L-cyclohexene with a yield of about 85 %. The beta zeolite was more active and selective for methoxylation of limonene than other catalysts described in the literature for this reaction type. For the methoxylation of alpha-pinene, however, the highest yield of the corresponding alpha-terpinyl methyl ether is only up to 50 % in both reactor types, as several bicyclic and double addition products are formed in parallel. The selectivity for all addition products together can reach values of about 85 % [7].

The α-pinene methoxylation has been carried out over beta zeolite [7], acidic cation exchange resins [8] and sulphonic-modified mesoporous silica [9], PVA with sulphonic acid groups [10] and heteropolyacids supported in silica [11].

Heteropolyacids (HPAs) are strong Brönsted acids, which have attracted importance as acid catalysts for the synthesis of fine chemicals. HPAs generally exhibit higher catalytic activities compared with mineral acids, ion-exchange resins and zeolites [12–15]. HPA catalysis lacks side reactions, such as sulphonation and chlorination, which frequently occur with mineral acids. However, HPAs have some disadvantages, such as low surface area ($1–10\ m^2\ g^{-1}$), separation problem from reaction mixtures and low thermal stability [12–14]. These disadvantages can be overcome immobilising the HPAs on different supports (zeolites, silica, activated carbons, polymers, etc.) [12–14]. HPAs have been reported as efficient catalysts for different reactions (Friedel-Crafts acylation, Fries

Scheme 7.1 Synthesis of 1-methyl-4-[alpha-alkoxy-isopropyl]-L-cyclohexenes from α- pinene, β-pinene and limonene

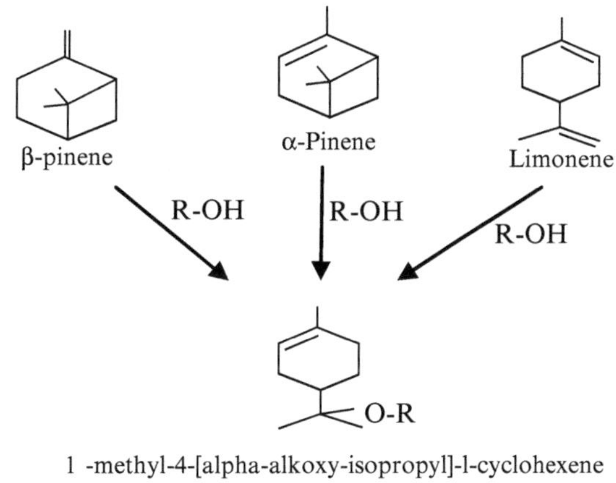

rearrangement, etherification, esterification, isomerisation, hydration and hydrolysis) [14, 15].

Heteropolyacids were also used as catalysts in different acid reactions, in which monoterpenes are involved, such as hydration of α-pinene [16–18], limonene [19] and camphene [20]; isomerisation of α-pinene [21]; acetoxylation of α-pinene [18], limonene [18, 19] and camphene [20]; and cyclisation of pseudoionone [22].

In this work, we report the application of silica-occluded $H_3PW_{12}O_{40}$ (PW/S) as solid catalyst for the alkoxylation of terpenes (α-pinene, β-pinene and limonene) with different alcohols.

2 Experimental Section

2.1 Catalyst Preparation and Characterisation

The silica-occluded tungstophosphoric acid (PW) was prepared by sol-gel method, as described elsewhere [11]. A mixture of water (2.0 mol), 1-butanol (0.2 mol) and tungstophosphoric acid (8.3×10^{-4} mol) was added to tetraethyl orthosilicate (0.2 mol) and stirred at 80 °C during 3 h. The hydrogel obtained was dehydrated slowly at 80 °C for 1.5 h in vacuo (25 Torr). The dried gel obtained was extracted in a soxhlet apparatus with methanol during 72 h and dried at 100 °C, overnight. The silica-occluded heteropolyacid was dried at 100 °C for 3 h prior to use in the catalytic reactions. The catalyst will be designed by PW/S.

The PW content was determined by ICP. The BET surface area was 254 $m^2\cdot g^{-1}$, the total pore volume was 0.12 $cm^3\cdot g^{-1}$ and the average pore volume was 0.12 $cm^3\cdot g^{-1}$. The integrity of Keggin structure of PW was verified by FT-IR (Fig. 7.1). The symmetric and asymmetric vibrations of different W–O bonds are observed in the following regions of the vibration spectra: W–O_d bonds (1,000–960 cm^{-1}), W–Ob–W bonds (890–850 cm^{-1}) and W–O_c–W bonds (800–760 cm^{-1}).

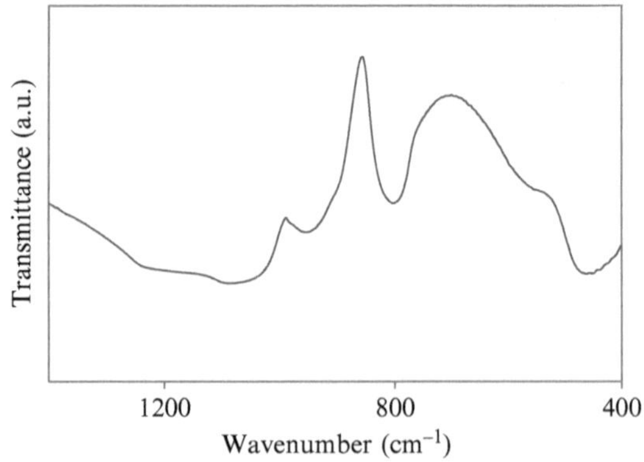

Fig. 7.1 FT-IR spectrum of PW/S

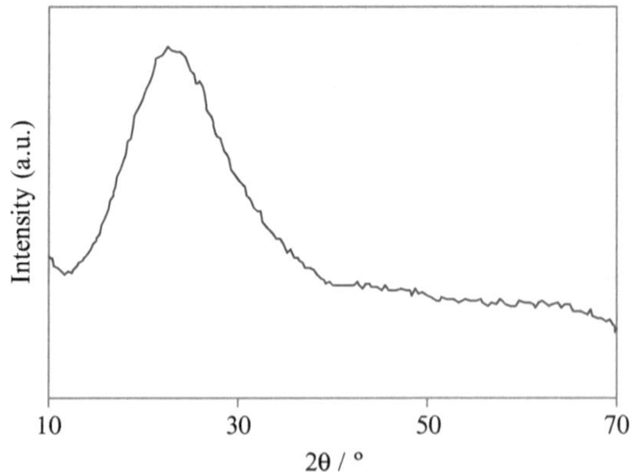

Fig. 7.2 XRD patterns of PW/S

From XRD (Fig. 7.2), the catalyst does not indicate the presence of any crystalline phases related to heteropolyacid [11].

2.2 *Catalytic Experiments*

The reactions were carried out in a batch reactor equipped with a magnetic stirrer at 60–80 °C. In a typical run, a mixture of the terpene (3 mmol), alcohol (30 mL) and solid PW/S catalyst (1.0 g) was intensively stirred under air at a specified temperature.

Stability tests of the catalyst were carried out by running four consecutive experiments, using the same reaction conditions. Between the catalytic experiments, the catalyst was separated from the reaction mixture by filtration, washed with ethanol and dried at 100 °C overnight.

It used nonane as the internal standard. The reaction progress was followed by gas chromatography (GC) using a KONIC HRGC-3000C instrument equipped with a 30 m × 0.25 mm DB-1 column and a flame ionisation detector.

The GC mass balance was based on the substrate charge. The difference was attributed to the formation of oligomers, which were not GC determinable. The products were identified by gas chromatography mass spectrometry (GC-MS) using a FISONS MD800 (Leicestershire, UK) instrument, equipped with a 30 m × 0.25 mm DB-1 column.

3 Results and Discussion

The main product of α-pinene, β-pinene and limonene alkoxylation was α-terpinyl alkyl ether, being also formed bicyclic products (endo-bornyl alkyl ether, β-fenchyl alkyl ether, exo-bornyl alkyl ether, bornylene and camphene) and monocyclic products (γ-terpinyl alkyl ether, β-terpinyl alkyl ether, terpinolene, limonene) as by-products. The reaction scheme of the alkoxylation of α-pinene, β-pinene and limonene is illustrated in Scheme 7.2.

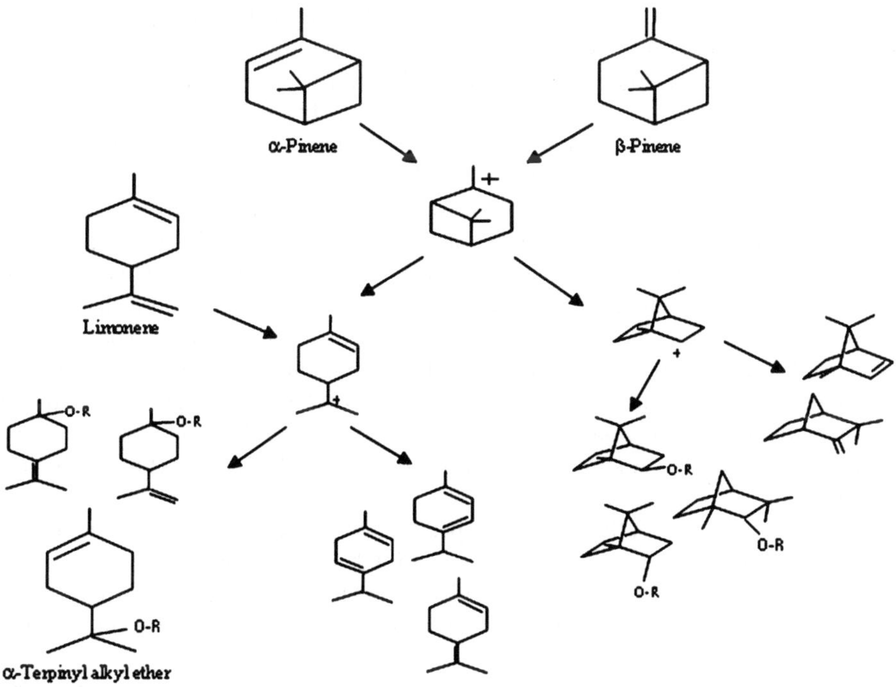

Scheme 7.2 Reaction scheme of the alkoxylation of α-pinene, β-pinene and limonene

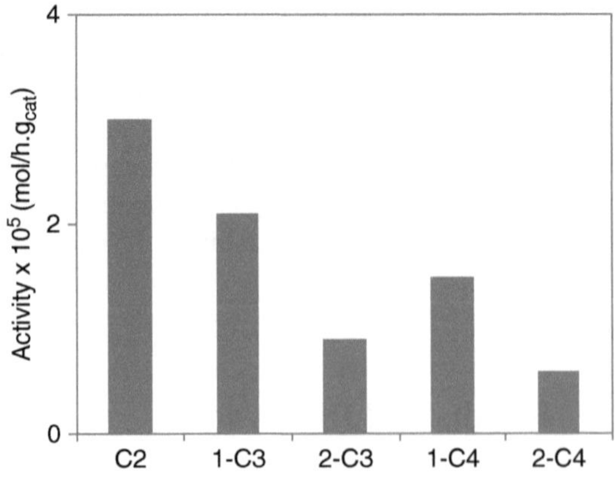

Fig. 7.3 Alkoxylation of α-pinene in the presence of PW/S. Effect of different alcohols (C2 ethanol, 1-C3 1-propanol, 2-C3 2-propanol, 1-C4 1-butanol, 2-C4 2-butanol)

3.1 Alkoxylation of α-Pinene

The main product of α-pinene alkoxylation with alcohols (methanol, ethanol, 1- propanol, 2-propanol, 1-butanol and 2-butanol) in the presence of silica-occluded heteropolyacid was α-terpinyl alkyl ether. The reaction scheme of the alkoxylation of α-pinene is illustrated in Scheme 7.2.

Figure 7.3 shows the catalytic activity of PW/S as catalyst in alkoxylation of α-pinene with different alcohols (methanol (C1), ethanol (C2), 1-propanol (1-C3), 2-propanol (2-C3), 1-butanol (1-C4) and 2-butanol (2-C4)). The alkoxylation of α-pinene with methanol was studied over different heteropolyacids supported in silica [11]. The catalytic activity of PW/S was 9.6×10^{-5} mol $h^{-1} \cdot g_{cat}^{-1}$ [11]. The catalytic activity decreased with the increase of the number of carbon atoms in the chain alcohol (Fig. 7.3). This behaviour can be explained due to the diffusive limitations of alcohols in structure of the catalyst, once it has been expected that in an electrophilic addition the longer-chain alcohols would be more reactive. This might be caused by sterical hindrance in the silica-occluded heteropolyacid pore structure in such a way that longer alcohols do not easily form the transition state. The catalytic activity decreased from linear alcohols (1-propanol and 1-butanol) to branched alcohols (2-propanol and 2-butanol) (Fig. 7.3). This behaviour can also be due to the presence of strong diffusion limitations of 2-propanol and 2-butanol inside porous system of the catalyst. A similar result was observed by K. Hensen et al. [7] in alkoxylation of limonene over beta zeolite.

Figure 7.4 shows the effect of the temperature on ethoxylation of α-pinene. It was observed that the catalytic activity increased with the temperature.

Table 7.1 shows the conversion of α-pinene and the selectivity to α-terpinyl alkyl ether obtained by reaction between α-pinene and various alcohols in a batch reactor.

By increase of chain length of the linear alcohols, a decrease of the selectivity of the α-terpinyl alkyl ether and a decrease of the α-pinene conversion were observed. These

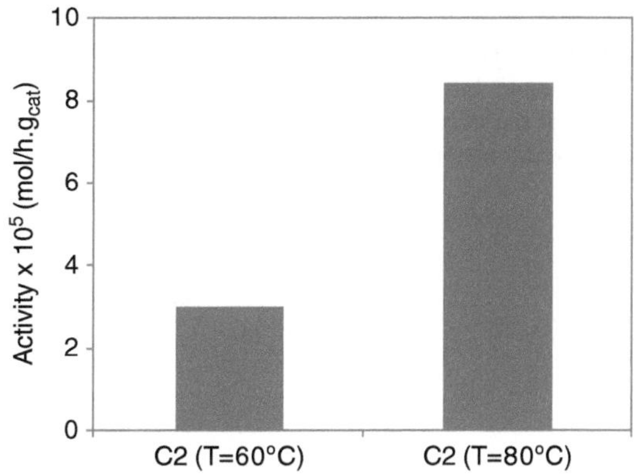

Fig. 7.4 Alkoxylation of α-pinene with ethanol in the presence of PW/S. Effect of temperature on the catalytic activity

Table 7.1 Conversion of α-pinene and selectivity of the α-terpinyl alkyl ether

Alcohol	Temperature (°C)	Conversion (%)[a] (α-pinene)	Selectivity (%) (α-terpinyl alkyl ether)	Yield (%)
Methanol	60	98	60	58.8
Ethanol	60	45	39	17.6
Ethanol	80	97	38	36.9
1-Propanol	60	38	36	13.7
2-Propanol	60	22	33	7.9
1-Butanol	60	33	35	11.6
2-Butanol	60	15	34	5.1

[a]Conditions: $t=50$ h; catalyst loading = 1.0 g PW-silica; 0.5 cm^3 α-pinene; 30 cm^3 alcohol

results can be caused by sterical hindrance in the silica-occluded heteropolyacid pore structure in such a way that longer alcohols do not easily form the transition state. Also, side reactions of α-pinene to its isomers occur.

In order to study the catalytic stability of the PW/S in the ethoxylation of α-pinene, consecutive batch runs with the same catalyst sample and the same reaction conditions were carried out. A small decrease of the catalytic activity from the first to the second use was observed. However, after the third use, a stabilisation of the catalytic activity was observed (Fig. 7.5).

3.2 *Alkoxylation of β-Pinene*

The main product of β-pinene alkoxylation with different alcohols in the presence of PW-silica was α-terpinyl alkyl ether. The reaction scheme of the alkoxylation of β-pinene is also illustrated in Scheme 7.2.

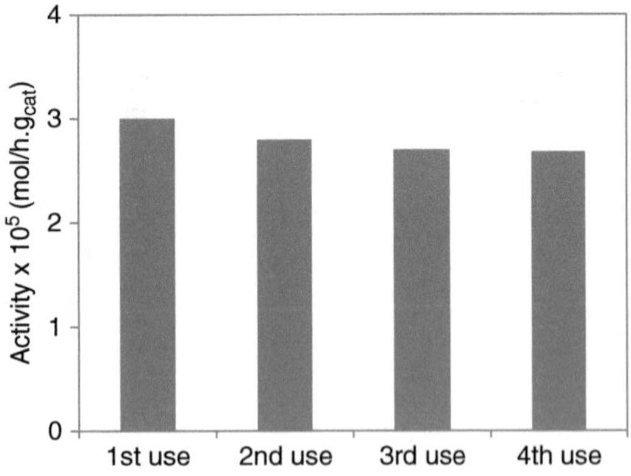

Fig. 7.5 Stability studies on PW/S catalyst in ethoxylation of α-pinene. Initial activities taken as the maximum observed reaction rate

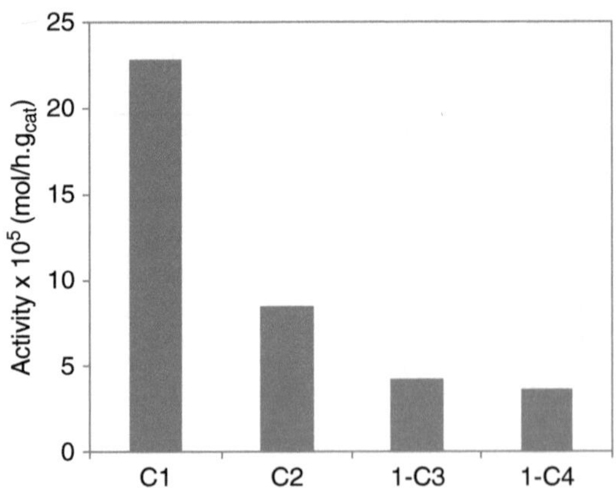

Fig. 7.6 Alkoxylation of β-pinene in the presence of silica-occluded PW. Effect of different alcohols (C1 methanol, C2 ethanol, 1-C3 1-propanol, 1-C4 1-butanol)

Figure 7.6 shows the catalytic activity of PW-silica as catalyst in alkoxylation of β-pinene with different alcohols (methanol (C1), ethanol (C2), 1-propanol (1-C3) and 1-butanol (1-C4)). It was observed that the catalytic activity decreased with the increases of the number of carbon atoms in the chain alcohol. As in the case of α-pinene alkoxylation, the sterical hindrance and diffusion restrictions in the porous system of catalyst might cause less reactivity.

Table 7.2 shows the conversion of β-pinene and the selectivity of α-terpinyl alkyl ether obtained by reaction between β-pinene and methanol, ethanol, 1-propanol and 1-butanol. A decrease of the selectivity of the α-terpinyl alkyl ether was observed, which can be explained due to some sterical hindrance in the silica.

The catalytic stability of the PW/S was also studied in the ethoxylation of β-pinene. Different batch runs with the same catalyst sample and in the same reaction conditions were carried out. A small decrease of the catalytic activity from the first to the fourth use is observed (Fig. 7.7).

Table 7.2 Conversion of β-pinene and selectivity of the α-terpinyl alkyl ether

Alcohol	Temperature (°C)	Conversion (%)[a] (α-pinene)	Selectivity (%) (α-terpinyl alkyl ether)	Yield (%)
Methanol	60	99[b]	57	56.4
Ethanol	60	95	40	38.0
1-Propanol	60	70	25	17.5
1-Butanol	60	64	23	14.7

[a]Conditions: $t = 50$ h; catalyst loading = 1.0 g PW-silica; 0.5 cm^3 α-pinene; 30 cm^3 alcohol
[b]$t = 25$ h

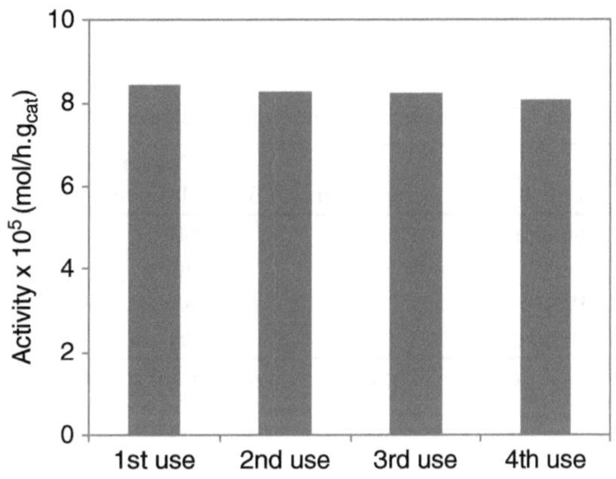

Fig. 7.7 Stability studies on PW/S catalyst in ethoxylation of β-pinene. Initial activities taken as the maximum observed reaction rate

3.3 *Alkoxylation of Limonene*

The main product of limonene alkoxylation with different alcohols in the presence of PW/S was α-terpinyl alkyl ether. Figure 7.8 shows the catalytic activity of PW/S as catalyst in alkoxylation of limonene with different alcohols (methanol (C1), ethanol (C2), 1-propanol (1-C3) and 1-butanol (1-C4)). The catalytic activity decreased with the increases of the number of carbon atoms in the chain alcohol (Fig. 7.8). The sterical hindrance and diffusion restrictions inside the catalyst can be responsible for the less reactivity, like it was observed in the case of α-pinene and β-pinene alkoxylation.

The conversion of limonene and the selectivity of α-terpinyl alkyl ether were shown in Table 7.3. A decrease of α-terpinyl alkyl ether selectivity with an increase of the chain length of alcohols was observed, which can be explained due to some sterical hindrance in the silica.

Figure 7.9 compares the initial activity of PW/S in alkoxylation of α-pinene, β-pinene and limonene at 60 °C. For all alcohols used in the alkoxylation reaction,

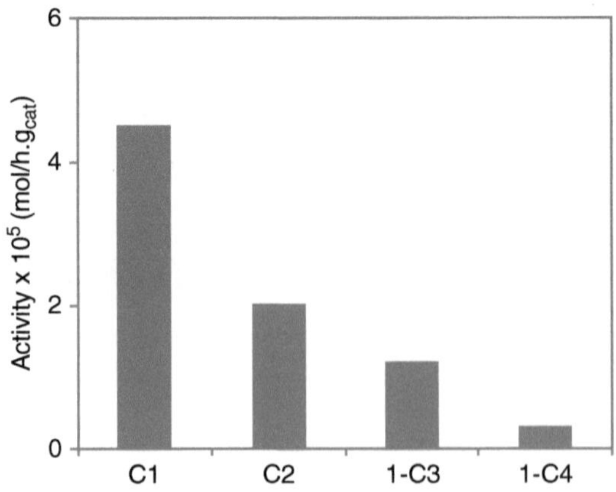

Fig. 7.8 Alkoxylation of limonene in the presence of PW/S. Effect of different alcohols (C1 methanol, C2 ethanol, 1-C3 1-propanol, 1-C4 1-butanol)

Table 7.3 Conversion of limonene and selectivity of the α-terpinyl alkyl ether

Alcohol	Temperature (°C)	Conversion (%)[a] (limonene)	Selectivity (%) (α-terpinyl alkyl ether)[b]	Yield (%)
Methanol	60	60	81	48.6
Ethanol	60	40	76	30.4
1-Propanol	60	35	72	25.2
1-Butanol	60	26	65	16.9

[a]Conditions: $t=50$ h; catalyst loading = 1.0 g PW-silica; 0.5 cm^3 limonene; 30 cm^3 alcohol
[b]$t=50$ h

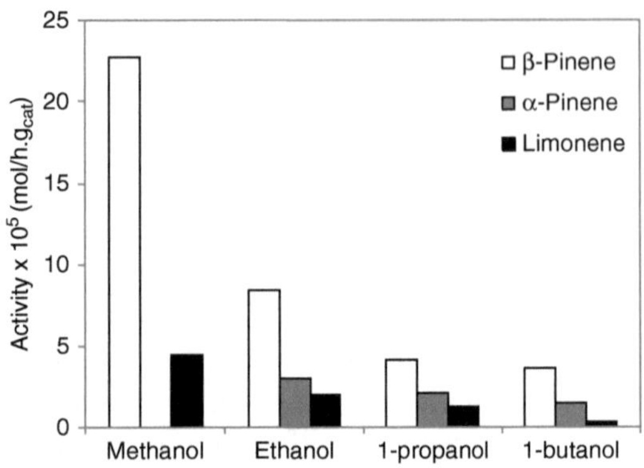

Fig. 7.9 Alkoxylation of α-pinene, β-pinene and limonene in the presence of PW/S. Effect of different alcohols (C1 methanol, C2 ethanol, 1-C3 1-propanol, 1-C4 1-butanol)

it was observed that β-pinene reacts faster than α-pinene and limonene. The reactivity of the terpenes decreases as follows: β-pinene > α-pinene > limonene. A possible explanation for this sequence is the higher stability of limonene molecule, while α-pinene and β-pinene have the angle strain of cyclobutane ring that makes them more reactive than limonene.

4 Conclusion

The acid-catalysed alkoxylation of α-pinene, β-pinene and limonene to α-terpinyl alkyl ether was carried out in the presence of silica-occluded heteropolyacid (tungstophosphoric acid). This ether is used as flavours and fragrances for perfume and cosmetic products, as additives for pharmaceuticals and agricultural chemicals as well as in the food industry for organic synthesis.

For the alkoxylation of α-pinene, β-pinene and limonene, the catalytic activity decreases with the increase of chain length of the linear alcohols. This behaviour could be explained due to the presence of sterical hindrance and diffusion restrictions in the porous system of catalyst.

The selectivity of silica-occluded heteropolyacid to α-terpinyl alkyl ether also decreased with the increase of chain length of the alcohols.

The catalytic stability of the PW/S was studied in the ethoxylation of β-pinene and α-pinene. It was observed that the catalyst can be easily recovered and reused without loss of activity.

It can also be concluded that the alkoxylation reactions with β-pinene substrate were faster than with α-pinene and limonene.

References

1. Erman WE (1985) Chemistry of the monoterpenes. An encyclopedic handbook. Marcel Dekker, New York
2. Bauer K, Garbe D, Surburg H (1997) Common fragrance and flavor materials: preparation, properties and uses. Wiley-VCH, New York
3. Pybus DH, Sell CS (eds) (1999) The chemistry of fragrances. RSC Paperbacks, Cambridge
4. Corma A, Iborra S, Velty A (2007) Chem Rev 107:2411–2502
5. Mäki-Arvela P, Holmbom B, Salmi T, Murzin DY (2007) Catal Rev 49:197–340
6. Monteiro JL, Veloso CO (2004) Top Catal 27:169–180
7. Hensen K, Mahaim C, Hölderich WF (1997) Appl Catal A Gen 149:311–329
8. Yoshiharu M, Masahiro M (1976) Jpn Kokai 75(131):948
9. Castanheiro JE, Guerreiro L, Fonseca IM, Ramos AM, Vital J (2008) Stud Surf Sci Catal 174: 1319–1322
10. Pito DS, Fonseca IM, Ramos AM, Vital J, Castanheiro JE (2009) Chem Eng J 147:302–306
11. Pito DS, Matos I, Fonseca IM, Ramos AM, Vital J, Castanheiro JE (2010) Appl Catal A Gen 373:140–146
12. Kozhevnikov IV (2002) Catalysts for fine chemicals, catalysis by polyoxometalates, vol 2. Wiley, Chichester
13. Kozhevnikov IV (1998) Chem Rev 98:171–198
14. Okuhara T, Mizuno N, Misono M (1996) Adv Catal 41:113–252
15. Timofeeva MN (2003) Appl Catal A Gen 256:19–35
16. Castanheiro JE, Ramos AM, Fonseca I, Vital J (2003) Catal Today 82:187–193
17. Castanheiro JE, Fonseca IM, Ramos AM, Oliveira R, Vital J (2005) Catal Today 104:296–304
18. Robles-Dutenhefner PA, da Silva KA, Siddiqui MRH, Kozhevnikov IV, Gusevskaya EV (2001) J Mol Catal A Chem 175:33–42

19. Avila MC, Comelli NA, Firpo NH, Ponzi EN, Ponzi MI (2008) J Chil Chem Soc 53:1460–1462
20. da Silva KA, Kozhevnikov IV, Gusevskaya EV (2003) J Mol Catal A Chem 192:129–134
21. da Silva KA, Robles-Dutenhefner PA, Kozhevnikov IV, Gusevskaya EV (2009) Appl Catal A Gen 352:188–192
22. Díez VK, Apesteguía CR, Di Cosimo JI (2008) Catal Lett 123:213–219

Chapter 8
Effect of Acidity, Structure, and Stability of Supported 12-Tungstophosphoric Acid on Catalytic Reactions

José Alves Dias, Sílvia Cláudia Loureiro Dias, and Julio Lemos de Macedo

Contents

1 Introduction 166
2 Experimental Section 168
  2.1 Preparation of Supported H_3PW 168
  2.2 Techniques of Characterization 169
  2.3 Experimental Setup of the Reactions 170
3 Results and Discussion 172
  3.1 H_3PW/SiO_2: Catalyst for Benzene Transalkylation with C9+ Aromatics for Production of Xylenes 172
  3.2 $H_3PW/SiO_2–Al_2O_3$: Catalyst for Esterification of Acetic Acid and Ethanol 174
  3.3 H_3PW/ZrO_2: Catalyst for Esterification of Oleic Acid and Ethanol 178
  3.4 H_3PW/MCM-41: Catalyst for Cyclization of (+)-Citronellal 182
4 Conclusion 185
References 185

Abstract Polyoxometalates are large metal–oxygen nanostructured compounds that can be modified at the molecular level. Catalysts based on 12-tungstophosphoric acid ($H_3PW_{12}O_{40}$) supported on different carries such as silica, silica–alumina, zirconia, and mesoporous MCM-41 have been successfully prepared and characterized by several physicochemical techniques (e.g., FTIR, FT-Raman, ^{31}P MAS-NMR, calorimetry, thermal analysis, TPD of pyridine). The materials were tested in different reactions: transalkylation of benzene with aromatics, esterification of acetic acid and ethanol, esterification of oleic acid with ethanol, and cyclization of

J.A. Dias (✉) • S.C.L. Dias • J.L. de Macedo
Campus Darcy Ribeiro – Asa Norte, Instituto de Química, Laboratório de Catálise (A1-62/21), Universidade de Brasília, caixa postal 4478, Brasília-DF 70904-970, Brazil
e-mail: jdias@unb.br; scdias@unb.br; julio@unb.br

A. Patel (ed.), *Environmentally Benign Catalysts: For Clean Organic Reactions*,
DOI 10.1007/978-94-007-6710-2_8, © Panch Tattva Publishers, Pune, India 2013

(+)-citronellal. The results, coupled with other applications of heteropoly acids in the literature, illustrate the potential for these catalysts to be used in greener processes, i.e., a more effective, selective, economical, and environmentally benign technique. The effects of support interaction, loading of $H_3PW_{12}O_{40}$, calcination temperature and stability, leaching, and acidity were addressed for all studied catalysts. These data were correlated to the conversion and/or selectivity of the main product and were used for recovery of the catalysts. Moreover, it was clear that the strength of interaction of the heteropoly acid with the surface support as well as the acidity define the capacity of utilization of these catalysts. Effective ways to recovery the initial activity is still a challenge to take full advantage of these materials.

1 Introduction

Polyoxometalates (POMs) are large metal–oxygen nanostructured compounds of variable size. They are divided into two classes according to chemical composition: isopolyanions ($[M_mO_y]^{p-}$) and heteropolyanions ($[X_xM_mO_y]^{q-}$, $x<m$), where M is the addenda atom and X is the heteroatom [1]. Among the various known POMs, the Keggin type is the most applied structure for catalytic purposes, and 12-tungstophosphoric acid ($H_3PW_{12}O_{40}$, H_3PW) is the strongest acid in that series [2]. The Keggin ion structure of H_3PW is well known and consists of a PO_4 tetrahedron surrounded by four W_3O_{13} triads formed by an edge-sharing octahedra. The triads are connected by corner-sharing oxygens [3]. These compounds can be tailored at the molecular level by either substituting the addenda atom or the counter cation, which lead them to be utilized in many fields including analytical chemistry, biochemistry, and materials science. This versatility is achieved by modifying properties such as size, mass, electron–proton transfer, thermal stability, lability of lattice oxygen, acidity, and solubility [4, 5].

Another way to change the properties of heteropoly acids (HPAs) is to prepare supported materials using a variety of matrices (e.g., silica, zirconia, alumina, carbon) [6]. Moreover, HPAs have been supported on different solids to increase the original, very low surface area. The enhancement of the specific surface area, development of higher dispersion of acidic protons, heterogenization of homogeneous systems, and fine control of acid strength are some of the goals for preparing supported HPAs. In addition, another goal is finding a suitable catalyst for nonpolar substrates with heteropoly compounds. The nature of reactants determines whether the reaction takes place in the bulk or on the surface of the catalyst [7]. It is known that HPAs can absorb polar molecules into their bulk, leading to a pseudoliquid-phase type of catalysis. Nonpolar molecules react only on the surface or between surface layers of the crystal [8]. Thus, increasing the low surface area by supporting HPAs may produce a catalyst with higher activity for reactions involving nonpolar substrates. According to the properties obtained for the supported HPAs, several examples demonstrate the potentiality of these systems.

For instance, aromatic alkylation reactions were studied [9] using 12-molybdophosphoric and 12-tungstophosphoric acids supported on silica. The alkylation of benzene and toluene was carried out with benzyl chloride, benzyl alcohol, cyclohexene, or cyclohexanol. The results show that the catalysts allowed rapid quantitative conversions with great yields in monoalkylation products and minimal formation of polyalkylation products. Regioselectivity in these reactions is similar to that described in literature for other catalysts. In formal cycloaddition [3+2] using benzyl alcohols catalyzed by $H_3PMo_{12}O_{40}/SiO_2$, the results were likewise to one catalyzed by $SnCl_4$.

In addition, methoxyacetophenones have been synthesized through Friedel–Crafts acylation of anisole with acetic anhydride using a H_3PW/SiO_2 catalyst [10]. High conversions and high *p*-selectivity were obtained at temperatures of 61–110 °C. However, the catalyst deactivated after 0.5 h at 61 and 83 °C due to strong but reversible adsorption of products. The catalyst could recover the activity without being fairly leached from the silica support, even though a progressive coking deactivated the catalyst.

Furthermore, supported $H_3PW_{12}O_{40}{\cdot}6H_2O$ and Pt on Zr–Ce mixed oxides have been used for the storage–reduction (absorption into H_3PW and adsorption on Zr–Ce support) of NO*x* [11]. Zr–Ce mixed oxides adsorb NO_x (mainly NO_2) as nitrates, which are desorbed thermally as NO_2 and NO; in the H_3PW–Pt/Zr–Ce system, NO and NO_2 were stored equimolarly by substitution of water molecules of the H_3PW structure. The optimum Zr:Ce support molar ratio was 0.5, and the presence of Pt had no influence on storage capacity. Nonetheless, Pt was essential for reducing by accelerating both the NO_x desorption and reduction processes.

Hydroisomerization of *n*-heptane was performed under a series of Pt-bearing dealuminated USY-supported Cs salt derivatives of $H_3PW_{12}O_{40}$ catalysts [12]. USY was found to be a poor support for Cs salts of H_3PW, leading to low catalytic activity in this reaction. In contrast, dealuminated USY-supported Cs salts of H_3PW showed high catalytic conversion and selectivity. Among the various Cs salts of H_3PW with different Cs:P ratios, $Cs_{2.0}HPW_{12}O_{40}$ with Pt loading of 0.4 wt% was the most effective, enhancing the conversion of *n*-heptane up to 76.2 % with a high selectivity to isomerized products of 92.2 %.

Liquid-phase esterification of acetic acid with *n*-butanol was also studied [13] using H_3PW and its cesium salts immobilized on dealuminated ultra-stable Y zeolite (DUSY). It was found that the catalytic activity could be remarkably improved by introducing H_3PW on dealuminated USY (49.5–86.4 %). The supported cesium salt of H_3PW on DUSY catalyst gave a very high conversion of *n*-butanol of 94.6 % and the 100 % selectivity for *n*-butyl acetate. The leaching test in water for the catalysts showed that the decrease of activity could be closely associated with the solubility of the active heteropoly components on the supports into the polar solvent, which is much lower for cesium salts than pure H_3PW.

Carbon-entrapped $H_3PW_{12}O_{40}$ catalyst has been tested for the intermolecular dehydration between butanol and *t*-butyl alcohol into butyl *t*-butyl ether [14]. No self-condensation product (dibutyl ether) or intramolecular dehydration product (isobutene) was observed under these reaction conditions. When the reaction was repeated several times using the recovered catalyst by renewing the reactants, no leakage of HPA was detected, and the catalytic activity remained unchanged.

Catalysts based on the impregnation of H_3PW on MCM-41 were prepared, characterized, and applied in detailed studies of Pechmann, esterification, and

Friedel–Crafts acylation reactions [15]. In these studies, H_3PW was incorporated on the mesochannels of MCM41, and acidity was controlled by different loadings of the polyacid. The sample with 60 % H_3PW showed the highest acidity and catalytic activity. In addition, tests of reutilization of the catalysts confirmed that they could be used several times without significant loss of activity.

Esterification of free fatty acids (palmitic acid of sunflower oil) with methanol was conducted with a series of catalysts containing 5–30 wt% of H_3PW supported on Nb_2O_5 [16]. Based on characterization of the catalysts by FTIR, XRD, and temperature programmed desorption of ammonia, as well as the activity results, it was observed that esterification activity depended on the structural variations of the catalysts. The 25%H_3PW/Nb_2O_5 was the best with palmitic acid conversion around 90 % at temperature of 65 °C, molar ratio of 13.7 (alcohol:acid), and calcinations of the catalyst at 400 °C.

Another study examined H_3PW supported on SBA-15 or obtained by direct synthesis and tested in the conversion of *n*-decane [17]. The material obtained by entrapping the polyacid in the ordered mesoporous silica was catalytically superior to the one obtained by impregnation. It achieved high yields of skeletal isomers (>70 %), multibranched skeletal isomers, and hydrocracking reactions limited to primary cracking. Compared to USY zeolites, the H_3PW/SBA-15 samples showed a broader optimum reaction temperature window for isomerization and higher *n*-decane dibranching selectivity.

Other researchers have carried out liquid-phase alkylation of phenol and cresols with *tert*-butyl alcohol using 30%H_3PW/ZrO_2 catalysts [18]. The catalysts showed high activity in terms of conversion and selectivity for the desired product under mild conditions. The proposed process could be economically viable, since the non-reacted phenol was separated by distillation and reused for further reaction. In addition, the regenerated catalyst could be used to achieve high yield transformation.

It is clear from the examples presented that HPAs have been used in many types of reactions. This chapter wishes to examine the relationship among the main characteristics of HPAs (acidity, structure, and stability) related to activity in reactions involving transalkylation, esterification, and cyclization. It is important to realize that the successful applicability of supported HPAs depends on these properties to carefully design the right material that can be well built enough to be reutilized at industrial level.

2 Experimental Section

2.1 Preparation of Supported H_3PW

The $H_3PW_{12}O_{40}$ used was obtained from Sigma-Aldrich. The studied supports (SiO_2, SiO_2–Al_2O_3, ZrO_2, and MCM-41) were obtained from Sigma-Aldrich except for MCM-41, which was synthesized according to the procedure described in [19]. The *x*% H_3PW/support samples were prepared by impregnation in aqueous acid solution

(HCl 0.1 mol L^{-1}), where x stands for the H_3PW loading (2–60 wt%). Adequate amounts of H_3PW and the support were placed in a glass round-bottom flask (ratio, 10 mL of the HCl solution per gram of support), and the suspension was kept at 80 °C under constant stirring until the solvent was completely dried. Then, the obtained solids were ground to fine particles and calcined in a muffle furnace (EDG model 3000) for 4–6 h. The supported catalysts were checked by elemental analysis and confirmed to be free of chloride after calcinations at 200 °C for at least 4 h (muffle furnace).

2.2 Techniques of Characterization

The materials were characterized by several methods. The spectroscopic methods were FTIR, FT-Raman, MAS-NMR, and XRD. Infrared spectra with Fourier transform were obtained by a Thermo Scientific spectrometer model Nicolet 6700 FTIR with a DTGS detector using 256 scans and a spectral resolution of 4 cm^{-1}. In the transmittance mode, each sample was pressed in dried 1 wt% KBr (Merck) pellets. FT-Raman data were collected in the pure samples with a Raman FRA 106/S module attached to the Bruker Equinox 55 spectrophotometer using a source laser (Nd–YAG) at 1,064 nm and power of 126 mW. The spectra were obtained at 2 cm^{-1} resolution, and 256 scans with the signal detected by a liquid N_2 cooled Ge detector. All FTIR and FT-Raman spectra were acquired for the samples previously calcined at 200 °C under ambient conditions at room temperature (25 °C). In addition, ^{31}P MAS-NMR spectra were acquired using a MAS probe of 7 mm in a Varian 7.05 T Mercury Plus spectrometer. The settings included single-pulse excitation (8.0 μs), recycle delay of 10 s, no ^{1}H decoupling, MAS rate at 5 kHz, and minimum of 256 scans. Signals were indirectly referenced to 85 wt% H_3PO_4. Powder XRD was obtained in a Bruker D8 FOCUS diffractometer between $5° < 2\theta < 70°$ (0.02° $step^{-1}$ and integration time of 1 s $step^{-1}$) with Cu-Kα radiation of 1.5418 Å (40 kV and 30 mA) using a graphite monochromator.

The thermal techniques included textural and thermogravimetric (TG) analysis. The specific surface area and pore volume of the samples were calculated from adsorption–desorption isotherms of nitrogen obtained at −196 °C on a micromeritics model ASAP 2020C instrument using BET and BJH models. TG analysis was conducted in a simultaneous TG–DSC model SDT 2960 from TA Instruments, with scan rate of 10 °C min^{-1}, from room temperature up to 800 °C under nitrogen (99.999 %) or synthetic air (99.999 %, $O_2 + N_2$, with 20 ± 0.5 % O_2) and flow of 100 mL min^{-1}. The acidity of the materials was calculated by thermal desorption of gaseous pyridine monitored by TG/DTG or by liquid-phase microcalorimetric adsorption of pyridine (Cal-Ad method). The number of acid sites was determined by quantitative analysis of TG/DTG curves of the materials after pyridine adsorption. The method involves the mass loss analysis of the materials before and after pyridine adsorption, taking into consideration the hydration degree of each sample, and thermal stability of the material as described elsewhere [20, 21].

The other acidity measurement was derived from the Cal-Ad method [22]. This method is based on liquid-phase microcalorimetric and adsorption of pyridine in cyclohexane slurries experiments. For calcined samples of supported H_3PW, a diluted pyridine solution in cyclohexane was added to slurry of the solid in anhydrous cyclohexane; each measurement of the heat evolved and the equilibrium amount of base in solution were determined by two independent experiments. These experiments should be parallel related to the ratio mass of solid to volume employed in each test. The solids were always handled in a nitrogen atmosphere glove box to ensure that no water was adsorbed in the samples. The volumetric apparatus had been calibrated previously. The heat of diluted pyridine added to cyclohexane, measured separately, was negligible. The temperature of both experiments was about the same. Both microcalorimetric and adsorption experiments were checked for equilibrium and diffusion constraints. For standing times above 3 min, there was no variation in the absorbance (adsorption measurements of base in solution) or in the heat measured. The heat measured was collected in a model ISC 4300 from Calorimetry Sciences Corporation using the incremental addition of pyridine. The absorbance measurements were obtained at 251 nm in a DU-650 UV–Vis spectrophotometer from Beckman. Data obtained by both experiments were used further for calculations, which were performed using a nonlinear least squares program with a simplex routine for parameter minimization. The program uses a Langmuir-type equation with a variable number of sites, which can be set by the user, according to Eq. 8.1:

$$\frac{h}{g} = \sum \frac{n_i K_i[\mathrm{B}]}{1 + K_i[\mathrm{B}]} \Delta H_i \tag{8.1}$$

Using this model, it is possible to calculate the number of each different type of site (n_i), the equilibrium constant (K_i), and the enthalpy (ΔH_i) from the analysis of the sum of the heat evolved per gram of the solid (h/g) and the concentration of base in solution ([B]). Details of the Cal-Ad method were recently published [23]. The Cal-Ad method was applied along with spectroscopic studies to analyze the acidity of different HPAs, including H_3PW [24], $Cs_xH_{3-x}PW$ [25], H_3PW/SiO_2 [26], H_3PW/Al_2O_3 [27], and H_3PW/Nb_2O_5 [23] as well as different solid acids such as zeolites [22, 28–30].

2.3 *Experimental Setup of the Reactions*

2.3.1 Transalkylation of Benzene with Trimethylbenzene Using H_3PW/SiO_2 Catalysts

The transalkylation reaction of benzene with trimethylbenzene was carried out in a tubular reactor at 1 atm and temperature of 470 °C. A catalyst mass of 0.5 g was used with a molar ratio H_2:hydrocarbon = 4, WHSV = 1 h^{-1} and a feed obtained by mixing benzene (99.8 % purity) and a C9+ stream in a benzene/C9+ (weight) = 70:30. The C9+ stream was collected in an industrial unit as a by-product of the naphtha catalytic reforming. The stream had the following mass composition: 2 % C10+ aromatics, 9 % propylbenzene, 43 % ethyltoluene, 45 % trimethylbenzene, and 0.6 % indane. The gaseous effluent was analyzed by gas chromatography connected to the reactor.

2.3.2 Esterification of Acetic Acid with Ethanol Using $H_3PW/SiO_2–Al_2O_3$ Catalysts

The esterification reaction of acetic acid with ethanol was carried out in a 5- or 10-mL custom-made glass reactor containing 2.00 g of acetic acid and the required amount of anhydrous alcohol for the reactions with 2:1, 1:1, and 1:2 (acetic acid:ethanol) molar ratios. Different amounts of catalysts were used (2, 5, 10, and 15 wt%, based on acetic acid); prior to the reactions, they were activated in a muffle furnace (EDG, model 3P-S) at 200 °C for 2 h. The system was stirred at four different reaction times (1, 2, 4, and 6 h) at 100 °C. After the reaction, each sample was cooled to room temperature, centrifuged to remove the catalyst, and analyzed by GC–FID (Shimadzu, model GC-17A) with an RTX®-5 cross bond® 5 % diphenyl/95 % dimethylpolysiloxane column from RESTEK (30 m × 0.25 mm × 0.25 μm).

2.3.3 Esterification of Oleic Acid with Ethanol Using H_3PW/ZrO_2 Catalysts

The esterification reactions were carried out in a reflux system. The mineral oil bath was held at 100 °C during the reaction time, while the cooling water was set to ca. 15–20 °C by a thermostatic bath. A 1:6 (oleic acid:ethanol) molar ratio and 10 wt% of the catalyst (related to oleic acid) were used. The reaction time was investigated for different xH_3PW/ZrO_2 in addition to pure zirconia, pure H_3PW, and the mechanical mixture with 20 % H_3PW in ZrO_2. At the end of the reaction period, the reactor system was shut down, and the reaction products were cooled to room temperature, filtered out, and washed in a 5 wt% NaCl solution. The obtained reaction products were kept in a closed glass flask with dry $MgSO_4$ in order to remove water. Quantitative analyses of the produced ester were carried out by ^1H NMR, and GC–MS analysis (Shimadzu, model GCMS QP5050A with a polydimethylsiloxane column CBPI PONA-M50-042 from Shimadzu, 100 m × 0.25 mm × 0.5 μm) confirmed that ethyl oleate was the only product of the reaction (100 % selectivity).

2.3.4 Cyclization of Citronellal Using H_3PW/MCM-41 Catalysts

First, each catalyst was activated at 300 °C for 4 h at 10 °C min^{-1} under air in a muffle furnace (EDG, model 3P-S) and cooled under N_2 atmosphere just before use in the

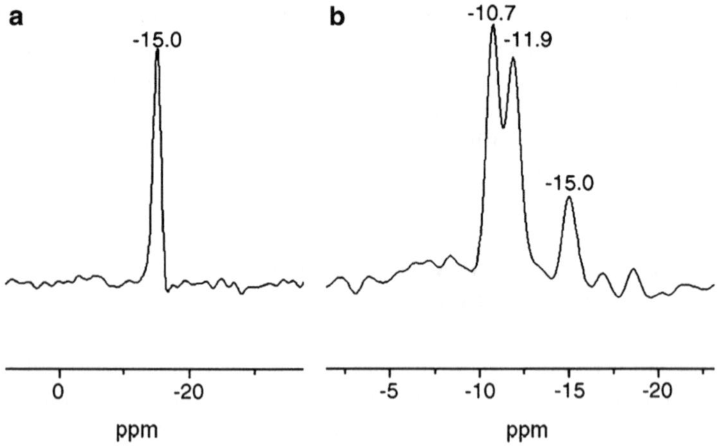

Fig. 8.1 ^{31}P MAS-NMR of 15 % H_3PW/SiO_2 before (**a**) and after transalkylation reaction (**b**)

cyclization reaction. Each catalyst was transferred to a round-bottom flask (10 wt% based on (+)-citronellal) with 5 mL of CH_2Cl_2. Next, 1 mmol of (+)-citronellal was added to that suspension, and the mixture was kept under magnetic stirring at room temperature. The time of the reactions varied from 30 min to 3 h. All products of the reaction were analyzed by ^{1}H NMR. The best catalyst was recycled.

3 Results and Discussion

3.1 *H_3PW/SiO_2: Catalyst for Benzene Transalkylation with C9+ Aromatics for Production of Xylenes*

Aromatic compounds are significant raw materials for intermediates of valuable petrochemicals and fine chemicals. Among them, benzene, toluene, and xylenes (BTX) are the most important. The major industrial sources of BTX (the reforming and gasoline pyrolysis) produce noticeable contents of trimethylbenzene and toluene. A convenient way to upgrade the low value C7 and C9 aromatics is to convert them to benzene and xylenes. However, the market reduction of benzene as a result of environmental issues underscores the importance of benzene transalkylation with C9+ aromatics as an important reaction for commercial applications [31–33]. Thus, the H_3PW/SiO_2 catalyst was tested to evaluate the benzene transalkylation with C9+ aromatics [34].

Catalysts based on H_3PW/SiO_2 have been widely prepared and characterized in the literature [26]. One of the best ways to determine the integrity of the Keggin anion is by ^{31}P MAS-NMR. Thus, the prepared catalysts (8–25 wt% H_3PW content) were characterized using this technique. In Fig. 8.1a, the signal that represents the

Table 8.1 Physical–chemical properties, acidity, and selectivity of the transalkylation reaction of trimethylbenzene with benzene using $x\%H_3PW/SiO_2$

Catalyst ($\%H_3PW$)	S^{f}_{BET} ($m^2 g^{-1}$)	S^{s}_{BET} ($m^2 g^{-1}$)	D_{XRD} (nm)	ΔH_{AVG} ($kJ mol^{-1}$)	T_{SITES} ($mmol g^{-1}$)	Sel_{XY} (%)	Coke (%)
8	74	23	12	85.9	0.078	0	1.6
15	63	60	15	100.4	0.159	3.9	0.3
20	57	43	17	108.8	0.204	2.8	1.0
25	40	20	18	116.7	0.252	0	2.4
MOR	n.a.	n.a.	n.a.	82.9	0.503	2.1	14.1

Keggin anion in the ^{31}P MAS-NMR spectrum is located at −15.0 ppm, which indicates that the structure was preserved after impregnation of H_3PW and calcination of the material.

Thermal stability is another important issue for the application of HPAs. Data of TG/DTG for the 15 % H_3PW/SiO_2 show that decomposition might occur after temperatures above 450 °C. The degree of decomposition is affected by the conditions for the utilization of the catalyst. It is well known that decomposition of H_3PW is dependent on the temperature as well as the time of exposition under such atmosphere [35, 36]. Thus, it is believed that partial decomposition of supported H_3PW could take place during the transalkylation reaction at 470 °C. Nonetheless, that temperature showed the highest activity for all catalysts and was compared to the commercial process, which uses mordenite (MOR) zeolite.

The main properties of xH_3PW/SiO_2 can be seen in Table 8.1. Silica gel (commercial Davisil®, grade 62) had a specific surface area of 260 $m^2 g^{-1}$. The supported samples decreased that area (Table 8.1, S^{f}_{BET}, column 2) to the range of 74–40 $m^2 g^{-1}$, depending on the loading of H_3PW. This decrease is attributed to the HPA deposit on the mesoporous surface of the silica [37]. These surface areas (S^{s}_{BET}) also decreased after the reaction (Table 8.1, column 3). In addition, the average size of nanocrystallites formed on the silica surface was in the range of 12–18 nm (Table 8.1, D_{XRD}, column 4). The acidity of these catalysts measured by the Cal-Ad method changed with H_3PW loading, from −85.9 to −116.7 kJ mol^{-1} (Table 8.1, $-\Delta H_{AVG}$, column 5). These later data are very relevant when coupled with the results of the benzene transalkylation with C9+ aromatics reaction.

All supported catalysts were active in the transalkylation of trimethylbenzene with benzene, but $15\%H_3PW/SiO_2$ was differentiated. Moreover, the relative deactivation on the stream was fast for all catalysts, except $15\%H_3PW/SiO_2$. The selectivity for the most import products (xylenes) was preponderant for that catalyst. Considering that $15\%H_3PW/SiO_2$ was not the strongest catalyst, a minimum acidity should be needed to activate the reaction (at least 80 kJ mol^{-1}). Nonetheless, the absence of selectivity for xylenes from $8\%H_3PW/SiO_2$ (only toluene was produced) indicates that the total amount (Table 8.1, T_{SITES}, column 6), and likely the distribution of acid sites, should be also mandatory. Thus, the higher density of Brønsted and Lewis acid sites in mordenite [29] and the presence of regular unidirectional pores probably favored the xylenes formation. The absence of those pores to stabilize the dialkylbenzenes led to the formation of coke (Table 8.1, column 7) on 8%H_3PW/SiO_2. In contrast, 15 and 20 % H_3PW are stronger acids and produce more xylenes than 8%H_3PW/SiO_2.

The decreased activity for the 20 % sample suggests that acid sites above 100 kJ mol^{-1} were not favorable to produce xylenes, because strong adsorption of the formed xylenes prevented desorption and produced a high amount of coke instead. This effect is confirmed by the 25%H_3PW/SiO_2, which does not produce any xylenes in the studied reaction conditions. It should be also noted that the amount of coke after the reaction was low in all supported catalysts and was associated with the loss of activity at the beginning of the reaction. Accordingly, almost no coke was produced on the catalyst with 15%H_3PW/SiO_2, which was the most stable. This sample also underwent the lowest loss on the specific surface area after reaction (Table 8.1, column 3). It is interesting to note that the Keggin structure of 15%H_3PW/SiO_2 had a clear degree of decomposition after reaction (Fig. 8.1b). The reason for the lowest deactivation on the stream of this catalyst can be attributed to the decomposed forms of H_3PW (−11.9 ppm attributed to $[\alpha\text{-}P_2W_{18}O_{62}]^{6-}$ and at −10.7 ppm assigned to $[\alpha\text{-}PW_{11}O_{39}]^{7-}$) that nonetheless are actives catalysts for the reaction.

In conclusion, supported H_3PW on silica was found to be active catalysts in the transalkylation of trimethylbenzene with benzene at 470 °C. An optimum value of strength, amount, and distribution of acid sites led to the best activity and selectivity for xylenes formation. The 15%H_3PW/SiO_2 was the best sample for this purpose with low deactivation by coke deposition and surface area reduction, thus offering promising properties to improve xylene production. Longer lifetime of the catalyst on the reaction stream may be reached by a lower temperature of reaction; this will decrease the degree of decomposition of the Keggin anion and also the activity of the catalyst, but may keep the selectivity level.

3.2 H_3PW/SiO_2–Al_2O_3: Catalyst for Esterification of Acetic Acid and Ethanol

Catalysts based on supported HPAs on silica–alumina have not been explored much in the literature, mainly due to the assumption that the polyacid would decompose on the surface. In order to verify this possibility, the supported H_3PW on silica–alumina was studied. The properties of the system H_3PW/SiO_2–Al_2O_3 have not yet been published, and discussion of its characterization will focus on the best catalyst. A series of catalysts containing 15, 20, 30, and 40 wt% loading of H_3PW on SiO_2–Al_2O_3 were characterized by leaching tests, FTIR, ^{31}P MAS-NMR, BET analysis, and pyridine adsorption.

Silica–alumina (Aldrich, 12 wt% Al_2O_3) had a surface area (S_{BET}) of 489 m^2 g^{-1}. The supported samples decreased that area linearly ($R^2=0.995$) as the loading of H_3PW increased (Table 8.2). This decrease is related to the deposit of H_3PW on the support surface. The interaction of heteropoly acids (HPAs) with several supports (e.g., silica gel, niobia, alumina, zirconia, MCM-41) have already been described in the literature [23, 26, 27, 38], evidencing that the nature of the support as well as the

Table 8.2 BET-specific surface area and ethyl acetate yield for *x*%H_3PW/SiO_2–Al_2O_3

Catalyst (%H_3PW)	S_{BET} ($m^2\ g^{-1}$)	Ethyl acetate yield (%)
0	489	42.3
15	402	70.2
20	372	73.5
30	292	78.3
40	240	63.0

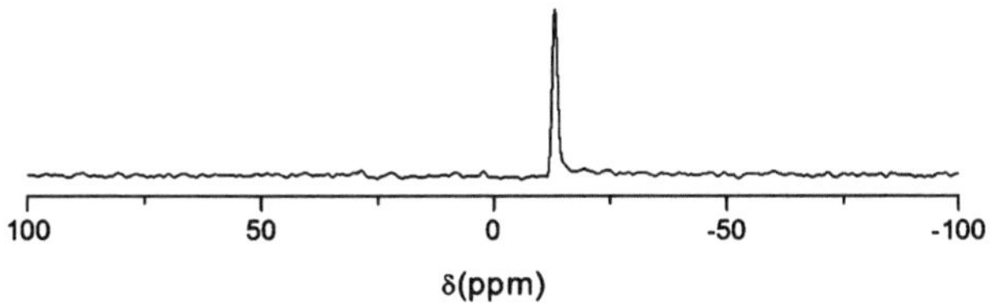

Fig. 8.2 ^{31}P MAS-NMR spectrum of 30%H_3PW/SiO_2–Al_2O_3

conditions used in the catalyst synthesis are important parameters in the development of supported HPA materials. Impregnation of H_3PW on γ-Al_2O_3 has caused partial degradation of the Keggin anion [39], attributed to the basicity of the support [3]. Nonetheless, other studies [27] show that successful impregnation procedures of H_3PW on γ-Al_2O_3 can be achieved using acetonitrile and 0.5 mol L^{-1} HCl aqueous solution as solvents. The 20%H_3PW/Al_2O_3 showed weaker acidity when compared to pure H_3PW and 20%H_3PW/SiO_2. Silica–alumina is expected to be a better support than silica gel due to the presence of both acid and basic sites [40], which can increase the strength of interaction between the HPA and the support without degradation of the Keggin anion. Indeed, grafting Al_2O_3 clusters at the silica gel surface increases H_3PW stability and interaction strength compared to alumina and silica gel [41].

Figure 8.2 presents the ^{31}P MAS-NMR spectrum of 30%H_3PW/SiO_2–Al_2O_3 material with a single signal at −14.8 ppm. The 20 and 40 % materials showed the same single signal, evidencing that there was no decomposition of the Keggin anion after preparation or calcination, as observed for H_3PW/Al_2O_3 [27]. The 15 wt% sample showed two signals at −13.5 and −14.8 ppm; the former might be related to the distorted Keggin structure of H_3PW.

Short-chain esters are extensively used in the chemical and pharmaceutical industry as solvents and for the production of other chemical substances [42]. The production of alkyl esters is most commonly achieved by the esterification reaction of carboxylic acids and alcohols with a homogeneous Brønsted acid catalyst (e.g., sulfuric acid). However, when conventional acid catalysts are used, an additional neutralization step is often required, which leads to catalyst degradation and generation of chemical wastes [43]. Thus, the development of heterogeneous catalysis to transform chemical reactions into more environmentally friendly processes is extremely important.

To determine the best H_3PW loading, catalytic activities in the acetic acid esterification were investigated using similar conditions. The catalytic runs were performed at 100 °C for 4 h using a 2:1 molar ratio (acetic acid:ethanol) and 10 wt%

Fig. 8.3 Effect of H_3PW loading in the esterification reaction of acetic acid with ethanol catalyzed by $x\%H_3PW/SiO_2–Al_2O_3$ (x = 15, 20, 30, and 40 wt%)

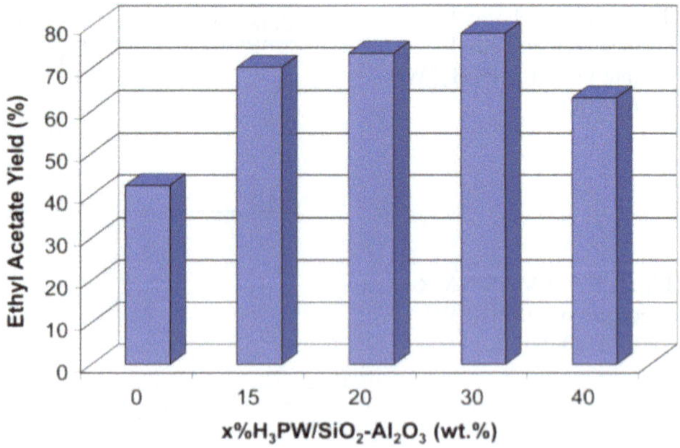

Fig. 8.4 Effect of acetic acid: ethanol molar ratio in the esterification reaction of acetic acid with ethanol catalyzed by $30\%H_3PW/SiO_2–Al_2O_3$

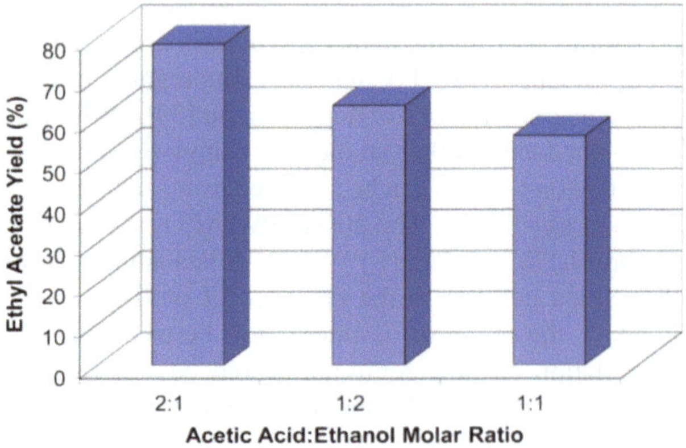

of the catalyst. The results (Fig. 8.3) show that the $30\%H_3PW/SiO_2–Al_2O_3$ material was the most active with about 78 % yield of ethyl acetate. Selectivity toward ester was 100 % for all studied catalysts.

The acetic acid:ethanol molar ratio (1:1, 1:2, and 2:1) in the esterification reaction was examined to determine its effect on the catalytic activity. The catalytic runs were performed at 100 °C for 4 h using 10 wt% of $30\%H_3PW/SiO_2–Al_2O_3$. Figure 8.4 shows that at 2:1 molar ratio, the $30\%H_3PW/SiO_2–Al_2O_3$ catalyst achieved the highest yield of ethyl acetate (78.3 %) and decreased to 63.4 and 56.0 % at 1:2 and 1:1 molar ratios, respectively. Selectivity was 100 % in all ratios studied. The esterification is an equilibrium reaction; the formation of the products was accomplished using one of the reactants in excess or by removing one of the products. The increase of ethyl acetate yield with acetic acid excess over ethanol excess may be related to the contribution of acetic acid to the self-catalyzed reaction. However, the self-catalyzed kinetic of esterification reactions is usually too slow, so a catalyst is used to improve kinetics and yield [44]. It has been suggested [45] that an excess of alcohol in the esterification reaction reduces the ester production by blocking the catalyst active sites [46].

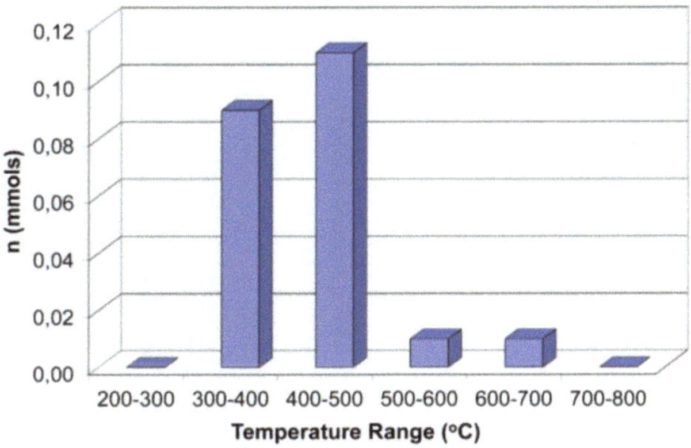

Fig. 8.5 Acid site distribution for 30%$H_3PW/SiO_2–Al_2O_3$ determined by TG/DTG after pyridine adsorption from 200 to 800 °C

The catalyst acidity was studied by gas-phase pyridine adsorption. The results show a total of 0.22 mmol g^{-1} acidic sites for the 30%$H_3PW/SiO_2–Al_2O_3$ material. The majority (0.20 mmol g^{-1}) of these sites were obtained by desorption at the 300–500 °C range with a peak maximum at 384 °C. Figure 8.5 presents the acid site distribution from 200 to 800 °C, which shows medium to strong acid sites with 0.09 mmol g^{-1} from 300 to 400 °C and 0.11 mmol g^{-1} from 400 to 500 °C. The same method of analysis applied to pure H_3PW showed 0.97 mmol g^{-1} of acid sites (theoretically, 1.03 mmol g^{-1}). Since the 30%$H_3PW/SiO_2–Al_2O_3$ sample possesses 0.30 g for each 1.00 g of catalyst, a total of 0.29 mmol g^{-1} acid sites would be expected for this material. The TG/DTA analysis reveals that 76 % of the H_3PW sites were accessible to pyridine, assuming that the $SiO_2–Al_2O_3$ sites were involved in acid–base interactions with the support and do not contribute to the acid site distribution of the catalyst.

Figure 8.6 shows the effect of the mass of the catalyst on the ethyl acetate production at 100 °C, 4 h, and 2:1 molar ratio (acetic acid:ethanol). For up to 0.1 g of catalyst (10 wt% related to acetic acid), a considerable increase from 59.7 to 76.0 % occurred in the ethyl acetate yield after which the yield remained practically constant. This indicates that the esterification reaction of acetic acid and ethanol had to some extent a dependency with the catalyst amount, i.e., the number of acid sites. According to the literature [45], the increase of ethyl acetate yield with the number of acid sites suggests that the reaction rate is dependent on the catalyst mass.

H_3PW is highly soluble in ethanol and moderately soluble in acetic acid and ethyl acetate [47]. To determine if any leaching occurred during the reaction, the catalyst stability on the most soluble solvent (ethanol) was tested after calcination procedures at 200, 300, and 400 °C during 1 h. In all tests, H_3PW leaching was not detected, evidencing that the esterification reaction proceeded by heterogeneous catalysis.

In conclusion, supported H_3PW on silica–alumina was found to be an active catalyst in the esterification of acetic acid and ethanol. It was developed an optimum value of *x*%H_3PW (30 %), reactants molar ratio (2:1 acetic acid:ethanol), and

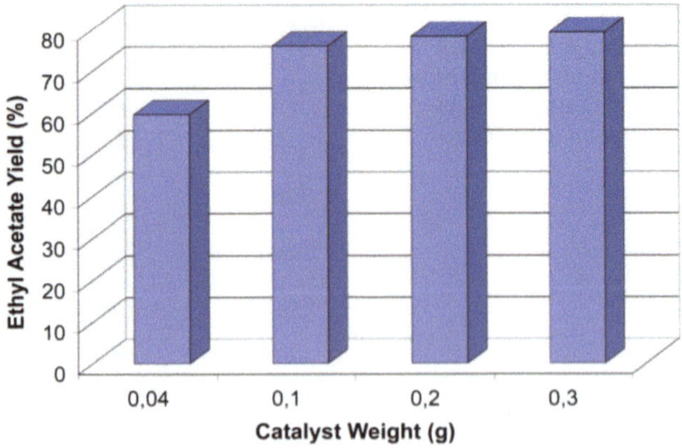

Fig. 8.6 Effect of the catalyst mass in the esterification reaction of acetic acid with ethanol catalyzed by 30%H_3PW/SiO_2–Al_2O_3

catalyst mass (10 wt%, related to acetic acid), which led to the best activity for ethyl acetate formation. In addition, 30%H_3PW/SiO_2–Al_2O_3 was also stable without any leaching under contact with ethanol for 1 h. The results obtained so far suggest that the reaction rate is dependent on the catalyst mass.

3.3 H_3PW/ZrO_2: Catalyst for Esterification of Oleic Acid and Ethanol

Esters are an industrially important class of substances with a variety of applications that are produced mainly by esterification of organic acids with alcohols [48]. Esterification usually has a low equilibrium constant and is a slow chemical reaction, especially at room temperature. The reactants, an alcohol and a carboxylic acid, yield an ester and water as the equilibrium shifts toward the products. Esterification as well as transesterification reactions of long-chain fatty acids has been largely applied to the production of biodiesel, which is considered a renewable fuel based on resources such as biomass or recycled feedstocks [49]. The esterification reaction of oleic acid is interesting in the context of biodiesel production, since this and other free fatty acids (FFA) may be present in different extensions in transesterification reactions of waste source of feedstocks. In that event, acid catalysts are more efficient than base ones, raising demand of active catalysts for both reactions simultaneously.

Supported H_3PW has been explored for the esterification and transesterification of a variety of reactants involving not only oleic acid but also a number of other FAA using mainly methanol and ethanol [16, 50–57]. Moreover, the system of H_3PW/ZrO_2 has been much tested and looks promising in many aspects. Thus, supported H_3PW/ZrO_2 was employed in the esterification of oleic acid with ethanol. The complete characterization of this system has been published elsewhere [58]. Nonetheless, the details of the better system correlated with acidity and activity will be discussed here.

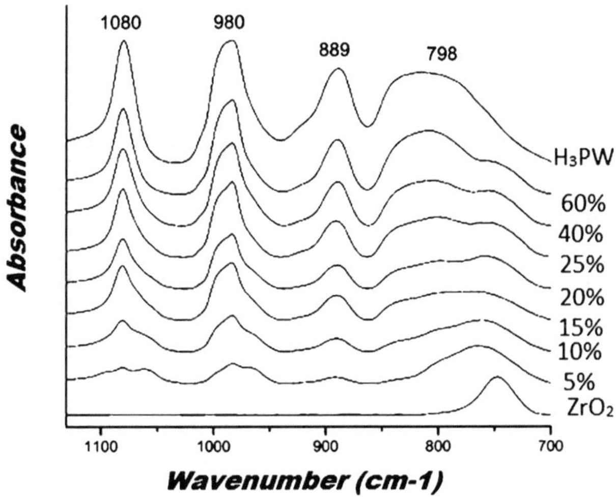

Fig. 8.7 Spectra of FTIR of $x\%H_3PW/ZrO_2$

In order to check the presence and structural integrity at the molecular level of the Keggin anion, FTIR, Raman, and ^{31}P MAS-NMR spectroscopies were used. All results demonstrate the preserved structure of H_3PW over ZrO_2 support in the 5–25 wt% range. As shown in Fig. 8.7, FTIR spectra illustrate these findings. Bands were assigned to ν_{as} (P–O$_a$) = 1,080 cm^{-1}, ν_{as} (W=O$_t$) = 982 cm^{-1}, ν_{as} (W–O$_c$–W) = 898 cm^{-1}, and ν_{as} (W–O$_e$–W) = 797 cm^{-1}, where a, t, c, and e stand for the specific positions (internal, terminal, corner, and edge-shared, respectively) of the oxygen atoms in the Keggin structure. The intensity of the bands, as expected, was proportional to the H_3PW loading. In addition, the results of ^{31}P MAS-NMR showed a strong interaction between H_3PW and ZrO_2, producing a two-signal spectra: one signal at −15.0 ppm (related to ^{31}P of the crystalline hexahydrate H_3PW) and the other at −13.0 ppm (assigned to ^{31}P bounded to zirconia toward surface bridges $\{[\equiv Zr{-}OH_2]_n^+[H_{3-n}PW_{12}O_{40}]^{n-3}\}$), respectively.

The textural properties of the support as well as the H_3PW/ZrO_2 catalysts were important features to determine their reactivity (Table 8.3). The commercial zirconia (Aldrich) displayed a monoclinic phase (XRD), a type II nitrogen adsorption isotherm, a low BET surface area of 6.1 m^2 g^{-1}, and an average pore size of 11.5 nm. Calcined H_3PW at 200 °C for 4 h showed a type II adsorption isotherm, a BET surface of 4.3 m^2 g^{-1}, and average pore size of 7.0 nm. Both solids showed an average distribution of mesoporous. An increase was observed on the surface area of the supported samples compared to pure zirconia and pure H_3PW, mainly within the range of 15–20 wt%. Above $20\%H_3PW/ZrO_2$, the pores started to collapse and/or clog, since the specific area decreased, and the pore volume shrinked considerably. This behavior, observed by different researchers, is explained by the formation of the surface overlayer from the H_3PW, added primarily on the mesoporous of zirconia. The supported phase interacts with ZrO_2, thereby increasing the total surface area of the final material. As the support by itself had a low specific surface area, it reflected the relative coverage of H_3PW, which led to materials with multiple

Table 8.3 Surface area (BET), pore size (BJH), and other important properties for $x\%H_3PW/ZrO_2$ in the esterification of oleic acid and ethanol

Catalyst	S_{BET} (m^2 g^{-1})	P_s^a (nm)	n_{H_3PW} (mmol g^{-1})	Coverage[b]	$TOF\left(mol_{EO}\ mol^{-1}_{H_3PW}\ h^{-1}\right)$
ZrO_2	6.1	10.2	0.0000	0.0	0.0
$5\%H_3PW/ZrO_2$	5.9	13.6	0.0174	2.0	183.5
$10\%H_3PW/ZrO_2$	6.6	13.1	0.0347	4.3	188.6
$15\%H_3PW/ZrO_2$	7.7	9.5	0.0521	6.8	193.7
$20\%H_3PW/ZrO_2$	9.1	9.5	0.0694	9.7	201.4
$25\%H_3PW/ZrO_2$	8.1	7.9	0.0868	12.9	195.8
$40\%H_3PW/ZrO_2$	8.1	7.1	0.1389	25.8	191.2
H_3PW[c]	4.3	7.0	0.3472	-	25.5

[a]BJH desorption average pore width (4 V/A)
[b]Coverage = $\%H_3PW \times 6.02 \times 10^{23} \times 1.13 \times 10^{-18}/(2{,}880 \times \%ZrO_2 \times BET_{ZrO_2})$
[c]TOF calculated based on a saturated solution of H_3PW in ethanol

surface layers at relative low loadings. These results are important when coupled to other data, such as coverage, which will be further detailed.

Initially, the reaction of acid oleic and ethanol was examined calculating the conversion for ethyl oleate versus different amounts of xH_3PW/ZrO_2 in a period of 4 h using 1:6 acid to alcohol molar ratio and 10 wt% of catalyst (based on acid oleic). Zirconia converted about 16 %, while H_3PW (homogeneous reaction) reached 98 %. The ratio of conversion increased quickly as the loading of H_3PW increased, forming a plateau. It was noted that at $20\%H_3PW$, the conversion started to saturate, and the amount of the polyacid could be optimum for that reaction [58].

In order to calculate the intrinsic catalytic activity of each supported material, the reaction kinetics was measured under initial conditions (i.e., low conversion range), which was chosen as 1 h. The actual catalytic activity of the supported catalysts was compared with turnover frequency (TOF) values. The TOF values, calculated based on the total amount of H_3PW (Table 8.3), confirmed the maximum conversion for $20\%H_3PW/ZrO_2$ catalyst. This catalyst had the highest surface area and is economically more attractive than pure H_3PW, since it reached in 4 h a similar conversion with less amount of the polyacid. In addition, the process could only be heterogeneous, which facilitates separation of the catalyst from the products.

The fast increase of conversion ratio toward a near saturation value, close to the ideal value expected for pure H_3PW, indicates effective coverage of the support active surface sites. This behavior may raise questions if the observed process is heterogeneous or has any homogeneous contribution. Thus, a test for leaching H_3PW was conducted to clarify the nature of the process. The test involved the determination of free H_3PW by UV–Vis spectroscopy [58]. About 8 % of H_3PW was leached after the reaction cycle (i.e., the actual loading of the 20 % catalyst was 18.4 wt%). Nonetheless, this was very different from similar tests with a proportional mechanical mixture of H_3PW and ZrO_2, which showed that H_3PW was completely leached, driving the process to a homogeneous one. Thus, the reaction in the presence of pure H_3PW, or when this polyacid is completely leached (e.g., mechanical mixture), is much faster, as expected for homogeneous process. Therefore, the obtained results demonstrate that under those studied conditions, the process with supported H_3PW on zirconia is heterogeneous with low contribution of the leached H_3PW, which reacts through a homogeneous reaction pathway. In the literature, similar reactions involving supported H_3PW on zirconia did not show any leaching. This is explained by the syntheses of the catalysts. The commercial ZrO_2 utilized had a very low specific surface area (about 6 m^2 g^{-1}); coverage above the monolayer might have facilitated the leaching. Other synthetic zirconia with surface areas as high as 150 m^2 g^{-1}, for example, would have better dispersion and probably a stronger interaction of H_3PW with the hydroxyl surface of ZrO_2.

Additional catalytic tests of 20%H_3PW/ZrO_2 were conducted in a variable period of time (1–12 h). These tests allowed a kinetic evaluation of the supported samples. A significant increase in the conversion ratio was observed in the first 4 h, after which the conversion reached a near saturation [58]. The reaction rate may have decreased due to competition between the substrate and water for adsorption on the active sites of the catalyst as well as the increase in the reverse hydrolysis reaction. Under the experimental conditions, the kinetics was compatible with a pseudo first-order reaction; i.e., the rate depended only on the oleic acid concentration. Nonetheless, ethanol might facilitate the reaction process by coming into the supported H_3PW structure in the direction of the protons, as polar molecules are able to penetrate the H_3PW secondary structure taking part in catalysis by the pseudoliquid phase (bulk type II) [59]. This hypothesis is supported by calculation of the monolayer coverage. Taking into account that the prepared 20%H_3PW/ZrO_2 had low surface area and was very active, the reaction kinetics must be driven by other than surface-type mechanism. Considering that the saturation to form monolayer coverage for these catalysts would be about 2.5 wt% (each Keggin unit occupies about 1.13 nm^2), the coverage of 20%H_3PW/ZrO_2 corresponds to 9.7 layers. Thus, a pseudoliquid process may better explain the high activity observed.

The most important parameter related to the activity of these catalysts is acidity. The nature and amount of the active sites may determine the activity. The nature was probed by pyridine adsorption and detected by FTIR, which attested to the formation of only Brønsted sites (absorption bands at 1,488, 1,532, and 1,540 cm^{-1}). Therefore, the observed activities were due to the presence of strong Brønsted protons provided by H_3PW. The amount of protons on the zirconia support was calculated by the loading, corresponding to 0.208 mmol g^{-1} of catalyst. The experimental result, based on pyridine desorption obtained by thermogravimetric analysis (TGA), showed 0.207 mmol g^{-1}. This demonstrates the efficient dispersion of H_3PW on the ZrO_2 surface and the high accessibility of the pyridine molecule to all protons of the structure.

Another important issue of any catalyst is stability under thermal treatments or reaction cycles. In order to evaluate the degradation of the supported H_3PW structure as a function of temperature, a series of esterification reactions were done with 20%H_3PW/ZrO_2 under the same experimental conditions, and the conversion and structure were evaluated. Initially, the conversion decreased slowly with the temperature (88 to 80 % under calcinations from 200 to 400 °C) but sharply with the heating treatment (80 to 20 % under calcinations from 400 to 700 °C). The structural data, based on ^{31}P MAS-NMR, FT-Raman, and XRD methods, show modifications of the original Keggin structure, which fully justify the decrease in reactivity [12]. These results are important for deactivation studies. The best treatment led to a decrease in the conversion from 88 to 40 % in the fourth cycle, in which the catalyst was recovered by filtration, washed with *n*-hexane, and calcined at 300 °C for 4 h. Even taking into consideration that leaching of H_3PW took place in the first run (not detected for the others), the deactivation was due to adsorption of both initial reagent (oleic acid) and product (ethyl oleate) detected from carbonaceous residues by elemental CHN analysis. Further studies of this issue should better clarify the question of deactivation. Note that research in the literature involving other reaction systems observed no deactivation for 4–6 cycles. This raises the question of whether preparation and structure of the zirconia support definitively affects the properties of the final catalyst.

3.4 H_3PW/MCM-41: Catalyst for Cyclization of (+)-Citronellal

The intramolecular cyclization of (+)-citronellal to produce (−)-menthol is composed by an acid-catalyzed isomerization of citronellal, resulting in a mixture of isopulegol intermediate isomers followed by a metal-catalyzed hydrogenation of the (−)-isopulegol [60, 61]. This is considered an important reaction in the fine chemical industry.

This study examined H_3PW supported on mesoporous MCM-41 for the cyclization of (+)-citronellal; the results were submitted for publication. The XRD of the supported catalysts (Fig. 8.8) showed that lower loadings (2–20 %) of H_3PW resulted in smaller crystallite size samples (23 nm) than pure MCM-41 (34 nm), which might be related to the hydrolysis of siloxane groups. On the other hand, higher loadings (40 %) of H_3PW produced larger particles (55 nm) due to agglomeration of MCM-41 particles through the formation of Keggin unit bridges [62].

As shown in Table 8.4, the mesoporous area contributed more to the total surface area (S_{BET}) of the material than the external surface. Comparison of pore volumes indicates that for the sample containing 2 %, the H_3PW species were mainly located inside the mesopores [63]. Between 5 and 20 %, a simultaneous distribution of H_3PW inside the pores and on the external surface resulted in a considerable decrease in the total surface area. For the 40 % sample, aggregates of H_3PW species were located preferentially on the external surface area, increasing the average pore size of the final material. Moreover, ^{31}P MAS-NMR studies indicated a characteristic signal at ca. −14.8 ppm related to Keggin structure, which is dependent on the hydration degree of the samples [64].

Thermal analysis showed the presence of three thermal events analog to pure H_3PW [34]. In addition, pyridine gaseous adsorption experiments indicated that at low loadings of H_3PW, the total amount of acid sites decreased, indicating that H_3PW protons were tightly bound to the stronger Si–OH groups of MCM-41. As the amount of H_3PW increased, a connected structure between terminal oxygens and water molecules hydrogen bonded to the protons resulted in Keggin anions

Fig. 8.8 XRD of $x\%H_3PW/$MCM-41: (**a**) H3PW; (**b**) 2%; (**c**) 5%; (**d**) 15%; (**e**) 20%; (**f**) 40%; (**g**) MCM-41

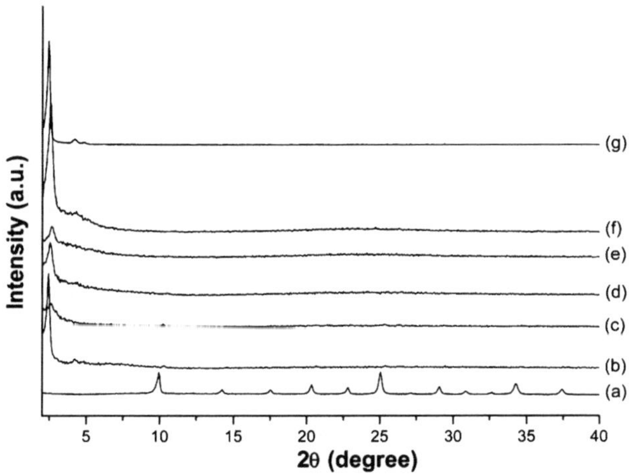

Table 8.4 Textural properties of $x\%H_3PW$/MCM-41

Catalyst	P_s^a (nm)	S_{BET} ($m^2\ g^{-1}$)	V_p^b ($cm^2\ g^{-1}$)
H_3PW	1.9	4.2	0.002
MCM-41	2.4	831.5	0.51
2%H_3W/MCM-41	2.0	743.1	0.37
5%H_3W/MCM-41	2.1	817.3	0.43
15%H_3W/MCM-41	2.4	647.9	0.38
20%H_3W/MCM-41	2.4	619.3	0.34
40%H_3W/MCM-41	2.8	740.9	0.53

[a]Average pore size diameter (4 V/A) calculated by BJH desorption method
[b]BJH desorption cumulative volume of pores

aggregates and larger clusters [26]. Higher loadings enhanced the overall acidity of the materials because of the higher amount of available protons. Two types of sites were identified by FTIR as Brønsted and hydrogen bonding (bands at 1,540, 1,490, and 1,444 cm^{-1}, which were different from pure MCM-41) due to the presence of H_3PW inside the channels and/or on the external surface of the MCM-41.

Table 8.5 presents the results of the cyclization reaction in 3 h for all studied catalysts. Isolated MCM-41 exhibited conversion around 35 % and selectivity to (−)-isopulegol around 37 %, related to the strength of its hydrogen bonding sites [19].

Four isomers were identified as products after complete conversion of the (+)-citronellal in 3 h: (−)-isopulegol, (+)-neoisopulegol, (+)-isoisopulegol, and (+)-neoisoisopulegol. For the desired isomer, 20%H_3PW/MCM-41 catalyst presented the highest selectivity (74.1 %). The selectivities to (−)-isopulegol using the supported mesoporous catalysts varied from 60 to 74 %, showing notable stereoselectivity compared to the literature [60, 61, 65, 66].

Table 8.5 Catalytic results for cyclization of (+)-citronellal over $x\%H_3PW$/MCM-41

Catalyst[a]	Stereoselectivity (−)-isopulegol (%)
MCM-41	37.3
$2\%H_3W$/MCM-41	65.0
$5\%H_3W$/MCM-41	65.9
$15\%H_3W$/MCM-41	62.2
$20\%H_3W$/MCM-41	74.1
$40\%H_3W$/MCM-41	61.7

[a]Conditions of the reaction: 1 mmol of (+)-citronellal, 10 wt% of catalyst, 5 mL CH_2Cl_2, and 3 h. Conversion was 35 % for MCM-41 and 100 % to xH_3PW/MCM-41

A proposed schematic mechanism of the intramolecular cyclization of (+)-citronellal starts with the activation of carbonyl group by hydrogen bonding through the supported H_3PW proton with the oxygen atom of (+)-citronellal molecule, increasing the electrophilic character of the carbonyl group. Next, the double bond nucleophilic attack occurs simultaneously with the formation of terminal olefin through the removal of hydrogen from the methyl group by oxygen. Among the four isomers, the (−)-isopulegol conformation is the more stable stereoisomer with carbocationic intermediates and a more energetic transition state.

Kinetic studies for the $20\%H_3PW$/MCM-41 catalyst revealed that the best reaction time was 60 min with conversion of 96.2 % and selectivity to the isomer (−)-isopulegol of 64.8 %. Under this time reaction, the catalyst was reused four times; a reduction of 25 % in the (+)-citronellal conversion was observed from the first cycle (96 %) to the last cycle (71 %). In addition, differences in the selectivity for the (−)-isopulegol isomer were small (13 %). The reduction of catalytic activity could be attributed to the catalyst activation before every cycle (300 °C for 4 h under air), which may have partially affected the Keggin structure of H_3PW [67]. Also, possible formation of coking on the surface of the catalysts cannot be ruled out. According to SEM studies, the $20\%H_3PW$/MCM-41 sample showed a good dispersion of particles compared with pure MCM-41.

In conclusion, catalysts based on H_3PW supported on MCM-41 were successfully prepared and tested in the cyclization of (+)-citronellal. All prepared catalysts were active in the cyclization of (+)-citronellal forming the main stereoisomer (−)-isopulegol, but a kinetic study demonstrated that $20\%H_3PW$/MCM-41 had the best performance with about 96 % conversion and 65 % selectivity for 1-h reaction. This catalyst was reutilized four times with a low degree of deactivation, keeping the selectivity about the same. The difference in the activity is attributed to the decrease in the dispersion degree into the pores of MCM-41 by formation of small agglomerates in the channels. Experimental characterization of the catalysts confirmed the integrity of the Keggin anion, which were preferentially anchored inside the mesochannels of the molecular sieve.

4 Conclusion

Supported H_3PW on silica, silica–alumina, zirconia, and mesoporous MCM-41 were prepared, characterized, and tested in different reactions: transalkylation of benzene with aromatics, esterification of acetic acid and ethanol, esterification of oleic acid with ethanol, and cyclization of (+)-citronellal. These are important reactions for industrial purposes. The studies show the potentiality of these materials to be greener catalysts than other traditional liquid acids used in those reactions. The controlled properties of these catalysts, in particular, acidity, demonstrate the importance of designing the right catalyst for the right reaction. This choice improves the activity and selectivity of the main desired products. The main challenge is still the capacity to recover the initial activity for supported HPAs. The effective elimination of carbonaceous deposits on the catalyst surface should be sought as the main goal. The other is to improve the strength of the interaction of H_3PW on the surface support in order to eliminate the possibility of deactivation by leaching.

Acknowledgments We would like to thank all of the graduate and undergraduate students from Laboratório de Catálise (LabCat family) that helped us to obtain the data presented in this manuscript. Also, we acknowledge CNPq for research scholarships and financial support provided by DPP/IQ/UnB, FINATEC, FAPDF, CAPES, MCT/CNPq, FINEP/CTInfra, FINEP/CTPetro, and PETROBRAS.

References

1. Greenwood NN, Earnshaw A (1994) Chemistry of the elements. Pergmon Press, Oxford, pp 1175–1185, reprinted
2. Okuhara T, Mizuno N, Misono M (1996) Adv Catal 41:113–252
3. Pope MT (1983) Heteropoly and isopoly oxometalates. Springer, Berlin, pp 58–80
4. Hill CL (1998) Ed Chem Rev 98:1–390
5. Kozhevnikov IV (2002) Catalysts for fine chemical synthesis – catalysis by polyoxometalates. Wiley, Chichester, pp 9–42
6. Moffat JB (2001) Metal-oxygen clusters: the surface and catalytic properties of heteropoly oxometalates. Springer, New York, pp 71–93
7. Jalil PA, Al-Daous MA, Al-Arfaj AA, Al-Amer AM, Beltramini J, Barri AI (2001) Appl Catal A 207:159–171
8. Misono M (1987) Catal Rev-Sci Eng 29:269–321
9. Pizzio LR, Vázquez PG, Cáceres CV, Blanco MN, Alesso EN, Torviso MR, Lantaño B, Moltrasio GY, Aguirre JM (2005) Appl Catal A 287:1–8
10. Cardoso LAM, Alves W Jr, Gonzaga ARE, Aguiar LMG, Andrade HMC (2004) J Mol Catal A 289:189–197
11. Gómez-García MA, Pitchon V, Kiennemann A (2005) Environ Sci Technol 39:638–644
12. Gu Y, Wei R, Ren X, Wang J (2007) Catal Lett 113:41–45
13. Fumin Z, Jun W, Chaoshu Y, Xiaoqian R (2006) Sci China Ser B Chem 49:140–147
14. Izumi Y (1998) Res Chem Intermed 24:461–471
15. Khder AERS, Hassan HMA, El-Shall MS (2012) Appl Catal A 411–412:77–86

16. Srilatha K, Lingaiah N, Devi BLAP, Prasad RBN, Venkateswar S, Sai Prasad PS (2009) Appl Catal A 365:28–33
17. Gagea BC, Lorgouilloux Y, Altintas Y, Jacobs PA, Martens JA (2009) J Catal 265:99–108
18. Bhatt N, Sharma P, Patel A, Selvam P (2008) Catal Commun 9:1545–1550
19. Braga PRS, Costa AA, Macedo JL, Ghesti GF, Souza MP, Dias JA, Dias SCL (2011) Microporous Mesoporous Mater 139:74–80
20. Garcia FAC, Braga VS, Silva JCM, Dias JA, Dias SCL, Davo JLB (2007) Catal Lett 119:101–107
21. Santos JS, Dias JA, Dias SCL, Garcia FAC, Macedo JL, Sousa FSG, Almeida LS (2011) Appl Catal A 394:138–148
22. Drago RS, Dias SC, Torrealba M, de Lima L (1997) J Am Chem Soc 119:4444–4452
23. Caliman E, Dias JA, Dias SCL, Garcia FAC, de Macedo JL, Almeida LS (2010) Microporous Mesoporous Mater 132:103–111
24. Dias JA, Osegovic JP, Drago RS (1998) J Catal 183:83–90
25. Dias JA, Caliman E, Dias SCL (2004) Microporous Mesoporous Mater 76:221–232
26. Dias JA, Caliman E, Dias SCL, Paulo M, de Souza ATCP (2003) Catal Today 85:39–48
27. Caliman E, Dias JA, Dias SCL, Prado AGS (2005) Catal Today 107–108:816–825
28. Dias SCL, de Macedo JL, Dias JA (2003) Phys Chem Chem Phys 5:5574–5579
29. de Macedo JL, Dias SCL, Dias JA (2004) Microporous Mesoporous Mater 72:119–125
30. de Macedo JL, Ghesti GF, Dias JA, Dias SCL (2008) Phys Chem Chem Phys 10:1584–1592
31. Hatch LF, Mater S (1979) Hydrocarbon Process 58:189–192
32. Tsai T, Liu S, Wang I (1999) Appl Catal A 181:355–398
33. Dimitriu E, Guimon C, Hulea V, Lutic D, Fechete I (2002) Appl Catal A 237:211–221
34. Dias JA, Rangel MC, Dias SCL, Caliman E, Garcia FAC (2007) Appl Catal A 328:189–194
35. Drago RS, Dias JA, Maier TO (1997) J Am Chem Soc 119:7702–7710
36. Dias JA, Dias SCL, Kob NE (2001) J Chem Soc Dalton Trans 3:228–231
37. Lefebvre F, Liu-Cai FX, Auroux A (1994) J Mater Chem 4:125–131
38. Vázquez P, Pizzio L, Cáceres C, Blanco M, Thomas H, Alesso E, Finkielsztein L, Lantaño B, Moltrasio G, Aguirre J (2000) J Mol Catal A 161:223–232
39. Pizzio LR, Cáceres CV, Blanco MN (1998) Appl Catal A 167:283–294
40. Tanabe K, Yamaguchi T (1966) J Res Inst Catal 14:93–100
41. Rao PM, Wolfson A, Kababya S, Vega S, Landau MV (2005) J Catal 232:210–225
42. Sakamuri R (2003) In: Kirk RE, Othmer DF (eds) Encyclopedia of chemical technology, vol 10. Wiley, London, pp 497–499
43. Bhorodwaj SK, Pathak MG, Dutta DK (2009) Catal Lett 133:185–191
44. Pereira CSM, Pinho SP, Silva VMTM, Rodrigues AE (2008) Ind Eng Chem Res 47:1453–1463
45. Das J, Parida KM (2007) J Mol Catal A 264:248–254
46. Chakraborty AK, Basak A, Grover V (1999) J Org Chem 64:8014–8017
47. Izume Y, Hasebe R, Urabe K (1983) J Catal 84:402–409
48. Parida KM, Mallick S (2007) J Mol Catal A 275:77–83
49. Ghesti GF, Macedo JL, Parente VCI, Dias JA, Dias SCL (2009) Appl Catal A 355:139–147
50. Kulkarni MG, Gopinath R, Meher LC, Dalai AK (2006) Green Chem 8:1056–1062
51. Pizzio L, Vázquez P, Cáceres C, Blanco M (2001) Catal Lett 77:233–239
52. Devassy BM, Lefebvre F, Halligudi SB (2005) J Catal 231:1–10
53. López-Salinas E, Hernández-Cortéz JG, Schifter I, Yorres-Garcia E, Navarrete J, Gutiérrez-Carrillo A, López T, Lottici PP, Bersani D (2000) Appl Catal A 193:215–225
54. Mallik S, Dash SS, Parida KM, Mohapatra BK (2006) J Coll Int Sci 300:237–243
55. Hatt NB, Shah C, Patel A (2007) Catal Lett 117:146–152
56. Devassy BM, Halligudi SB (2006) J Mol Catal A 253:8–15
57. Patel S, Purohit N, Patel A (2003) J Mol Catal A 192:195–202
58. Oliveira CF, Dezaneti LM, Garcia FAC, de Macedo JL, Dias JA, Dias SCL, Alvim KSP (2010) Appl Catal A 372:153–161
59. Lee KY, Nakata TAS, Asaoka S, Okuhara T, Misono M (1992) J Am Chem Soc 114:2836–2842
60. Mertens P, Verpoort F, Parvulescu A-N, Vos D (2006) J Catal 243:7–13

61. Neatu F, Coman S, Pârvulescu VI, Poncelet G, De Vos D, Jacobs P (2009) Top Catal 52:1292–1300
62. Llanos A, Melo L, Avendaño F, Montes A, Brito JL (2008) Catal Today 133–135:1–8
63. Nandhini KU, Mabel JH, Arabindoo B, Palanichamy M, Murugesan V (2006) Microporous Mesoporous Mater 96:21–28
64. Kozhevnikov IV (1998) Chem Rev 98:171–198
65. da Silva KA, Robles-Dutenhefner PA, Sousa EMB, Kozhevnikov EF, Kozhevnikov IV, Gusevskaya EV (2004) Catal Commun 5:425–429
66. Mäki-Arvela PM, Kumar N, Nieminen V, Sjöholm R, Salmi T, Murzin DY (2004) J Catal 225:155–169
67. Milone C, Gangemi C, Neri G, Pistone A, Galvagno S (2000) Appl Catal A 199:239–244

Chapter 9
Biodiesel Production over 12-Tungstosilicic Acid Anchored to Different Mesoporous Silica Supports

Varsha Brahmkhatri and Anjali Patel

Contents

1 Introduction 190
  1.1 What Is Biodiesel? 190
  1.2 Biodiesel Feedstocks 191
  1.3 Catalysts for Transesterification and Esterification Reactions 191
2 Experimental Section 193
  2.1 Materials 193
3 Results and Discussion 195
  3.1 Characterization 195
  3.2 Esterification of Oleic Acid over TSA_3/MCM-41 199
  3.3 Recycling of the Catalyst 202
  3.4 Transesterification of Triglycerides (*Jatropha* Oil) 202
  3.5 Effect of Supports on Activity of TSA in Biodiesel Production 205
4 Conclusion 206
References 207

Abstract This chapter describes the use of heterogeneous catalysts comprising 12-tungstosilicic acid and mesoporous silicas, MCM-41 and SBA-15, for esterification of free fatty acid and oleic acid with methanol. The influence of various reaction parameters such as catalyst concentration, acid/alcohol molar ratio, and temperature was studied to optimize the conditions for maximum conversion of oleic acid. The catalysts show potential of being used as recyclable catalyst materials after simple regeneration. As an application, studies were carried out for biodiesel production from *Jatropha* oil, as feedstocks without any pretreatment, with methanol over the present catalysts. The catalytic activity of both the catalysts was correlated with the nature of the support.

V. Brahmkhatri • A. Patel (✉)
Department of Chemistry, Faculty of Science, M. S. University of Baroda,
Vadodara 390 002, India
e-mail: aupatel_chem@yahoo.com

A. Patel (ed.), *Environmentally Benign Catalysts: For Clean Organic Reactions*,
DOI 10.1007/978-94-007-6710-2_9, © Panch Tattva Publishers, Pune, India 2013

1 Introduction

Biodiesel has gained tremendous interest in recent years as an alternative, renewable liquid transportation fuel. Compared with conventional diesel, biodiesel has the advantages of being biodegradable, renewable, nontoxic, and less polluting, especially SO_x emission [1, 2]. The major obstacle to the commercialization of biodiesel from vegetable oil, in comparison to petroleum-based diesel, is primarily the high raw material cost [3–6].

1.1 What Is Biodiesel?

Biodiesel is a renewable fuel comprised of monoalkyl esters of fatty acids. Biodiesel is recognized as a "green fuel" that has several advantages over conventional diesel. Biodiesel is simple to use, biodegradable, nontoxic, and essentially free of sulfur and aromatics.

1.1.1 How Is Biodiesel Produced?

Biodiesel is usually produced from vegetable oils and animal fats via their trans-esterification reaction or from free fatty acids via their esterification.

Transesterification reaction

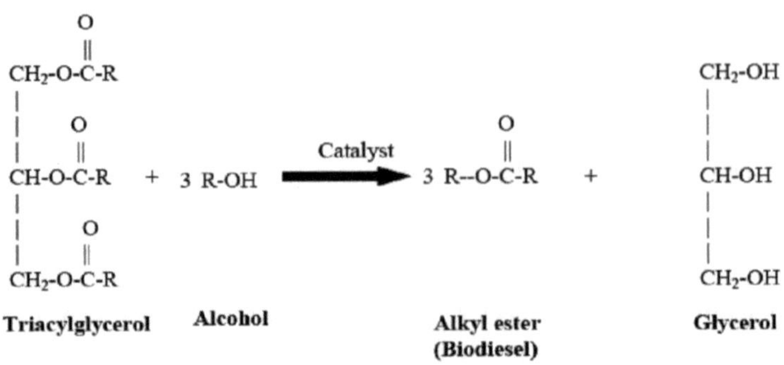

Esterification of free fatty acids

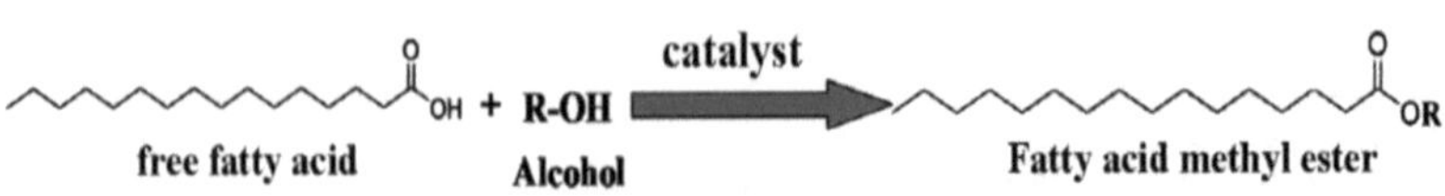

1.2 Biodiesel Feedstocks

In biodiesel production process, raw materials account for almost 75 % of total biodiesel cost. Basically all vegetable oils and animal fat can be used as feedstock for biodiesel production. Most of these oils and fats have a similar chemical composition; they consist of triglycerides with different amounts of individual fatty acids. The major fatty acids are those with a chain length of 16 and 18 carbons, whereas the chain could be saturated or unsaturated. Methyl esters produced from these fatty acids have very similar combustion characteristics in a diesel engine, because the major components in fossil diesel fuel are also straight-chain hydrocarbons with a chain length of about 16 carbons (hexadecane, "cetane"). The major feedstocks for the biodiesel production today are rapeseed oil (Canola), soybean oil, and palm oil.

Especially in Asian countries like India and China, the use of non-edible seed oils for biofuel production is very popular; in that case there would be no competition with the food production, especially when these oil plants are grown on marginal areas not suitable for food production. Especially *Jatropha curcas* L. has attracted enormous attention in the last few years, especially in India, Indonesia, and the Philippines. As there would be no competition with the food production and also with the traditional agricultural areas, *Jatropha* could fill the gap between actual vegetable oil production and demand for biofuels.

An interesting alternative for low-cost biodiesel production is the utilization of low-quality raw materials as feedstocks such as waste cooking oil obtained from canteens restaurants and from houses which are rich in free fatty acids.

1.3 Catalysts for Transesterification and Esterification Reactions

1.3.1 Homogeneous Catalysts (Alkaline/Acid)

The conventional biodiesel production technology involves the use of alkaline homogeneous catalysts such as NaOH and KOH, but sometimes $NaHCO_3$ or $KHCO_3$ is also employed mainly in large-scale production plants. These are not compatible for feedstocks with large amounts of free fatty acids (FFAs) and moisture due to the formation of soaps that strongly affect the feasibility of glycerol separation which is an important coproduct of transesterification reaction.

The traditional liquid acids such as HCl and H_2SO_4 were found to be more efficient, but they need a very long reaction time and a very high molar ratio of methanol to oil. Also corrosion of reaction vessels and problem of recycling are the key issues with traditional liquid acids. Therefore, commercialization of biodiesel production is difficult due to the technological drawbacks such as separation and purification steps that increase the cost factor to maximum.

1.3.2 Enzyme Catalysts

Enzymatic catalysts are able to effectively catalyze the transesterification of triglycerides in either aqueous or nonaqueous systems, which can overcome the problems mentioned above. In particular, the by-products, glycerol, can be easily removed without any complex process, and also that free fatty acids contained in waste oils and fats can be completely converted to alkyl esters. Enzymatic catalysts are very selective, but they are very expensive and have unstable activities and slow reaction kinetics.

High energy consumption and costly separation of the catalyst from the reaction mixture have inspired the development of heterogeneous catalysts for biodiesel.

1.3.3 Heterogeneous Catalysts

Heterogeneous catalysts have the benefit of easy separation from the product formed without requirement of washing. Reusability of the catalyst is another advantage of the heterogeneous catalysts. Heterogeneous catalysts are categorized as solid acid and solid base. Considering solid acid catalysts, a literature survey shows that there are fewer reports than those on solid bases. Compared with solid base catalysts, solid acid catalysts have lower activity but higher stability; thus, they can be applied for feedstock with large amounts of free fatty acids without catalyst deactivation.

Various solid acid catalysts such as ion exchange resins [7], superacids such as tungstated and sulfated zirconia [8], polyaniline sulfate [9], metal oxides[10, 11], zeolites [12, 13], acidic ionic liquids[14], and supported heteropoly acids [15–17] have been reported for transesterification of triglycerides.

Catalysis by supported HPAs has been greatly expanded during the past few years from the viewpoint of their variety of structures and compositions. They provide the opportunities for tuning their chemical properties, such as acidities, and reactivities by choice of appropriate support. Considering acidic properties, they have found enormous applications in various industrially important classes of reactions such as alkylations, acylations, and esterifications [18–23]. Recently, they have gained tremendous interest, in the synthesis of biodiesel.

A literature survey shows that 12-tungstophosphoric acid supported onto different supports is well studied for various organic transformations [24–27]. At the same time, even though 12-tungstosilicic acid is the next acidic and stable [18, 19, 28, 29] heteropoly acid in the Keggin series, not much work has been carried out for the same. T. Dogu et al. have reported silicotungstic acid impregnated MCM-41-like mesoporous solid acid catalysts for dehydration of ethanol [30]. Halligudi et al. reported silicotungstic acid/zirconia immobilized on SBA-15 for esterification [31]. Thus, it will be interesting to explore the catalytic activity of 12-tungstosilicic acid anchored to different mesoporous silica supports such as MCM-41 and SBA-15 for esterification and transesterification of free fatty acids and triglycerides.

In this chapter, we describe esterification of oleic acid for biodiesel production over anchored 12-tungstosilicic acid (TSA). It is known that "support" does not play always merely a mechanical role but it can also modify the catalytic properties of the HPAs. The nature of the support is an important factor for a successful reaction toward the better catalytic activity. Hence, in this chapter, we have also investigated the use of different mesoporous silica materials for anchoring TSA.

A series of catalysts comprising 12-tungstosilicic acid (TSA) and mesoporous silica, MCM-41 and SBA-15, were synthesized. The supports and the synthesized catalysts were characterized by various physicochemical techniques. The catalytic activity was evaluated for biodiesel production by esterification of free fatty acids, oleic acid. Influence of various reaction parameters (such as catalyst concentration, acid/alcohol molar ratio, and reaction temperature) on catalytic performance was studied. The catalytic activity of both the catalysts was correlated with the nature of support. This chapter also describes the application of present catalysts in biodiesel production from *Jatropha* oil, as feedstocks without any pretreatment and with methanol over the present catalyst.

2 Experimental Section

2.1 Materials

All chemicals used were of A.R. grade. $H_4SiW_{12}O_{40}{\cdot}nH_2O$ (Loba Chemie, Mumbai), pluronic123 (Aldrich), tetraetxhyl orthosilicate (TEOS), CTAB, oleic acid, and methanol were used as received from Merck.

2.1.1 Synthesis of the Supports

Synthesis of MCM-41

MCM-41 was synthesized using reported procedure [32] with slight modification. Surfactant (CTAB) was added to the very dilute solution of NaOH with stirring at 60 °C. When the solution became homogeneous, TEOS was added drop wise and the obtained gel was aged for 2 h. The resulting product was filtered, washed with distilled water, and then dried at room temperature. The obtained material was calcined at 555 °C in air for 5 h and designated as MCM-41.

Synthesis of SBA-15

The synthesis of SBA-15 was carried out as reported in literature [33]. The SBA-15 synthesis was carried out without hydrothermal conditions. In a typical preparation, 4 g of Pluronic P123 was dissolved in 30 mL water and 120 mL 2 M HCl with stirring at 35 °C. Then 8.5 g of TEOS was added into the solution with stirring at 35 °C for 20 h. Then the mixture was aged at 80 °C for 48 h without hydrothermal conditions. The solid product was recovered, washed, and air-dried at room temperature. The calcination was carried out at 500 °C for 6 h.

2.1.2 Synthesis of the Catalysts

TSA Anchored to MCM-41

A series of catalysts containing 10–30 % of TSA anchored to MCM-41 were synthesized by impregnation. 1 g of MCM-41 was impregnated with an aqueous solution of TSA (0.1 g/10 mL to 0.3 g/30 mL of double distilled water) and dried at 100 °C for 10 h. The obtained materials were designated as TSA_1/MCM-41, TSA_2/MCM-41, and TSA_3/MCM-41.

TSA Anchored to SBA-15

A series of catalysts containing 10–30 % of TSA anchored to SBA-15 were synthesized by impregnation. 1 g of SBA-15 was impregnated with an aqueous solution of TSA (0.1 g/10 mL to 0.3 g/30 mL of double distilled water) and dried at 100 °C for 10 h. The obtained materials were designated as TSA_1/SBA-15, TSA_2/SBA-15, and TSA_3/SBA-15.

2.1.3 Characterization

The FT-IR spectra of the samples were obtained by using KBr pellets on PerkinElmer. Adsorption–desorption isotherms of N_2 for samples were recorded on a Micromeritics ASAP 2010 surface area analyzer at −196 °C. From adsorption–desorption isotherms, surface area was calculated using BET method. The XRD pattern was obtained by using PHILIPS PW-1830. The conditions used were Cu Kα radiation (1.5417 Å), scanning angle from 0° to 60°. TEM was done on JEOL (JAPAN) TEM instrument (model-JEM 100CX II) with accelerating voltage 100 kV. The samples were dispersed in Ethanol and ultrasonicated for 5–10 min. A small drop of the sample was then taken in a carbon-coated copper grid and dried before viewing.

n-Butylamine Acidity Determination

The total surface acidity for all the materials has been determined by *n*-butylamine titration [34]. A 0.025 M solution of *n*-butylamine in toluene was used for estimation. The catalyst weighing 0.5 g was suspended in this solution for 24 h, and excess base was titrated against trichloroacetic acid using neutral red as an indicator. This gives the total surface acidity of the material.

2.1.4 Catalytic Activity: Esterification and Transesterification

Esterification of Oleic Acid

The esterification of oleic acid (0.01 mol) with methanol (0.4 mol) was carried out in a 100 mL batch reactor provided with a double-walled air condenser, magnetic stirrer, and a guard tube. The reaction mixture was heated at 40 °C for 4 h. Samples were taken periodically and analyzed by gas chromatograph (Nucon-5700) equipped with a 30 m×0.25 mm BP1 capillary column. Products were identified by comparison with the authentic samples and finally by gas chromatography–mass spectroscopy (GC–MS).

Transesterification of Triglyceride Feedstocks (*Jatropha* Oil)

The typical reaction of transesterification was carried out in a 100 mL batch reactor, provided with thermometer, mechanical stirring, and condenser. Jatropha oil and methanol were added in 1:8 wt /wt ratios and followed by catalyst addition, and then reaction mixture was held at 65 °C for respective time with stirring in order to keep system uniform in temperature and suspension. After the reaction is completed, the mixture was rotary evaporated at 50 °C to separate the methyl esters. The conversion of FFA in the JO to biodiesel was calculated by means of the acid value (AV) of the oil layer with the following equation [35]:

$$\text{Conversion }(\%) = \left(1 - \frac{\text{AV}_{\text{OL}}}{\text{AV}_{\text{JO}}}\right)$$

where OL and JO refer to oil layer and Jatropha oil, respectively.

3 Results and Discussion

3.1 Characterization

Figure 9.1 shows FT-IR spectra of TSA, MCM-41, TSA_3/MCM-41, SBA-15, and SA_3/SBA-15. FT-IR band assignments of supports as well as catalyst are shown in Table 9.1.

The presence of these bands (Table 9.1) strongly reveals that the primary structure of TSA Keggin anion is preserved even after anchoring to supports. The absence of vibration band at 783 cm^{-1} of TSA may be because of the low concentration of TSA or the TSA bands may be superimposed with those of support.

The XRD patterns of pure MCM-41 TSA_3/MCM-41 are shown in Fig. 9.2. The XRD pattern of the calcined MCM-41 showed a sharp peak around $2\theta = 2°$ and few weak peaks in $2\theta = 3 \sim 5°$, which indicated well-ordered hexagonal structure of MCM-41. No separate HPA crystal phase of TSA was observed in the TSA_3/MCM-41. Further, the absence of characteristic peaks of crystalline phase of TSA indicates

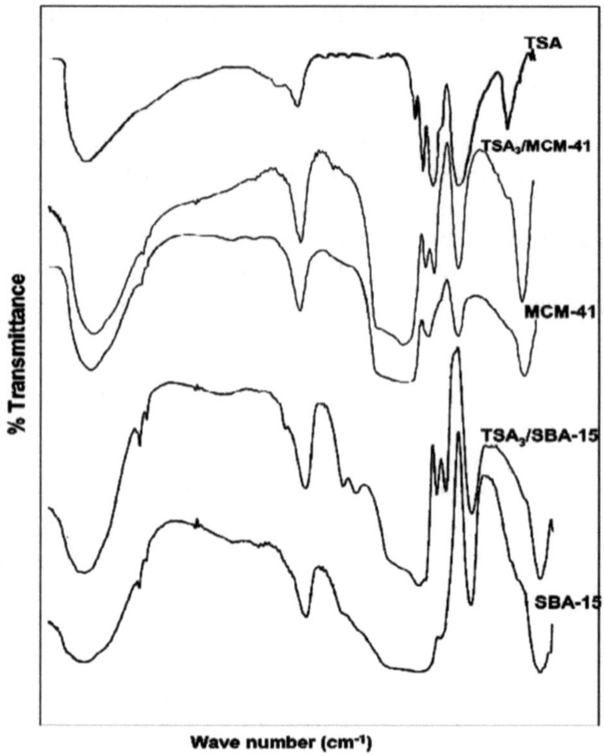

Fig. 9.1 FT-IR spectra of supports and catalysts

Table 9.1 FT-IR band assignments of supports and catalysts

Material	Band frequency (cm^{-1}) Si–OH	ν_{sym} Si–O–Si	ν_{asym} Si–O–Si	Sym bending Si–O–Si	W–O–W	W=O	Si–O
MCM-41	3,448	801	1,300–1,000	458	–	–	–
SBA-15	3,448	801	1,165	458	–	–	–
TSA	–	–	–	–	783	979	923
TSA_3/ MCM-41	3,448	801	1,300–1,000	458	–	977	922
TSA_3/ SBA-15	3,448	801	1,165	458	–	978	924

that TSA is finely dispersed inside the hexagonal channels of MCM-41 and there must be some chemical interaction between the host MCM-41 and the guest TSA.

XRD patterns of TSA, SBA-15, and TSA_3/SBA-15 are shown in Fig. 9.2. The XRD patterns of parent SBA-15 show three well-resolved peaks at 0.89°, 1.50°, and 1.72° which are indexed to (1 0 0), (1 1 0), and (2 0 0) reflections of ordered hexagonal mesophase. It is interesting to note that, in TSA_3/SBA-15, the intensity of peaks corresponding to (1 1 0) and (2 0 0) planes of SBA-15 decreases. Further, the comparison of the XRD patterns of SBA-15 and TSA_3/SBA-15 reveals that the

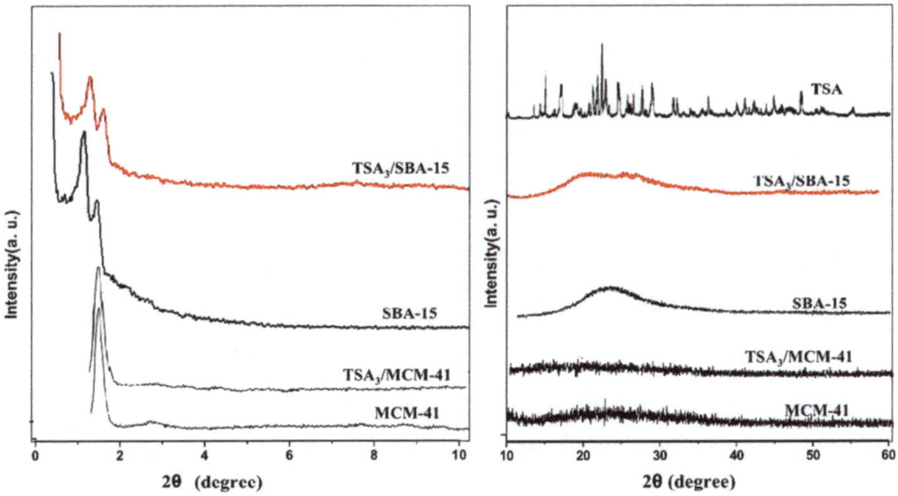

Fig. 9.2 XRD patterns of TSA, MCM-41, SBA-15, TSA$_3$/MCM-41, TSA$_3$/SBA-15

Table 9.2 Textural properties of supports and catalysts and total acidity

Catalyst	Surface area (m^2/g)	Pore diameter *d* (Å)	Mesopore volume (cm^3/g)	Total acidity (mmol/g)
MCM-41	659	47.9	0.79	0.82
TSA$_1$/MCM-41	539.29	29.62	0.39	1.14
TSA$_2$/MCM-41	464.16	29.45	0.30	1.21
TSA$_3$/MCM-41	349.26	29.23	0.26	1.33
SBA-15	834	68	1.26	1.11
TSA$_1$/SBA-15	689	62.5	0.9741	1.39
TSA$_2$/SBA-15	677	61.38	0.7341	1.52
TSA$_3$/SBA-15	645	60.0	0.6418	1.65

mesoporous structure is rather intact even after the TSA loading. As compared to MCM-41 support, SBA-15 could accommodate more discrete TSA species because of its large pore volume and high surface area. The XRD patterns of TSA$_3$/SBA-15 show the appearance of very small peaks of TSA, but all the typical characteristic diffraction peaks of crystalline phase of TSA were not observed as compared to the XRD of bulk phase of TSA. Further, the absence of characteristic peaks of crystalline phase of TSA indicates that TSA is highly dispersed inside the hexagonal channels of SBA-15 [36].

The values of surface area, pore size, and pore volumes are presented in Table 9.2. Specific surface area, porosity, and pore diameter all strongly decreased for the catalyst relative to the support. As the TSA loading increases, surface area, pore diameter, and pore volume all strongly decrease. The reason being, as the TSA species will enter the mesopores, it decreases the pore diameter, and also probably some TSA species will appear in the mesoporous channels that decrease the average pore volume as well as the surface area.

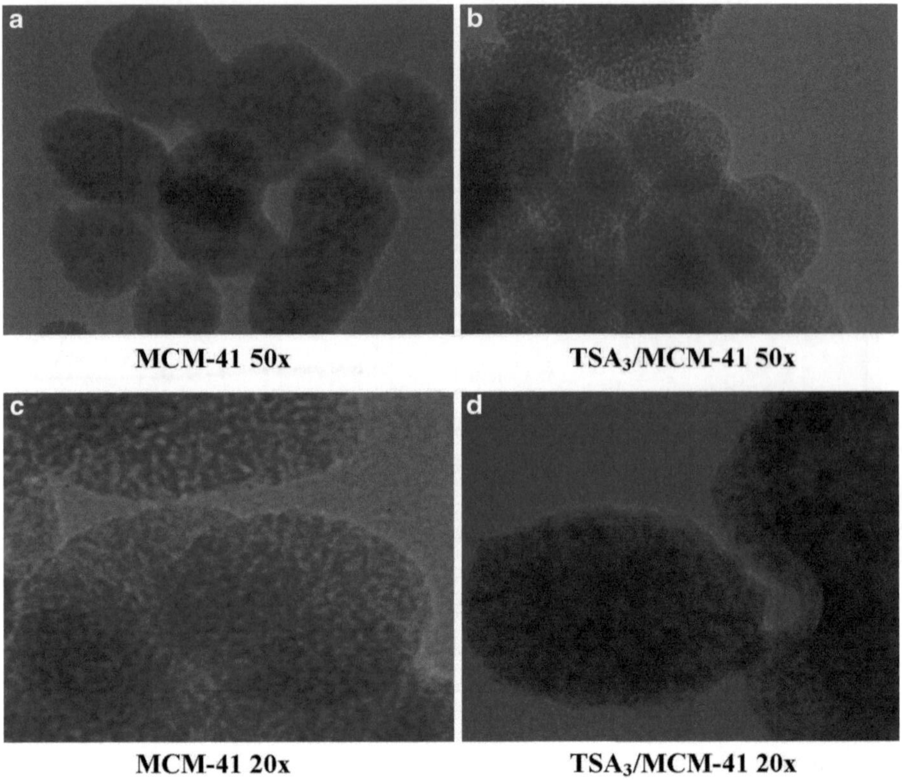

Fig. 9.3 TEM images of MCM 41 and TSA_3/MCM41

Total acidity values for all the catalysts are presented in Table 9.2. Total acidity values indicate that as the TSA loadings increases, acidity values also increase and the results are as expected.

Figure 9.3a, c and b, d shows the TEM image of MCM 41 and TSA_3/MCM41, respectively. Figure 9.3a, c clearly shows hexagonal mesopores in MCM-41. The TEM images of TSA_3/MCM41 (Fig. 9.3b, d) show that most of the hexagonal pores are covered with dark-colored fine particles. This indicates uniform dispersion of TSA inside the hexagonal pores of MCM-41.

Figure 9.4 shows TEM images of SBA-15 and TSA_3/SBA-15. The TEM images of SBA-15 (Fig. 9.4a) show the morphology of 2D hexagonal arrays of channels with uniform pore size. Figure 9.4c clearly shows hexagonal mesopores. The TEM images of TSA_3/SBA-15 (Fig. 9.4b, d) show well-ordered nanochannels and the nanochannels were arranged on 2D hexagonal structure over very large scales. The uniform structure of SBA-15 was well maintained throughout the structure even after TSA loading, which reveals that the TSA species were well dispersed inside the hexagonal channels. Other possibility is that TSA has formed very small (nm) crystals in these channels.

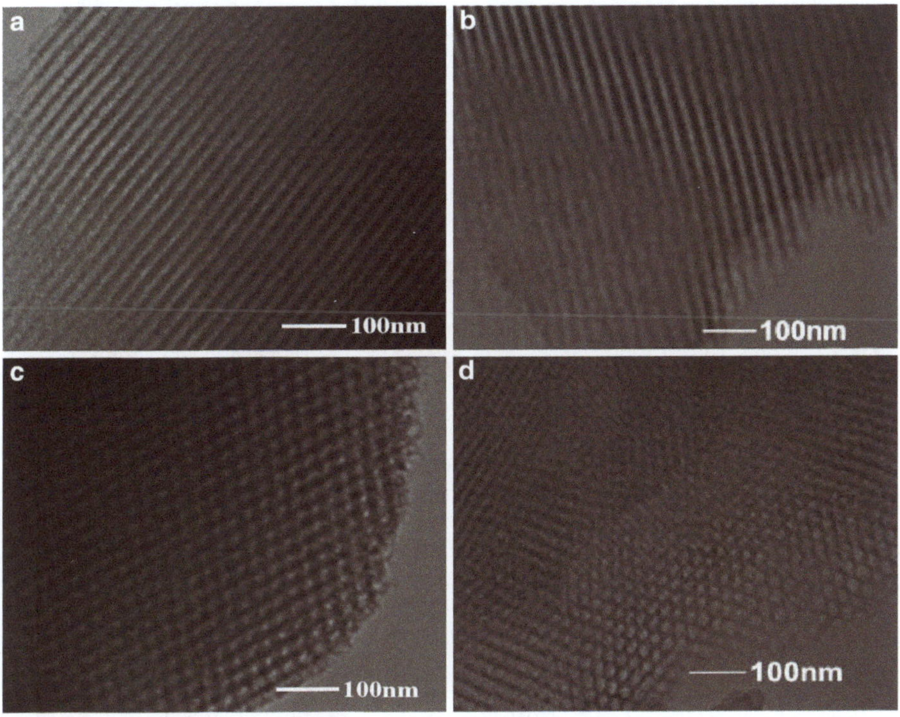

Fig. 9.4 TEM images of SBA-15 (**a**, **c**) and TSA_3/SBA-15 (**b**, **d**)

Scheme 9.1 Esterification of oleic acid with methanol

3.2 *Esterification of Oleic Acid over TSA_3/MCM-41*

The esterification of free fatty acids is an equilibrium-limited reaction. In order to overcome the equilibrium limitation, generally esterification of free fatty acids is carried out by taking alcohol in excess in order to favor the forward reaction. The esterification of oleic acid with alcohol is shown in Scheme 9.1.

The effect of various reaction parameters such as acid/alcohol molar ratio, amount of catalyst, reaction time, and temperature was studied over TSA_3/MCM-41 to optimize the conditions for getting maximum conversion of oleic acid.

The reaction was carried out by varying mole ratio of oleic acid to methanol, with 0.1 g of the catalyst for 10 h at 60 °C. It can be observed from Fig. 9.5 that the oleic acid conversion increases with increase oleic acid/methanol ratio and reaches maximum 99 % at oleic acid/methanol 1:40. With further increase in molar ratio,

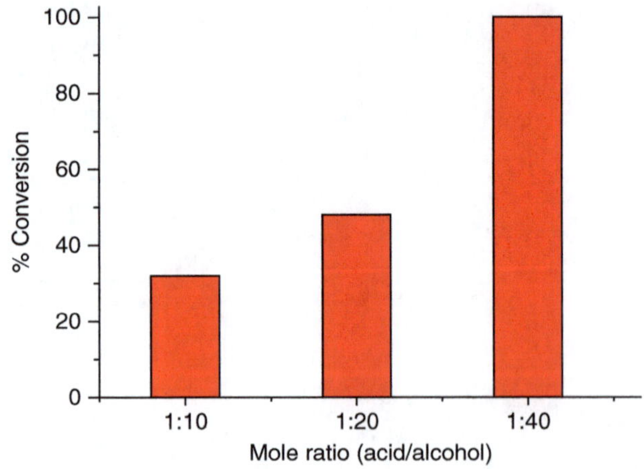

Fig. 9.5 Effect of mole ratio. Reaction conditions: amount of catalyst 0.1 g; reaction time 10 h, reaction temperature 60 °C

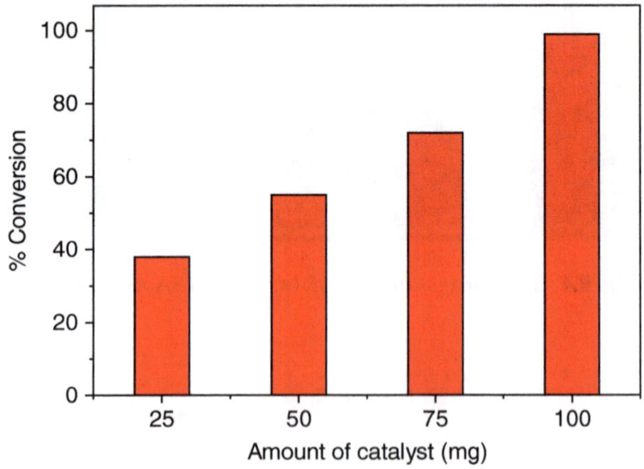

Fig. 9.6 Effect of amount of catalyst. Reaction conditions: mole ratio 1:40; reaction time 10 h, reaction temperature 60 °C

increase in conversion was not that much appreciable. Hence molar ratio of 1:40 was selected for obtaining high conversions.

Effect of amount of catalyst on oleic acid conversion was investigated. The catalyst amount was varied in the range of 25–100 mg. As shown in Fig. 9.6, with increase in amount of catalyst, conversion of oleic acid also increases. The maximum conversion obtained was 90 % with 100 mg of catalyst.

3.2.1 Reaction Time (h)

Effect of reaction time on conversion of oleic acid was investigated. It was observed (Fig. 9.7) that the oleic acid conversion increases with increase in reaction time. After 10 h maximum 99 % oleic acid was achieved for TSA_3/MCM-41.

Effect of reaction temperature on oleic acid conversion was studied, and it was found that with as the reaction temperature increases conversion of oleic acid also

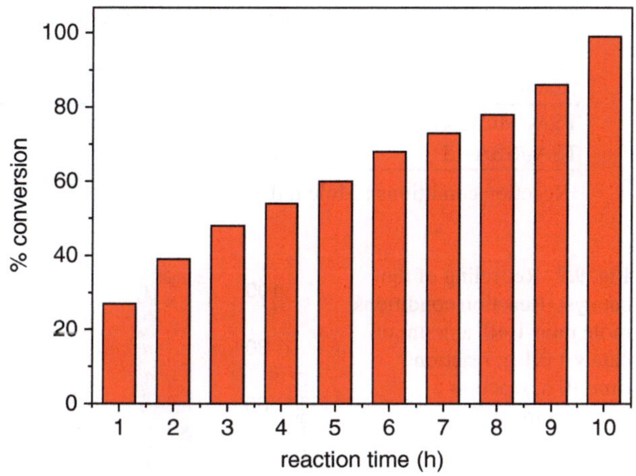

Fig. 9.7 Effect of reaction time. Reaction conditions: mole ratio 1:40; amount of catalyst 0.1 g; reaction temperature 60 °C

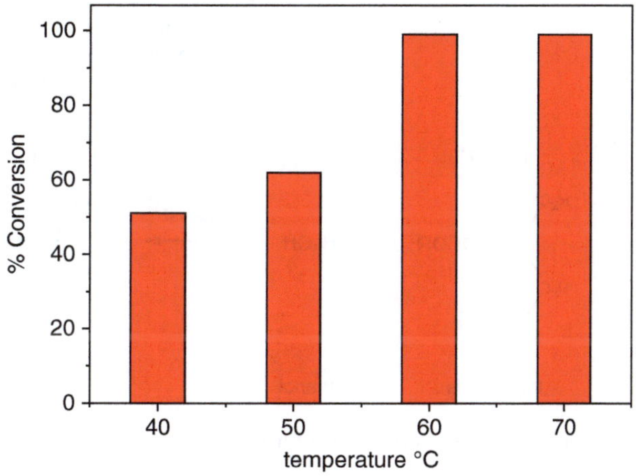

Fig. 9.8 Effect of reaction temperature. Reaction conditions: mole ratio 1:40, amount of catalyst 0.1 g; reaction time 10 h

increases (Fig. 9.8). At 60 °C maximum 99 % conversion was achieved for TSA_3/MCM-41. But with further increasing the temperature up to 70 °C, % conversion remains the same.

The optimized conditions for esterification of oleic acid over TSA_3/MCM-41 are mole ratio of acid to alcohol 1:40, amount of catalyst 0.1 g, reaction temperature 60 °C, and reaction time 10 h.

Esterification of oleic acid over TSA_3/SBA-15 was carried out under optimized conditions and the results are shown in Table 9.3. It can be observed from Table 9.3 that both the catalysts can efficiently catalyze esterification of oleic acid, but the TSA_3/SBA-15 shows much higher activity at much lower temperature and in very short reaction time as compared to TSA_3/MCM-41.

Table 9.3 Esterification of oleic acid over TSA_3/MCM-41 and TSA_3/SBA-15 under optimized conditions

[a]Catalyst	% conversion (at 40 °C)	% conversion (at 60 °C)
TSA_3/MCM-41	51 (10 h)	99 (10 h)
TSA_3/SBA-15	99 (5 h)	99 (4 h)

[a]Reaction conditions : Mole ratio of acid to alcohol 1:40; Amount of catalyst 0.1 g

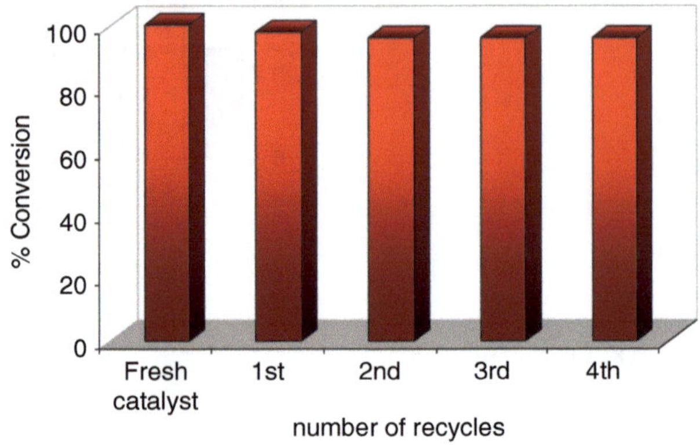

Fig. 9.9 Recycling of the catalyst. Reaction conditions: mole ratio 1:40, amount of catalyst 0.1 g; reaction time 10 h, reaction temperature 60 °C

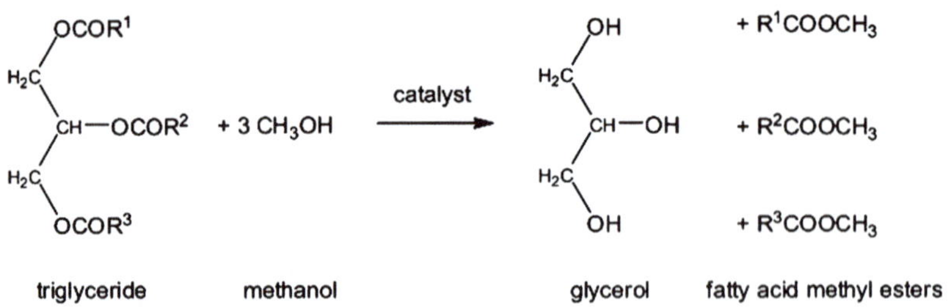

Scheme 9.2 Transesterification of triglycerides with methanol

3.3 *Recycling of the Catalyst*

The catalyst was recycled in order to test its activity as well as stability. The catalyst was separated from the reaction mixture only by simple filtration, first washed with methanol till the filtrate is free from the acid (unreacted oleic acid if any), followed by conductivity water and then dried at 100 °C, and the recovered catalyst was charged for the further run. There is no appreciable change in the % conversion of oleic acid using regenerated catalyst up to four cycles (Fig. 9.9).

3.4 *Transesterification of Triglycerides* (**Jatropha** *Oil*)

Transesterification of triglycerides (TGs) with low molecular weight alcohols (Scheme 9.2) produces biodiesel. *Jatropha* oil is non-edible and does not compromise

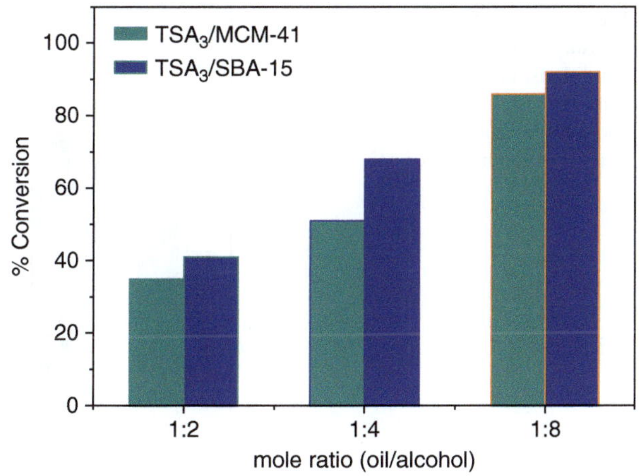

Fig. 9.10 Effect of weight ratio of oil/alcohol. Reaction conditions: amount of catalyst 0.3 g, reaction time 20 h for TSA_3/MCM-41 and 8 h for TSA_3/SBA-15, reaction temperature 65 °C

the edible oil, which are mainly used for food consumption. Non-edible oils are not suitable for human consumption because of the presence of toxic components. Further, *Jatropha* seed has a high content of oil and the biodiesel produced has similar properties to that of petroleum-based diesel.

The effect of various reaction parameters such as oil/alcohol weight ratio, amount of catalyst, reaction time, and temperature was studied over TSA_3/MCM-41 and TSA_3/SBA-15 to optimize the conditions for maximum conversion.

An important variable which can effect on the conversion of triglycerides is the molar ratio of oil to alcohol. According to the reaction stoichiometry, three moles of alcohol per mole of triglyceride are required to yield 3 mol of fatty esters and 1 mol of glycerin. But in practice a higher oil/alcohol ratio is employed in order to obtain a higher conversion for ester production. We selected weight ratio of oil to methanol (wt/wt) between 1:2 and 1:8. The ratio 1:8 is suitable for obtaining high yields of products as summarized in Fig. 9.10. It can be seen in Fig. 9.10 that oil conversion increased with the increases of the oil/methanol ratio and reached a maximum at 1:8, with a ratio of less than 1:8 the reaction was incomplete. When the oil/methanol ratio was higher than 1:8, the conversion decreased. So the ratio 1:8 is suitable for obtaining higher conversion.

The effect of the catalyst amount on the oil conversion is shown in Fig. 9.11. Experiments were carried out by varying the amount of the catalyst between 0.1 and 0.5 g, keeping the oil to methanol ratio at 1:8 at 65 °C. It can be seen from Fig. 9.11 that an increase in the conversion of oil was noticed when the amount of the catalyst increased from 0.1 to 0.3 g and acid-catalyzed process attains a maximum conversion at 0.3 g of the catalyst. The increase in the conversion with an increase in the catalyst amount can be attributed to an increase in the availability and number of catalytically active sites. With further increasing the amount of catalyst from 0.3 to 0.5 g, decrease in the oil conversion was observed which may be due to the blocking of active sites.

Effect of reaction time on conversion of triglycerides was studied. It was observed (Fig. 9.12) that the conversion increases with increase in reaction time. After 8 h maximum 91 and 86 % conversion was achieved in 8 h and in 20 h for *Jatropha* oil over TSA_3/MCM-41 and TSA_3/SBA-15, respectively.

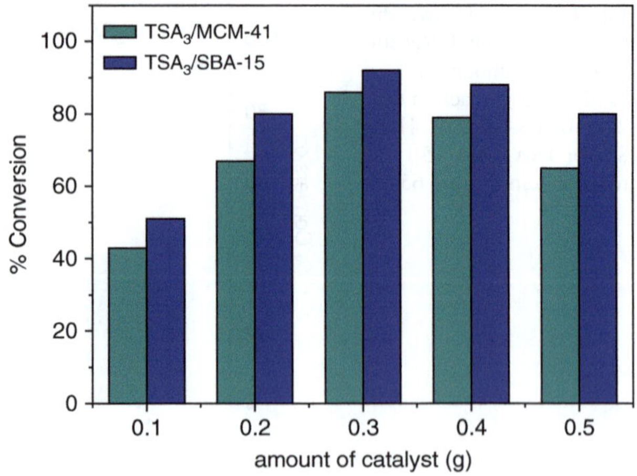

Fig. 9.11 Effect of amount of catalyst. Reaction conditions: ratio 1:8, reaction time 20 h for TSA_3/MCM-41 and 8 h for TSA_3/SBA-15, reaction temperature 65 °C

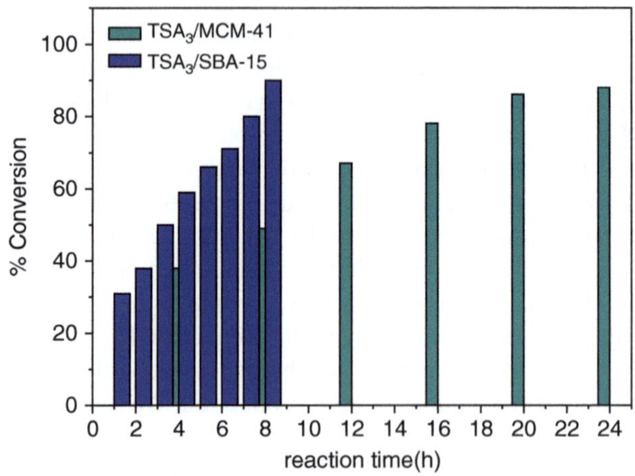

Fig. 9.12 Effect of reaction time. Reaction conditions: ratio 1:8, amount of catalyst 0.3 g, reaction temperature 65 °C

Methanolysis of oils is normally performed near the boiling point of the alcohol. The effect of reaction temperature on conversion was studied, and it was found that as the reaction temperature increases, conversion of oleic acid also increases (Fig. 9.13). At 65 °C maximum 92 and 86 % conversion was achieved for *Jatropha* oil over TSA_3/MCM-41 and TSA_3/SBA-15, respectively. But with further increasing the temperature up to 70 °C, % conversion decreases. Usually the increase in reaction temperature should lead to usual increase of reaction extent, but it could also cause acceleration of secondary reactions which reduce the conversion toward desired product.

The optimized conditions for transesterification of and *Jatropha* oil over TSA_3/SBA-15 are weight ratio of oil to alcohol 1:8, amount of catalyst 0.3 g, reaction temperature 65 °C, and reaction time 8 h. For TSA_3/MCM-41 maximum conversion was achieved in 20 h and the rest of the parameters were same as that of TSA_3/SBA-15.

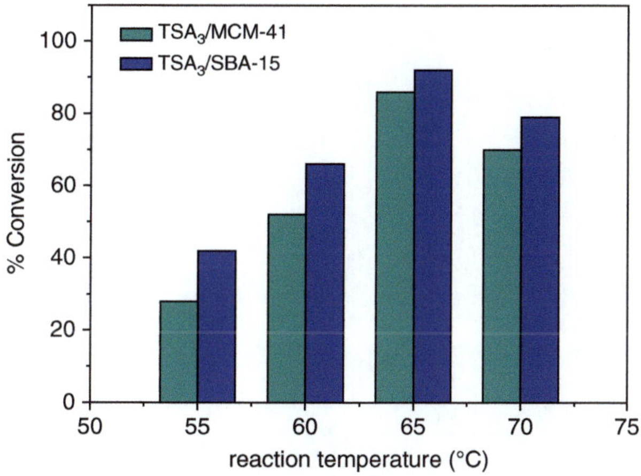

Fig. 9.13 Effect of reaction temperature. Reaction conditions: mole ratio of oil/alcohol 1:8, amount of catalyst 0.3 g, reaction time 20 h for TSA_3/MCM-41 and 8 h for TSA_3/SBA-15

Table 9.4 Effect of supports on activity of TSA in esterification and transesterification reactions

Catalysts	% Conversion (oleic acid) (at 60 °C)	TON/TOF (min^{-1})	% Conversion (*Jatropha* oil)	Surface area (m^2/g)	Pore diameter (Å)	Total acidity	Keggin ion density (TSA) (nm^{-2})
TSA_3/ MCM-41	99 (10 h)	2,232/3.72[a]	86	349	29.23	1.33	0.17979
TSA_3/SBA-15	99 (4 h)	2,232/9.3[a]	92	645	60.0	1.65	0.0973

[a] TOF calculated for oleic acid conversion

3.5 *Effect of Supports on Activity of TSA in Biodiesel Production*

Esterification of oleic acid and transesterification of *Jatropha* oil with methanol was used to probe the catalytic activity of TSA onto different supports, MCM-41 and SBA-15. It was observed from Table 9.4 that higher conversions were observed with TSA_3/SBA-15 than that of TSA_3/MCM-41 indicating the former is more active. The obtained difference in catalytic activity may be due to the nature of supports and this can be explained as follows.

First of all as mentioned earlier, support may also play a role in modifying the catalytic activity of HPAs. The surface area of SBA-15 is higher than that of MCM-41, and the same trend was observed for the catalysts. Increase in surface area may be due to the nature of support. The higher value of surface area for TSA_3/SBA-15 as compared to that of TSA_3/MCM-41 is responsible for higher catalytic activity.

The total acidity of TSA_3/SBA-15 is higher than TSA_3/MCM-41 catalyst, even though the % loading of TSA is same for both the supports. It is well known that the SBA-15 is more acidic support as compared to MCM-41. Therefore, the higher value of total acidity for TSA_3/SBA-15 was observed collectively from TSA as well as support SBA-15. As the total acidity of TSA_3/SBA-15 is higher as compared to TSA_3/MCM-41, superior catalytic activity from TSA_3/SBA-15 is as expected.

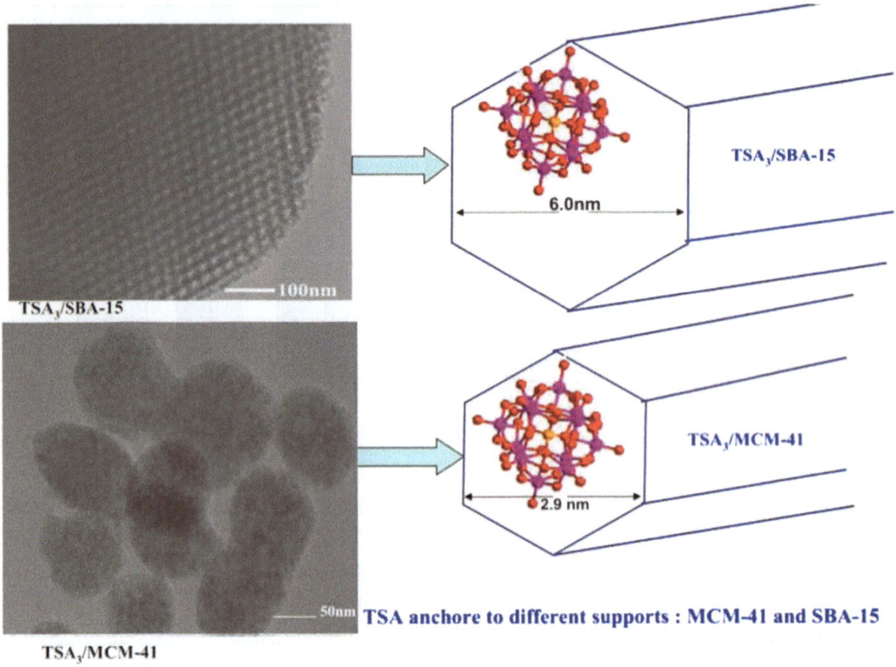

Scheme 9.3 Effect of support pore diameter on catalytic activity of TSA

The density of Keggin ion (TSA nm^{-2}), expressed as number of Keggin anion per square nanometer, was calculated according to actual TSA loading and catalyst surface area [37]. As seen from Table 9.4, the Keggin ion density of TSA_3/MCM-41 is higher as compared to TSA_3/SBA-15, suggesting multilayer formation of TSA when anchored to MCM-41. In other words fewer Bronsted acid sites are available, whereas Keggin ion density of TSA_3/SBA-15 is low, suggesting monolayer formation of TSA, and, hence, more number of Bronsted acid sites are available for catalytic reaction.

In case of mesoporous support, textural properties of support also affect the catalytic activity. In the present case, for SBA-15 support, pore diameter is 6.8 nm and after anchoring TSA it decreases to 6.0 nm; still there is enough space available for reactants to enter, whereas for MCM-41 support, pore diameter is 4.7 nm which decreases to 2.9 nm after anchoring TSA; hence, the available space is much more suitable for small molecules rather than bigger molecules as reactants (Scheme 9.3).

Hence, TSA_3/SBA-15 exhibits higher activity as compared to TSA_3/MCM-41.

4 Conclusion

The present chapter describes heterogeneous acid catalysts comprising 12-tungstosilicic acid anchored to different mesoporous silica supports, MCM-41 and SBA-15, for biodiesel production by esterification oleic acid and transesterification of

Jatropha oil with methanol. The higher conversions for both esterification and transesterification were obtained with TSA_3/SBA-15 than that of TSA_3/MCM-41, indicating the former is more active. The large surface area, high value of total acidity, and availability of more Bronsted acid sites all together imply the excellent catalytic activity of TSA_3/SBA-15 as compared to that of TSA_3/MCM-41. Catalytic results demonstrated that the size of the support channels as well as pore diameter plays a crucial role in catalyst performance.

Acknowledgements Ms. Varsha Brahmkhatri is thankful to the Department of Science and Technology, New Delhi, for DST-INSPIRE fellowship. Anjali Patel thanks the University Grants Commission (Project No.39-837/2010 (SR)) for financing a part of this work. Authors are also thankful to Mr. Nilesh Narkhede for assistance in preparation of chapter.

References

1. Angina S, Ram P (2000) Renew Sustain Energy Rev 4:111
2. Gerpen JV (2005) Fuel Process Technol 86:1097
3. Haas MJ (2005) Fuel Process Technol 86:1087
4. Haas MJ, McAloon AJ, Yee WC, Foglia TA (2006) Bioresour Technol 97:671
5. Kulkarni MG, Dalai AK (2006) Ind Eng Chem Res 45:2901
6. Zhang Y, Dube MA, McLean DD, Kates M (2003) Bioresour Technol 90:229
7. Feng Y, He B, Cao Y, Li J, Liu M, Yan F et al (2010) Bioresour Technol 101:1518
8. Tanabe K, Yamaguchi T (1994) Catal Today 20:185
9. Zieba A, Drelinkiewicz A, Konyushenko EN, Stejskal J (2010) Appl Catal A Gen 383:169
10. Abreu FR, Alves MB, Macedo CCS, Zara LF, Suarez PAZ (2005) J Mol Catal A Chem 227:263
11. Guo HF, Yan P, Hao XY, Wang ZZ (2008) Mater Chem Phys 112:1065
12. Shu Q, Yang B, Yuan H, Qing S, Zhu G (2007) Catal Commun 8:2159
13. Benson TJ, Hernandez R, French WT, Alley EG, Holmes WE (2009) J Mol Catal A Chem 303:117
14. Han M, Yi W, Wu Q, Liu Y, Hong Y, Wang D (2009) Bioresour Technol 100:2308
15. Alsalme A, Kozhevnikova EF, Kozhevnikova IV (2008) Appl Catal A Gen 349:170
16. Brahmkhatri V, Patel A (2011) Appl Catal A Gen 403:161
17. Brahmkhatri V, Patel A (2011) Ind Eng Chem Res 50:6620
18. Mizuno N, Misono M (1998) Chem Rev 98:199
19. Kozhevnikov IV (1998) Chem Rev 98:171
20. Cavani F (1998) Catal Today 41:73
21. Bhatt N, Sharma P, Patel A, Selvam P (2008) Catal Commun 9:1545
22. Bhatt N, Patel A (2009) Kin Catal 50:401
23. Bhatt N, Patel A, Taiwan J (2011) Inst Chem Eng 42:356
24. Kamalakar G, Komura K, Kubota Y, Sugi Y (2006) J Chem Technol Biotechnol 81:981
25. Costaa VV, Rochaa, da Silva KA, Kozhevnikov IV, Gusevskayaa EV (2010) Appl Catal A: Gen 383:217
26. Srilatha K, Issariyakul T, Lingaiah N, Sai Prasad PS, Kozinski J, Dalai AK (2010) Energy Fuel 24:4748
27. Kamiya Y, Okuhara T, Misono M, Miyaji A, Tsuji K, Nakajo T (2008) Catal Surv Asia 12:101
28. Okuhara T, Mizuno N, Misono M (1996) Adv Catal 41:113
29. Moffat JB, Twigg MV, Spencer MS (eds) (2001) Fundamental and applied catalysis. Kluwer Academic Press, New York
30. Varisli D, Dogu T, Dogu G (2008) Ind Eng Chem Res 47:4071

31. Sawant DP, Vinub A, Mirajkar SP, Lefebvre F, Ariga K, Anandanb S, Mori T, Nishimura C, Halligudi SB (2007) J Mol Catal A Chem 271:46
32. Cai Q, Luo ZS, Pang WQ, Fan YW, Chen XH, Cui FZ (2001) Chem Mater 13:258
33. Zhao D, Huo O, Feng J, Chamelka BF, Stucky SD (1998) J Am Chem Soc 120:6024
34. Sahu HR, Rao GR (2000) Bull Mater Sci 23:349–354
35. Wang Y, Ou S, Liu P, Zhang Z (2007) Energy Convers Manage 48:184
36. Liu QY, Wu WL, Wang J, Ren XQ, Wang YR (2004) Micropor Mesopor Mater 76:51
37. Chai SH, Wang HP, Liang Y, Xu BQ (2008) Green Chem 10:1087

Chapter 10
Solvent-Free Selective Oxidation of Benzyl Alcohol to Benzaldehyde over Monolacunary Phosphomolybdate Supported onto Hydrous Zirconia

Soyeb Pathan and Anjali Patel

Contents

1 Introduction 210
2 Experimental 211
2.1 Synthesis of Hydrous Zirconia (ZrO_2) 211
2.2 Synthesis of Monolacunary Phosphomolybdate (PMo_{11}) 211
2.3 Synthesis of the Catalyst 212
3 Characterization 212
3.1 Catalytic Reaction 212
4 Results and Discussion 213
4.1 Analytical Techniques 213
4.2 Catalytic Studies 216
4.3 Heterogeneity Test 221
4.4 Recycling of the Catalyst 221
4.5 Characterization of the Regenerated Catalyst 222
5 Conclusion 223
References 223

Abstract The present chapter describes synthesis and characterization of supported lacunary phosphomolybdate as well as its use as heterogeneous catalyst for solvent-free oxidation of benzyl alcohol. Influence of the reaction parameters (molar ratio of substrate to H_2O_2, amount of the catalyst, reaction time, and reaction temperature) was studied. The catalyst was reused three times without any significant loss in the catalytic activity. Present catalyst gives single selective product, benzaldehyde, as well as high TON for the same, under mild reaction conditions.

S. Pathan • A. Patel (✉)
Department of Chemistry, Faculty of Science, M. S. University of Baroda, Vadodara 390002, India
e-mail: aupatel_chem@yahoo.com

A. Patel (ed.), *Environmentally Benign Catalysts: For Clean Organic Reactions*, DOI 10.1007/978-94-007-6710-2_10,

1 Introduction

Benzaldehyde is a very valuable chemical which has widespread applications in perfumery, dyestuff, and agrochemical industries [1]. It is the second most important aromatic molecule (after vanillin) used in the cosmetics and flavor industries. Generally, it is obtained by oxidation of styrene, benzyl alcohol, and toluene as well as hydrolysis of benzyl chloride, important transformations in organic synthesis [2–4]. The oxidation of styrene gives a variety of products such as carbonyl compounds, epoxides, diols, and products (aldehydes) of oxidative cleavage of C–C bond. Production of benzaldehyde is the first step in oxidation of benzyl alcohol, and further oxidation leads to the production of benzoic acid with side reactions leading to the other products like toluene, dibenzyl acetal, and benzyl benzoate [5]. Oxidation of toluene is usually carried out in organic solvents which are environmentally undesirable, and benzaldehyde produced from hydrolysis of benzyl chloride often contains traces of chlorine impurities and copious waste is generated in this process [6, 7]. These procedures suffer from lack of selectivity, use of organic solvents, toxicity of the reagents, and waste production which require environmentally benign alternative.

As a result selective and solvent-free oxidation reactions involving catalytic amount of the oxidizing agents or liquid phase catalytic oxidation have come into prominence. In this context, heteropoly acids/polyoxometalates (HPAs/POMs) have been gaining importance due to their redox properties [8–13]. It is well known that the redox properties of POMs can be tuned at the atomic/molecular levels without affecting their structure [14–19], by removing one or more addenda atoms. These types of POMs are called lacunary POMs (LPOMs) and lead to the development of active and selective catalysts. However, like POMs, the main drawback of LPOMs is their low thermal stability, which results in a more or less degradation of its Keggin structure under catalytic conditions, resulting in a drop in catalytic performance. The mentioned problem can be overcome by modifying their properties via supporting onto the supports. In our laboratory, during our systematic research on supported POMs, we surprisingly found that stability as well as catalytic performance of the POMs are improved by supporting it onto different supports [20, 21].

A careful investigation of the literature shows that reports on catalytic aspect of mono LPOMs are very scanty [22]. The first report on the catalytic evaluation of monolacunary phosphotungstates as well as silicotungstates was made in 1988, by M. Schwegler et al., for oxidation of cyclohexene [22]. Then in 1995, Hill and coworkers studied the role of lacunary phosphotungstates for oxidation of various organic substrates such as alkenes, alkanes, and sulfides [17]. The same group reported the use of monolacunary silicotungstate as well as phosphotungstate for aerobic oxidation of H_2S to obtain elemental sulfur [23]. They have also evaluated the catalytic activity for alkene epoxidation over the quaternary ammonium salt of monolacunary phosphotungstate [24]. The protonated $[PW_{11}O_{39}]^{7-}$ species was also used for alkene epoxidation by Kuznetsova et al. [25]. Z. Weng et al. reported the use of $[PW_{11}O_{39}]^{7-}$ for oxidation of benzyl alcohol using H_2O_2 as an oxidant [26]. Recently, studies on the mechanisms of radical reactions for generation of monolacunary phosphotungstate species and its reactivity with methyl radicals for production of propene and 2-methylpropene was demonstrated by Dan Meyerstein's group [27].

All the reported articles describe the use of LPOMs which forms the bulk of homogeneous catalysts. It was also found that no literature is available on the catalytic aspects of supported LPOMs. To the best of our knowledge, there are few articles on the catalytic activity of supported LPOMs and that was by our research group only [28–32].

In the present chapter, we report the synthesis and characterization of supported sodium salt of monolacunary phosphomolybdate supported onto hydrous zirconia as well as its use as catalyst for selective oxidation of benzyl alcohol. A series of catalysts containing 10–40 % of monolacunary phosphomolybdate (PMo_{11}) onto hydrous zirconia (ZrO_2) were synthesized. The synthesized catalyst was characterized by various thermal, spectral, and surface techniques.

The catalytic activity was evaluated for the non-solvent liquid phase oxidation of benzyl alcohol with aqueous 30 % H_2O_2. The conditions for maximum conversion as well as selectivity for desired product was optimized by varying different parameters such as molar ratio of substrate to H_2O_2, amount of the catalyst, reaction time, and reaction temperature. A catalytic property for recycled catalyst was also evaluated for the oxidation of benzyl alcohol under optimized condition.

2 Experimental

All the chemicals used were of AR grade. Sodium molybdate (Merck), disodium hydrogen phosphate (Merck), and zirconium oxychloride (Loba Chemie, Mumbai) were used as received. Nitric acid, acetone, dichloromethane, and benzyl alcohol were obtained from Merck and used as received.

2.1 Synthesis of Hydrous Zirconia (ZrO_2)

Hydrous zirconia was prepared by adding an aqueous ammonia solution to an aqueous solution of $ZrOCl_2 \cdot 8H_2O$ up to pH 8.5. The resulted precipitates were aged at 100 °C over a water bath for 1 h, filtered, washed with conductivity water until chloride-free water was obtained, and dried at 100 °C for 10 h. The obtained material is designated as ZrO_2.

2.2 Synthesis of Monolacunary Phosphomolybdate (PMo_{11})

Monolacunary phosphomolybdate was synthesized by following method reported by us earlier [31]. Sodium molybdate dihydrate (0.22 mol, 5.32 g) and anhydrous disodium hydrogen phosphate (0.02 mol, 0.28 g) were dissolved in 50–70 ml of conductivity water and heated to 80–90 °C followed by the addition of concentrated nitric acid in order to adjust the pH to 4.3. The volume was then reduced to half by evaporation, and the heteropolyanion was separated by liquid–liquid extraction with 50–60 ml of acetone. The extraction was repeated until the acetone extract showed absence of NO_3^- ions (ferrous sulfate test). The extracted sodium salt was dried in air. The obtained material was designated as PMo_{11}. Anal Calc %: Na, 7.65; Mo, 50.12; P, 1.47; O, 39.52. Found %: Na, 7.60; Mo, 49.99; P, 1.44; O, 39.92.

2.3 *Synthesis of the Catalyst*

A series of catalysts containing 10–40 % of PMo_{11} were synthesized by impregnating 1 g of ZrO_2 with an aqueous solution of PMo_{11} (0.1–0.4 g/10–40 ml of conductivity water) and dried at 100 °C for 10 h. The obtained materials were designated as 10 % PMo_{11}/ZrO_2, 20 % PMo_{11}/ZrO_2, 30 % PMo_{11}/ZrO_2, and 40 % PMo_{11}/ZrO_2.

3 Characterization

The synthesized materials were characterized by FT-IR, TGA, DRS, powder XRD, and BET surface area. FT-IR spectra of the samples were recorded as the KBr pellet on the PerkinElmer instrument. The total weight loss was calculated by the TGA method on the Mettler Toledo Star SW 7.01 up to 600 °C. The UV-visible-DR spectrum was recorded at ambient temperature on PerkinElmer 35 LAMBDA instrument using the 1 cm quartz cell. The spinning rate was 4–5 kHz. The powder XRD pattern was obtained by using the instrument Philips Diffractometer (Model PW 1830). The conditions used were Cu Kα radiation (1.5417 A). Elemental analysis was carried out using JSM 5610 LV combined with INCA instrument for EDX-SEM. The BET specific surface area was calculated by using the standard Brunauer, Emmett, and Teller method on the basis of the adsorption data. Adsorption–desorption isotherms of samples were recorded on a micromeritics ASAP 2010 surface area analyzer at −196 °C.

3.1 *Catalytic Reaction*

The oxidation reaction was carried out in a borosilicate glass reactor provided with a double-walled condenser containing catalyst, benzyl alcohol, and H_2O_2 at 90 °C with constant stirring for 24 h. Reaction temperature was maintained constant by immersing the reactor in a constant temperature oil bath. Same reaction was carried out by varying different parameters such as effect of % loading, molar ratio of substrate to H_2O_2, amount of the catalyst, reaction time, and reaction temperature. After completion of the reaction, the catalyst was removed and the product was extracted with dichloromethane. The product was dried with magnesium sulfate and analyzed on gas chromatograph using BP-1 capillary column. Product was identified by comparison with the authentic samples and finally by gas chromatography–mass spectroscopy (GC–MS).

4 Results and Discussion

4.1 Analytical Techniques

Any leaching of the active species from the support makes the catalyst unattractive and hence it is necessary to study the stability as well as leaching of PMo_{11} from the support. Heteropoly acids can be quantitatively characterized by the heteropoly blue color, which is observed when it reacted with a mild reducing agent such as ascorbic acid. In the present study, this method was used for determining the leaching of PMo_{11} from the support. Standard samples containing 1–5 % of PMo_{11} in water were prepared. To 10 ml of the above samples, 1 ml of 10 % ascorbic acid was added. The mixture was diluted to 25 ml. The resultant solution was scanned at a λ_{max} of 785 cm^{-1} for its absorbance values. A standard calibration curve was obtained by plotting values of absorbance against % concentration. 1 g of supported catalyst with 10 ml conductivity water was refluxed for 18 h. Then 1 ml of the supernatant solution was treated with 10 % ascorbic acid. Development of blue color was not observed, indicating that there was no leaching. The same procedure was repeated with benzyl alcohol and the filtrate of the reaction mixture after completion of reaction in order to check the presence of any leached PMo_{11}. The absence of blue color indicates no leaching of PMo_{11}.

The FT-IR spectra of PMo_{11} and 20 % PMo_{11}/ZrO_2 are presented in Fig. 10.1. The bands at 1,048 and 999 cm^{-1}, 935 and 906 cm^{-1}, and 855 cm^{-1} in parent PMo_{11} (Fig. 10.1) attributed to asymmetric stretching of P–O, Mo–O_t, and Mo–O–Mo, respectively. The FT-IR spectra of 20 % PMo_{11}/ZrO_2 (Fig. 10.1) shows a band at 557 cm^{-1} assigned to Zr–O–H stretching.

Bands at 1,039, 990, and 910 cm^{-1} correspond to asymmetric stretching of P–O and Mo–O–Mo, respectively. The shifting and broadening of bands as well as disappearance of Mo–O_t band (935 cm^{-1}) may be due to interaction of terminal oxygen of PMo_{11} with ZrO_2.

The TGA of PMo_{11} (Fig. 10.2a) shows the initial weight loss of 16 % up to 30–200 °C. This may be due to the removal of water molecules. The final weight loss is observed at 235 °C indicates the decomposition of PMo_{11}. The TGA of 20 % PMo_{11}/ZrO_2 (Fig. 10.2b) shows initial weight loss up to 150 °C may be due to the removal of adsorbed water molecules. No significant loss occurs up to 300 °C. DTA of PMo_{11} shows (Fig. 10.2a) endothermic peaks at 80 and 140 °C, due to the loss of adsorbed water and water of crystallization, respectively. In addition, DTA of PMo_{11}

Fig. 10.1 FT-IR of PMo_{11} and 20 % PMo_{11}/ZrO_2

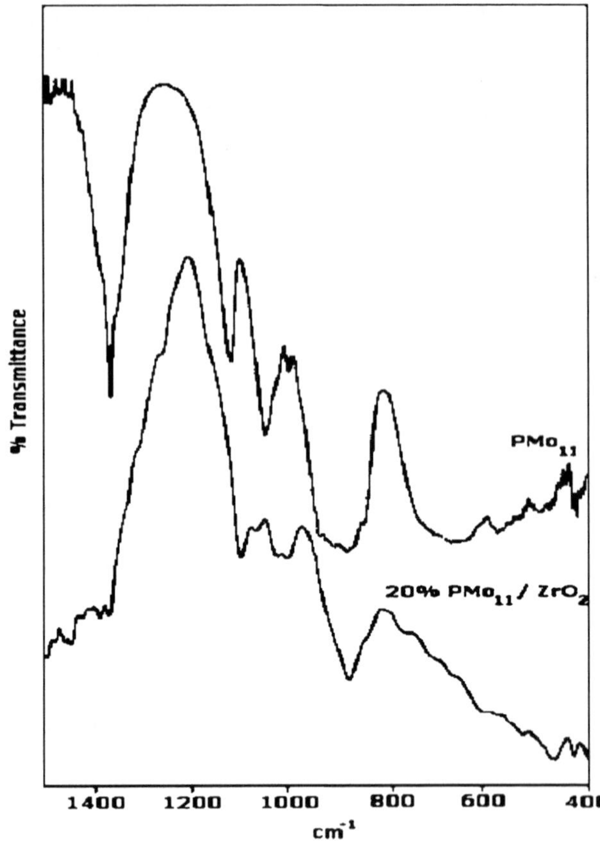

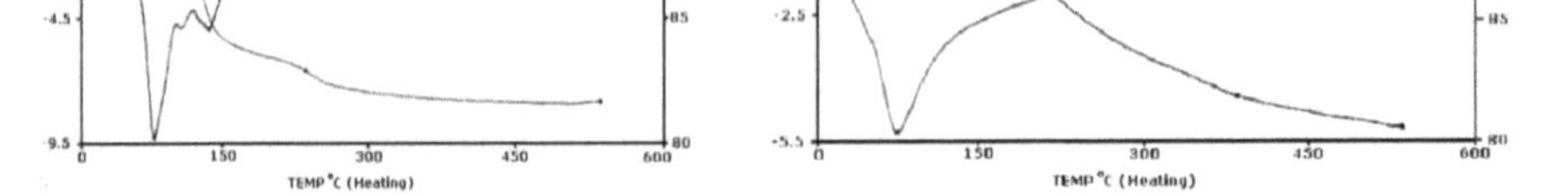

Fig. 10.2 TG-DTA for (a) PMo_{11} and (b) 20 % PMo_{11}/ZrO_2

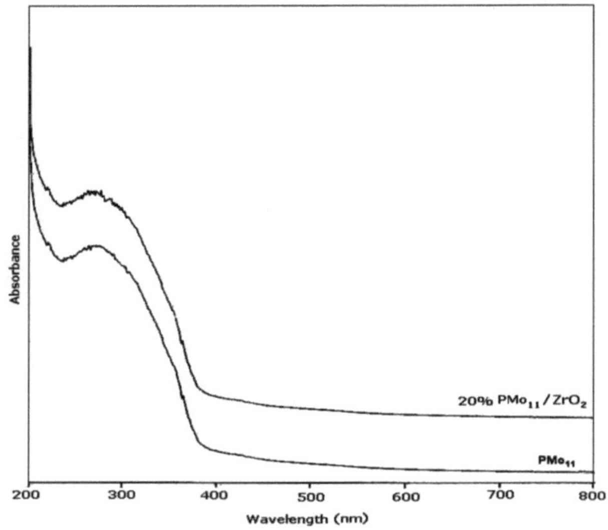

Fig. 10.3 DRS of PMo_{11} and 20 % PMo_{11}/ZrO_2

also shows a broad exothermic peak in the region of 270–305 °C. This may be due to the decomposition of PMo_{11}. DTA of 20 % PMo_{11}/ZrO_2 (Fig. 10.2b) shows an endothermic peak at 80 °C which may be due to adsorbed water. It also shows a broad exothermic peak in the region 285–325 °C, which may be due to decomposition of PMo_{11} on to the surface of the support. From the TG-DTA, it can be concluded that supporting of PMo_{11} on the surface of ZrO_2 increases the thermal stability of PMo_{11}. An increase in thermal stability indicates the presence of chemical interaction between PMo_{11} and ZrO_2.

The DRS gives information about the non-reduced heteropolyanion due to charge transfer from oxygen to metal. The band at 300 (λ_{max}) is attributed to O → Mo charge transfer.

The λ_{max} of 20 % PMo_{11}/ZrO_2 is same as that for PMo_{11} (Fig. 10.3). This confirms the presence of the undegraded PMo_{11} on the surface of ZrO_2. In other words, the Keggin structure remains unaltered after supporting onto ZrO_2.

XRD of PMo_{11} (Fig. 10.4) that shows sharp peaks indicates its crystalline nature. XRD pattern of 20 % PMo_{11}/ZrO_2 shows no characteristic peaks of PMo_{11}, indicating a very high dispersion of PMo_{11} in a noncrystalline form on the surface of ZrO_2.

The data for BET surface area measurements are shown in Table 10.1. The larger surface area of the 20 % PMo_{11}/ZrO_2 as compared to that of the support was because of supporting of PMo_{11} and it is as expected. With increase in the percent loading of PMo_{11}, the surface area decreases. This may be due to the stabilization/blocking of the sites.

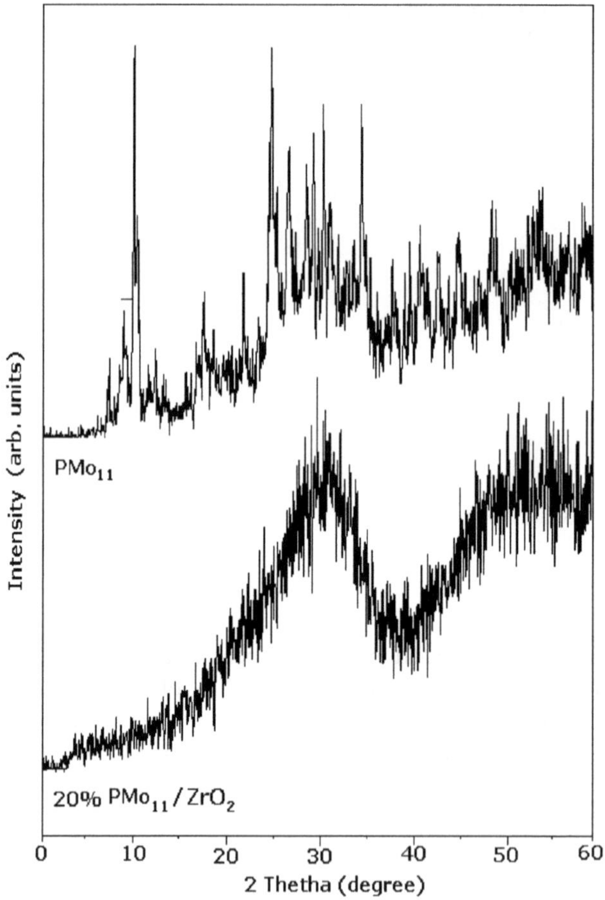

Fig. 10.4 Powder XRD of PMo_{11} and 20 % PMo_{11}/ZrO_2

Table 10.1 Total surface area for ZrO_2, 10 % PMo_{11}/ZrO_2, 20 % PMo_{11}/ZrO_2, 30 % PMo_{11}/ZrO_2, and 40 % PMo_{11}/ZrO_2

Catalyst	Surface area ($m^2\ g^{-1}$)
ZrO_2	170
10 % PMo_{11}/ZrO_2	191
20 % PMo_{11}/ZrO_2	197
30 % PMo_{11}/ZrO_2	188
40 % PMo_{11}/ZrO_2	187

4.2 Catalytic Studies

A detail study was carried out on the oxidation of benzyl alcohol to optimize the conditions. To ensure the catalytic activity, all reactions were carried out without catalyst. It was found that no oxidation takes place. The support, ZrO_2, was also used as catalyst for the oxidation of benzyl alcohol and no conversion was found, indicating PMo_{11} are only active species.

Scheme 10.1 Oxidation of benzyl alcohol

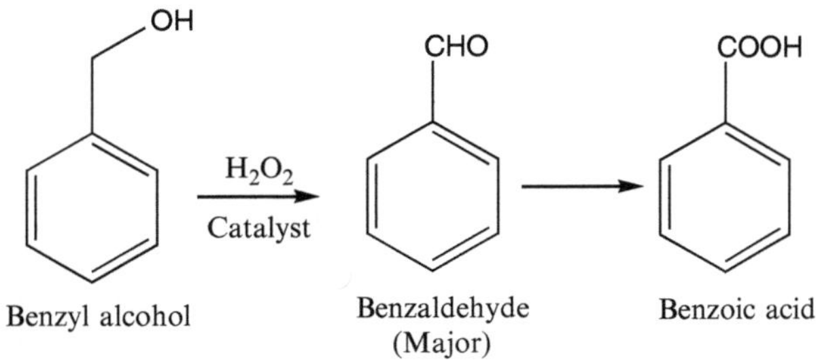

The reaction was carried out by varying the mole ratio of benzyl alcohol to H_2O_2 with 25 mg of the catalyst for 24 h at 90 °C. Generally, benzyl alcohol on oxidation gives benzaldehyde and benzoic acid (Scheme 10.1). However, benzaldehyde was characterized as the major oxidation product in the present case. With 1:3 molar ratio the conversion of benzyl alcohol was 24.5 % and the selectivity for benzaldehyde was 93 %. While with 1:4 molar ratio, the conversion was 28 % and the selectivity for benzaldehyde was 82 %. Hence, further optimization of the conditions was carried out with a 1:3 molar ratio.

The oxidation of benzyl alcohol was carried out with 1:3 molar ratio of benzyl alcohol to H_2O_2 by using 25 mg of fresh catalysts for 24 h at 90 °C are presented in Fig. 10.5. Figure 10.5 shows an increase in the conversion with increase in the % loading of PMo_{11} from 10 to 20 %.

Further, with increase in % loading from 20 to 40 %, decrease in conversion is found. This may be due to the blocking of the active sites. The decrease in % conversion may be correlated with the values of surface area. Hence, a decrease in % conversion of benzyl alcohol is as expected.

Thus, the loading of PMo_{11} on the support was fixed at 20 %. A detail study was carried out at 90 °C over 20 % PMo_{11}/ZrO_2.

The effect of temperature on the oxidation of benzyl alcohol was investigated at three different temperatures, namely, 70, 90, and 110 °C, keeping the other parameters fixed, namely, benzyl alcohol (10 mmol), 30 % H_2O_2 (30 mmol), catalyst (25 mg), and reaction time (24 h). The results are shown in Fig. 10.6, which reveals that 5.8, 24.5, and 40.7 % conversion were found corresponding to 70, 90, and 110 °C, respectively. On increasing the temperature from 90 to 110 °C, improvement in conversion was observed, but selectivity for benzaldehyde decreases from 93 to 67 %. This may be due to oxidation of benzaldehyde to benzoic acid at elevated temperature. Hence, further optimization of the conditions was carried out with a 90 °C temperature.

In order to determine the effect of H_2O_2 on the oxidation of benzyl alcohol to benzaldehyde, we studied six different benzyl alcohol–H_2O_2 molar ratios (1:1, 1:2, 1:3, 1:4, 2:1, and 3:1) keeping other parameters fixed, namely, catalyst (25 mg), temperature (90 °C), and reaction time (24 h). The results are shown in Fig. 10.7.

Benzyl alcohol to H_2O_2 molar ratio of 1:1 and 1:2 resulted in 4.6 and 19.1 % conversion, respectively, and when benzyl alcohol to H_2O_2 molar ratio was changed to 1:3, conversion increased to 24.5 %, keeping all other conditions similar. However, conversion was found to be almost the same at 24.8 % when the benzyl

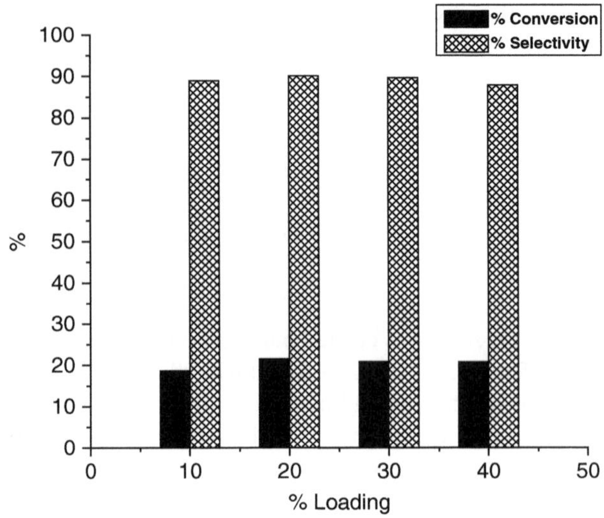

Fig. 10.5 Effect of % loading; % conversion is based on benzyl alcohol; time = 24 h; temperature = 90 °C; amount of catalyst = 25 mg; molar ratio of benzyl alcohol to H_2O_2 = 1:3

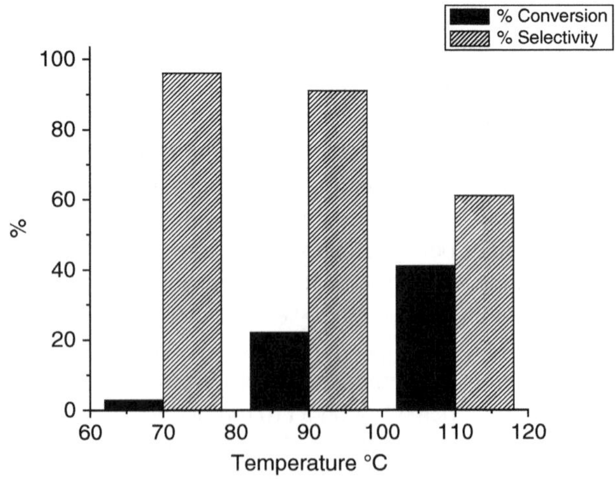

Fig. 10.6 Effect of temperature; % conversion is based on benzyl alcohol; time = 24 h; amount of catalyst = 25 mg; molar ratio of benzyl alcohol to H_2O_2 = 1:3

alcohol to H_2O_2 molar ratio was further changed to 1:4. Therefore, 1:3 molar ratio of benzyl alcohol to H_2O_2 was found to be the optimum in terms of conversion of benzyl alcohol.

The amount of catalyst has a significant effect on the oxidation of benzyl alcohol. Five different amounts of 20 % PMo_{11}/ZrO_2, namely, 10, 15, 20, 25, and 30 mg, were used, keeping all other reaction parameters fixed, namely, temperature (90 °C), benzyl alcohol (10 mmol), 30 % H_2O_2 (30 mmol), and reaction time (24 h). The results are shown in Fig. 10.8, indicating 12.2, 16.3, 18.3, 24.5, and 24.7 % conversion corresponding to 10, 15, 20, 25, and 30 mg catalyst, respectively.

Lower conversion of benzyl alcohol into benzaldehyde with 10–20 mg catalyst may be due to fewer catalytic sites.

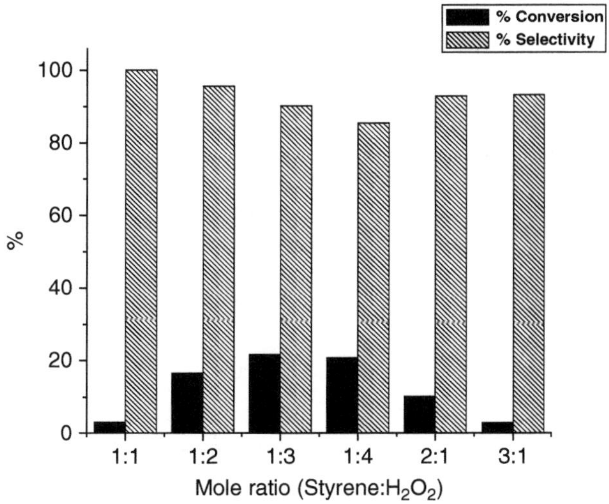

Fig. 10.7 Effect of mole ratio; % conversion is based on benzyl alcohol; time = 24 h; temperature = 90 °C; amount of catalyst = 25 mg

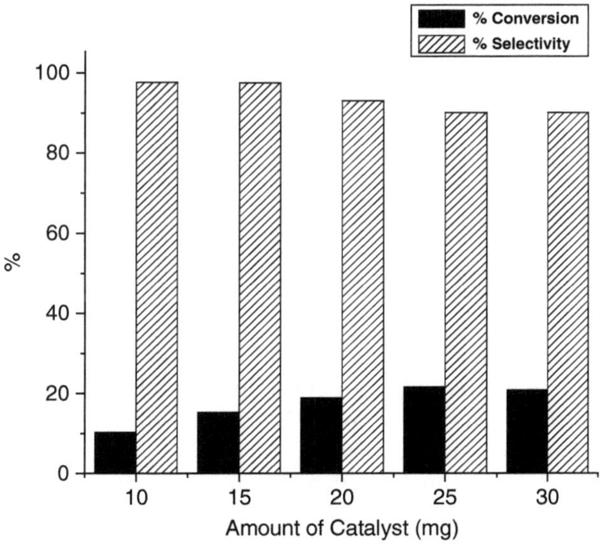

Fig. 10.8 Effect of amount of catalyst; % conversion is based on benzyl alcohol; time = 24 h; temperature = 90 °C; molar ratio of benzyl alcohol to H_2O_2 = 1:3

The maximum percentage conversion was observed with 25 mg catalyst, but there was no remarkable difference in the progress of reaction when 25 or 30 mg of catalyst was employed. Therefore, 25 mg amount of catalyst was taken to be optimal.

The time dependence of catalytic solvent-free oxidation of benzyl alcohol was studied by performing the reaction of benzyl alcohol (10 mmol) with 30 % H_2O_2 (30 mmol) in the presence of 25 mg of catalyst at 90 °C with constant stirring. The percentage of conversion was monitored at different reaction times. It is seen from Fig. 10.9 that with increase in reaction time, the % conversion also increases. Initial conversion of benzyl alcohol increased with the reaction time. This is due to the reason that more time is required for the formation of reactive

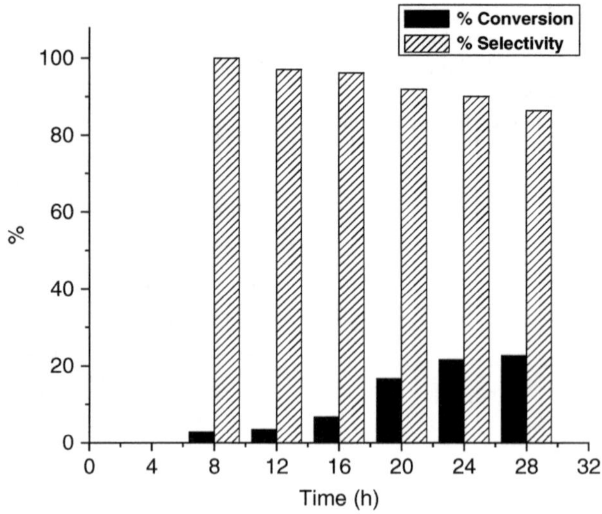

Fig. 10.9 Effect of reaction time; % conversion is based on benzyl alcohol; amount of catalyst = 25 mg; temperature = 90 °C; molar ratio benzyl alcohol to H_2O_2 = 1:3

Table 10.2 Control experiments for the oxidation of benzyl alcohol

		% Selectivity
Material	% Conversion	Benzaldehyde
ZrO_2	–	–
PMo_{11}	24.3	94

% Conversion is based on benzyl alcohol; amount of PMo_{11}, 4.2 mg; amount of ZrO_2 21.8 mg; molar ratio of substrate to H_2O_2; time, 24 h; temperature 90 °C

intermediate (substrate + catalyst) which is finally converted into the products. It is seen that 24.5 % conversion was observed at 24 h; when the reaction was allowed to continue for 28 h, 26.2 % conversion was observed, but selectivity of benzaldehyde decreases.

The optimum conditions for maximum % conversion of benzyl alcohol to benzaldehyde are the mole ratio of benzyl alcohol to H_2O_2 as 1:3 with 25 mg of catalyst and 24 h reaction time at 90 °C.

The control experiments with ZrO_2 and PMo_{11} were also carried out under optimized conditions with benzyl alcohol and H_2O_2. It can be seen from Table 10.2 that ZrO_2 is inactive toward oxidation of benzyl alcohol, indicating the catalytic activity is due only to PMo_{11}. The same reaction was carried out by taking the active amount of PMo_{11} (4.2 mg).

It was found that (Table 10.2) the active catalyst gives 24.3 % conversion of benzyl alcohol with 94 % selectivity of benzaldehyde. Almost the obtained same activity for supported catalyst indicates that PMo_{11} is the real active species. Thus, we are successful in supporting PMo_{11} onto ZrO_2 without any significant loss in activity and hence in overcoming the traditional problems of homogeneous catalysis.

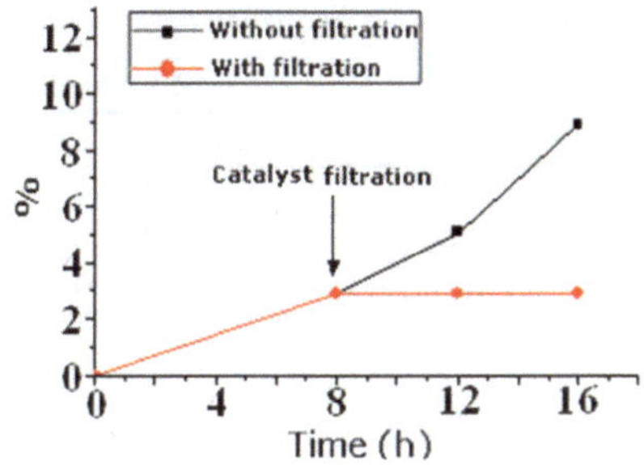

Fig. 10.10 % Conversion is benzyl alcohol; amount of catalyst = 25 mg; molar ratio of substrate to H_2O_2 = 1:3; temperature = 90 °C

Table 10.3 Oxidation of benzyl alcohol with fresh and recycled catalysts

Catalyst	Cycle	% Conversion	% Selectivity	TON
			Benzaldehyde	
20 % PMo_{11}/ZrO_2	Fresh	24.5	93	1,253
	1	24.4	93	1,248
	2	24.4	93	1,248
	3	24.2	93	1,238

% Conversion is based on benzyl alcohol; amount of catalyst = 25 mg; molar ratio of substrate to H_2O_2 = 1:3; time = 24 h; temperature = 90 °C

4.3 Heterogeneity Test

Heterogeneity test was carried out for oxidation of benzyl alcohol as an example. For the rigorous proof of heterogeneity, a test [33] was carried out by filtering catalyst from the reaction mixture at 90 °C after 8 h and the filtrate was allowed to react up to 16 h. The reaction mixture of 8 h and the filtrate were analyzed by gas chromatography. No change in the % conversion as well as % selectivity was found, indicating the present catalyst fall into category C [33]. The results are presented in Fig. 10.10.

4.4 Recycling of the Catalyst

Oxidation of benzyl alcohol was carried out with the recycled catalyst, under the optimized conditions. The catalyst was removed from the reaction mixture after completion of the reaction by simple filtration, washed with dichloromethane, and dried at 100 °C.

The catalyst was recycled in order to test its activity as well as stability. The obtained results are presented in Table 10.3. As seen from the table, there was no appreciable change observed in selectivity; however, little decrease in conversion was observed which shows that the catalysts are stable and can be regenerated for repeated use.

Table 10.4 FT-IR frequency data for fresh and regenerated catalysts

	FT-IR frequency, cm^{-1}				
Catalyst	H–O–H	Zr–O–H	P–O	Mo=Ot	Mo–O–Mo
PMo_{11}	1,614; 1,384	–	1,048; 999	935	906; 855
20 % PMo_{11}/ZrO_2	1,623; 1,403	557	1,039; 990	–	910
R-20 % PMo_{11}/ZrO_2	1,621; 1,399	560	1,038; 990	–	910

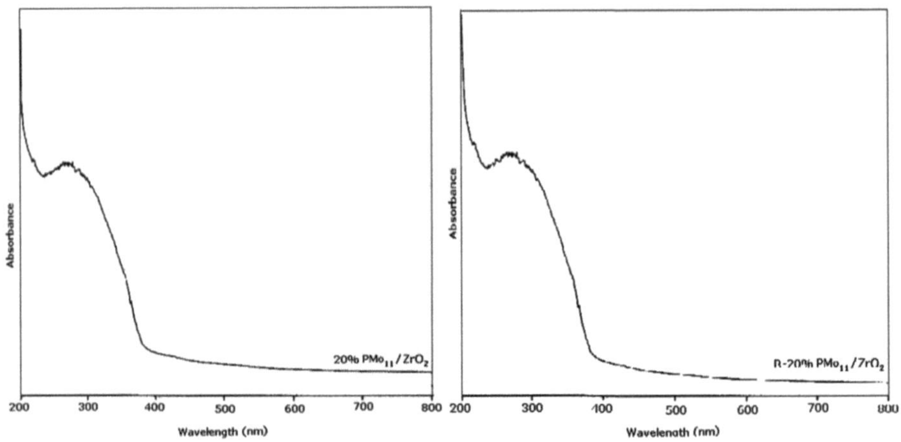

Fig. 10.11 DRS of 20 % PMo_{11}/ZrO_2 and R-20 % PMo_{11}/ZrO_2

4.5 Characterization of the Regenerated Catalyst

The catalyst, 20 % PMo_{11}/ZrO_2, was regenerated in order to test its stability. The regenerated catalyst (R-20 % PMo_{11}/ZrO_2) was characterized for FT-IR and DRS in order to confirm the retention of the catalyst structure, after the completion of the reaction.

The FT-IR data for the fresh as well as the regenerated catalysts are represented in Table 10.4. The characteristic FT-IR bands for PMo_{11} appear at 1,048 and 999 cm^{-1}, 935 and 906 cm^{-1}, and 855 cm^{-1}. No appreciable shift in the FT-IR band position of the regenerated catalyst compared to 20 % PMo_{11}/ZrO_2 indicates the retention of Keggin-type PMo_{11} onto ZrO_2.

The DRS gives information about the non-reduced heteropolyanion due to charge transfer from oxygen to metal. The λ_{max} of 20 % PMo_{11}/ZrO_2 and R-20 % PMo_{11}/ZrO_2 is same as that for PMo_{11} (Fig. 10.11). This confirms the presence of the undegraded PMo_{11} species in catalysts. In other words, the Keggin structure remains unaltered even after recycling of catalyst.

5 Conclusion

In conclusion, a heterogeneous catalyst comprising of lacunary phosphomolybdate and zirconia was proved to be successful for solvent-free selective oxidation of benzyl alcohol with H_2O_2 as an oxidant under mild reaction conditions. Superiority of the catalyst lies in obtaining good conversion of benzyl alcohol to benzaldehyde (93 %) with high TON (1253). FT-IR and DRS of regenerated catalyst showed that the catalyst is stable under present reaction conditions. Moreover, removal of the catalyst consists of the single filtration and catalyst can be reused after a simple workup.

Acknowledgments Mr. Soyeb Pathan is thankful to the University Grants Commission (UGC-Maulana Azad National Fellowship) New Delhi. Anjali Patel thanks the Department of Science and Technology (DST, No. SR/S5/GC-01/2009) for financing a part of this work.

References

1. Zhihuan Weng, Jinyan Wang, Xigao Jian (2008) Catal Commun 9:1688–1691 and reference No. 1 therein
2. Pybus DH, Sell CS (eds) (1999) The chemistry of fragrances. RSC Paperbacks, Cambridge
3. Singh RP, Subbarao HN, Dev S (1979) Tetrahedron 35:1789
4. Fey T, Fischer H, Bachmann S, Albert K, Bolm C (2001) J Org Chem 66:8154
5. Li G, Enache DI, Edwards JK, Carley AF, Knight DW, Hutchings GJ (2006) Catal Lett 110:7
6. McGrath DV, Grubbs RH, Ziller JW (1991) J Am Chem Soc 113:3611
7. Knight DA, Schull TL (2003) Synth Commun 33:827
8. Ding Y, Ma B, Gao Q, Li G, Yan L, Suo J (2005) J Mol Catal A Chem 230:121
9. Peng G, Wang Y, Hu C, Wang E, Feng S, Zhou Y, Ding H, Liu Y (2001) Appl Catal A Gen 218:91
10. Li M, Shen J, Ge X, Chen X (2001) Appl Catal A Gen 206:161
11. Yang JH, Lee DW, Lee JH, Hyun JC, Lee KY (2000) Appl Catal A Gen 195:123
12. Li W, Lee K, Oshihara K, Ueda W (1999) Appl Catal A Gen 182:357
13. Mizuno N, Tateishi M, Iwamoto M (1996) J Catal 163:87
14. Mizuno N, Yamaguchi K, Kamata K (2005) Coord Chem Rev 249:1944
15. Mizuno N, Misono M (1998) Chem Rev 98:199
16. Neumann R (1998) Prog Inorg Chem 47:317
17. Hill CL, Chrisina C, Prosser-McCartha M (1995) Coord Chem Rev 143:407
18. Hill CL (2003) Compr Coord Chem II 4:679
19. Mizuno N, Hikichi S, Yamaguchi K, Uchida S, Nakagawa Y, Uehara K, Kamata K (2006) Catal Today 117:32
20. Sharma P, Patel A (2009) Appl Surf Sci 255:7635
21. Sharma P, Patel A (2009) J Mol Catal A: Chem 299:37
22. Schwegler M, Floor M, Van Bekkuw H (1988) Tetrahedron Lett 29:823
23. Harrup MK, Hill CL (1994) Inorg Chem 33:5448
24. Khenkin A, Hill C (1993) Mendleev Commun 3:140
25. Kuznetsova NI, Detusheva LG, Kuznetsova LI, Fedotov MA, Likholobov VA (1992) Kinet Katal 33:516

26. Weng Z, Wang J, Jian X (2008) Catal Commun 9:1688
27. Popivker I, Zilbermann I, Maimon E, Matana Y, Cohen H, Meyerstein D (2010) Inorg Chimca Acta 363:4202
28. Shringarpure P, Patel A (2008) Dalton Trans 3953
29. Shringarpure P, Patel A (2010) Dalton Trans 39:2615
30. Pathan S, Patel A (2011) Dalton Trans 40:348
31. Shringarpure P, Patel A (2011) Ind Eng Chem Res 50:9069
32. Pathan S, Patel A (2012) Ind Eng Chem Res 51:732
33. Sheldon A, Walau M, Arends IWCE, Schuchurdt U (1998) Acc Chem Res 31:485

Chapter 11
$H_3PMo_{12}O_{40}$ Heteropolyacid: A Versatile and Efficient Bifunctional Catalyst for the Oxidation and Esterification Reactions

Márcio José da Silva, Raquel da Silva Xavier, and Lidiane Faria dos Santos

Contents

1 Introduction 226
1.1 Heteropolyacid-Catalysed Oxidation and Esterification Reactions 226
2 Experimental Section 231
2.1 Esterification Reaction Conditions 231
2.2 Oxidation Reaction Conditions 232
3 Results and Discussion 233
3.1 $H_3PMo_{12}O_{40}$-Catalysed Esterification Reactions: Effect of the Fatty Acids Nature 233
3.2 $H_3PMo_{12}O_{40}$-Catalysed Oleic Acid Esterification: Effect of the Alcohol Nature 234
3.3 $H_3PMo_{12}O_{40}$-Catalysed Camphene Oxidation Reactions by Hydrogen Peroxide 235
3.4 Molybdenum-Catalysed Oxidative Desulfurisation Reactions Using DBT as Model Compound and Hydrogen Peroxide as Oxidant 238
4 Conclusion 241
References 242

Abstract This work summarises our recent developments for H_2O_2-based green oxidation reactions and the results obtained in esterification reactions, both catalysed by $H_3PMo_{12}O_{40}$ and their salts. We have found that these catalysts are versatile bifunctional catalysts, highly active in esterification reactions as well as oxidation reactions. Herein, their performance was assessed in three different kinds of reactions: natural olefin oxidation, gasoline oxidative desulfurisation reactions and acid-catalysed esterification of fatty acids. The $H_3PMo_{12}O_{40}$ catalyst has two key features which allow it to act efficiently in all these reactions: Brønsted and Lewis acid sites and redox sites. At first, $H_3PMo_{12}O_{40}$-catalysed oxidation reaction of

M.J. da Silva (✉) • R. da Silva Xavier • L.F. dos Santos
Department of Chemistry, Exacts Science Center, Federal University of Viçosa, Avenue P.H. Rolfs, Viçosa, Minas Gerais 3670-000, Brazil
e-mail: silvamj2003@ufv.br

A. Patel (ed.), *Environmentally Benign Catalysts: For Clean Organic Reactions*,
DOI 10.1007/978-94-007-6710-2_11, © Panch Tattva Publishers, Pune, India 2013

dibenzothiophene (DBT) into dibenzosulphone (DBTO) by hydrogen peroxide was investigated in a biphasic system (i.e. isooctane/acetonitrile). The DBT is soluble only in the non-polar layer (isooctane); thus, after their oxidation into DBTO, this later migrates to polar layer (acetonitrile). This is very useful process for the removal of sulfur compounds, usually found in gasoline samples. Surprisingly, the $AlPMo_{12}O_{40}$ heteropolyacid salt was found to be more efficient than $H_3PMo_{12}O_{40}$ catalyst in DBT oxidation reactions. To the best of our knowledge, this is the first application of $AlPMo_{12}O_{40}$ catalyst for mentioned oxidation reactions without using phase-transfer agent. High removal rates (higher than 90 %) were achieved using $AlPMo_{12}O_{40}$ as catalyst and hydrogen peroxide as oxidant. On the other hand, studying the second oxidative process, we have found that $H_3PMo_{12}O_{40}/H_2O_2$ system was also highly effective in oxidation reactions of monoterpenes. Monoterpenes are abundant and renewable raw materials and were converted into oxygenates products which are valuable ingredient for fragrance, agrochemicals and pharmaceutical industries. Camphene was selectively oxidised into epoxy and allylic products with conversion rates on the range 80–90 %. Hydrogen peroxide, an environmentally benign reactant, was employed as oxidant in the camphene oxidation reactions. Finally, we also have described novel results of $H_3PMo_{12}O_{40}$-catalysed fatty acid esterification reactions with different alcohols.

1 Introduction

1.1 Heteropolyacid-Catalysed Oxidation and Esterification Reactions

In recent years, several works have reported new catalytic applications of heteropolyacids in organic reactions [1]. Some of the most relevant properties for technological applications rely on solubility, acidity and redox potentials [2]. Heteropolyacids (HPAs) are well-defined molecular clusters that are recognised for their molecular and electronic structural diversity and have been used in homogeneous or heterogeneous catalysis [3]. Among several HPAs studied, those Keggin-type are the most outstanding class. In special, those containing tungsten as addenda atom (i.e. $H_3PW_{12}O_{40}$) are the strongest Brønsted acids and have high thermal stability [4]. These HPAs are commercially available and more easily handled when used in solid state. In comparison with the conventional liquid acid catalysts, HPAs as solid heterogeneous catalysts have the advantages such as less corrosive action and, consequently, are environmentally benign catalysts because their use avoids neutralisation steps resulting on lower generation effluents [5]. Besides the strong Brønsted acidity of Keggin-type heteropolyacids, the softness of the heteropoly anions is also responsible for their high catalytic activity in the esterification reactions [6]. On the other hand, redox properties of the Keggin-type HPAs are more noticeable when the addenda atom is molybdenum, (i.e. $H_3PMo_{12}O_{40}$) as it has low oxidation potential [7, 8].

Among the stoichiometric oxidants, hydrogen peroxide is regarded as desirable "greener oxidant" because of their high atom efficiency and high active oxygen content; the only by-product is water [9]. Oxidation catalytic processes are becoming the most attractive methods when used with clean and inexpensive oxidants especially dioxygen and hydrogen peroxide, which results in water as by-product. These oxidants can be used when HPAs are the only catalysts or when they are employed in combination with other metal catalyst. However, it is important to note that when dioxygen is the oxidant, a two electron-transfer process occurs in almost all cases; conversely, when hydrogen peroxide is the oxidant, a one electron-transfer process is most observed. For instance, the use of the heteropolyacids/palladium combination in a multicomponent catalytic system was successfully reported and dioxygen was employed as oxidant [10]. Palladium(II)-copper(II)-heteropolyacid combinations were found to be an efficient modified Wacker-type catalyst system for the oxidation of cyclohexene into cyclohexanone [11]. Various kinds of metal-substituted polyoxometalates have been synthesised and applied in selective oxidation processes, such as oxidative alkyne homocoupling reactions [12]. Actually, about 30 years ago, Matveev and Kozhevnikov demonstrated that $H_3PV_xMo_{12-x}O_{40}$ ($x = 1–6$) phosphovanadomolybdate Keggin-type compounds could be used as co-catalysts in place of copper salts in palladium-catalysed oxidations [13]. Since then, the most often used polyoxometalate for catalytic oxidation involving two electrons (i.e. with dioxygen as oxidant) has been based on the acidic polyanion containing two vanadium atoms (i.e. $H_5PV_2Mo_{10}O_{40}$ or $H_5PV_2Mo_{10}O_{40}$) which have the α-Keggin structure [14].

On the other hand, heteropolyacid complexes containing Mo or W as central atom and the ligand-cetylpyridynium chloride were used in two-phase processes of olefin epoxidation and allylic alcohols with hydrogen peroxide [15]. Indeed, Misono stated that in spite of extensive attempts using molybdenum HPAs as catalysts for alkane oxidation, the performance was considerably below the level of commercialisation [16]. However, the use of $H_3PMo_{12}O_{40}$ as catalysts in olefin oxidation reactions by hydrogen peroxide seems to be extremely attractive, as we show in the next topics.

1.1.1 Oxidation Reactions of Monoterpenes

The oxidation of monoterpenes is an important method for valorisation of these cheap, abundant and renewable natural olefins [17]. Oxygenated terpenes are largely applied in organic synthesis as chiral building blocks [18] and are valuable ingredients for flavour, fragrance [19] and pharmaceutical industries [20]. However, their production via non-catalytic oxidation reactions generally takes place with low yield and poor selectivity, where the use of toxic and expensive stoichiometric metallic oxidants is also required. Thus, to develop catalytic processes based on clean oxidants, such as hydrogen peroxide, becomes strategically important because they result in a reduction of the environmental impact and contribute to reduced production costs [21]. Peroxide oxidants have been employed in terpene oxidation with transition metal porphyrin complexes catalysts [22]. However, the main disadvantage of these catalysts is the laborious synthesis as well as the requirement of the co-catalyst. However, among the several catalysts employed in the olefin oxidation, deserves highlights the Wacker system [23]. Originally, the Wacker catalyst (i.e. $PdCl_2$/$CuCl_2$) employs dioxygen as final oxidant and is industrially used in the production of acetaldehyde from ethylene. However, this process has serious drawbacks (i.e. high Lewis acidity of Cu(II) ions and also the need for high concentration of ions chloride) which make its use less efficient in the olefin oxidation as monoterpenes. An alternative is to replace dioxygen by hydrogen peroxide, which is able of directly oxidise palladium without the presence of an auxiliary metal reoxidant ($CuCl_2$) [24]. Mimoun and co-workers described one of the first works employing hydrogen peroxide as the final oxidant in Pd(II)-catalysed olefin oxidation reactions [25]. However, comparatively to dioxygen oxidant, there is yet little information available in literature concerning the use of hydrogen peroxide as a palladium oxidant in Wacker-type processes [26].

The current goal of our research group is to develop catalysts that can effectively activate these greener oxidants and/or renewable substrates under mild reaction conditions and oxidise the different substrates with high efficiency and selectivity.

In this work, we are describing for the first time the results obtained in $H_3PMo_{12}O_{40}$-catalysed oxidation reactions of camphene by hydrogen peroxide.

1.1.2 Desulfurisation Oxidative Reactions

Sulfur oxides (SOx) emissions are generated during fuel combustion from the oxidation of sulfur compounds. However, the dramatic environmental impact provoked by sulfur oxides present in these emissions is becoming sulfur content specifications in fuels very more stringent worldwide [27]. These environmental concerns have driven the need to remove sulfur-containing compounds from fuels. The most important industrial process is called hydrodesulfurisation (HDS) that treats the fuel under high temperatures and high pressures with hydrogen. In the petroleum refining industry, HDS is the conventional process to reduce the sulfur levels of the fuels. However, in this process, to achieve the low sulfur content higher hydrogen pressure, higher reaction temperature and more active catalysts are needed [28]. This is because dibenzothiophene (DBT) and its derivatives are very hard to remove by HDS, owing to their broad stereo hindrance [29].

In recent years, various alternative desulfurisation processes operated under moderate conditions without requiring hydrogen have been extensively assessed, among which is the oxidative desulfurisation (ODS) process [30]. ODS appears to be particularly promising and is currently receiving increasing attention because it avoids the use of hydrogen and allows the process to be conducted under mild reaction conditions [31]. Although the process is effective, there are still some problems such as cross-contamination between extractant and fuel and loss of fuel. In the ODS process, organic sulfur compounds, such as DBT and its derivatives, can be oxidised selectively to their corresponding sulfoxides and sulfones, which can be then removed by extraction or adsorption [32].

The most widely used hydroperoxide in ODS processes where sulfur compounds are extracted by solvent is the *tert*-butyl hydroperoxide (TBHP) [33]. Nevertheless, the best option for an oxidising agent is hydrogen peroxide, because it presents a high amount of active oxygen by mass unit, is an inexpensive commercial product often used at industrial scale and gives only water as a by-product [34]. The first papers of the ODS with hydrogen peroxide described the use of radiation aiming generating of hydroperoxide groups [35]. However, the longer photo-irradiation time (*ca.* 48 h) hampers that this process may be used industrially.

Undoubtedly, the use of heterogeneous recyclable catalyst is the better option, and a large number of solid catalysts have been developed [36]. However, the main drawbacks are related to laborious work synthesis, leaching and stability of catalyst, the longer reaction time and the oxidant stability, which may undergo decomposition under reaction temperature [37]. Moreover, although efficient, catalyst based on noble metals (i.e. Ag, Au) makes the ODS process very expensive [38].

An alternative that allows a drastic reduction of the reaction time is soluble Lewis acid catalysts, based on transition metal salts [39]. Several works describe the use of transition metal catalysts in acetic acid solutions or in presence of others organic acids [40]. In this case, they are formed radicals from organic peracids which are highly reactive in the oxidation reactions [41]. Nevertheless, the main drawback is related to the presence of two reaction phases: a non-polar phase with the sulfur compound and a polar phase that contains the oxidant and the catalyst. This serious inconvenience may be circumvented by employing phase-transfer catalysts [42]. However, this requires introduction to additional limitation related to rate of mass transfer between two phases, which may affect the efficiency of the removal of sulfur compounds.

The active site of these homogeneous or heterogeneous catalysts is a transition metal, frequently in a high oxidation state and consequently with high Lewis acidity. Molybdenum catalysts are commonly used, and frequently a specie Mo(IV) is an active site [43]. Frequently, molybdenum catalysts are supported on acid oxides; however, the catalyst leaching is yet a main challenge to be overcome [44].

An attractive alternative is the use of polyoxometalates as catalysts in ODS reactions. Recently, by using the selenium-containing dinuclear peroxotungstate catalysts, various kinds of sulfides could be converted into the corresponding sulfoxides or sulfones in excellent yields with stoichiometric amounts of H_2O_2 [45]. Another example of the polyoxometalates-catalysed ODS reactions in HOAc solutions using H_2O_2 as oxidant was recently described; tetrabutylammonium salts of $[V(VW_{11})O_{40}]^{4-}$ anion show higher activity than $(VO)_2P_2O_7$, reaching removal rates equal to 90 % in the removal of sulfur model compounds added to fuel (i.e. benzothiophene, dibenzothiophene, 4,6-dimethyldibenzothiophene [46]). Recently, a promising methodology for ultra-deep desulfurisation of diesel fuel using oxidation followed by phase transfer of oxidised sulfur compound was described [47]. In those works, to remove sulfur compounds, tetraoctylammonium bromide, $H_3PW_{12}O_{40}$ and hydrogen peroxide were efficiently used as phase-transfer agent, catalyst and oxidant, respectively.

1.1.3 Heteropolyacid-Catalysed Fatty Acids Esterification Reactions for Biodiesel Production

The biodiesel is a "green" alternative fuel that has arisen as an attractive option, mainly because it is less pollutant than its counterpart fossil and can be obtained from renewable sources [48]. Nowadays, most of the biodiesel consumed worldwide is produced via transesterification edible oils, carried out under alkaline homogeneous catalysis conditions [49]. In this process, it is not possible to recycle the catalyst, besides resulting in a greater generation of effluents and salts from neutralisation steps of the products and wastes [49]. Moreover, the raw materials commonly used in those processes have high prices and need large land reserves for its cultivation [50]. Some of the natural oils or fats contain high amounts of free fatty acids (FFA), which are undesirable for the alkaline transesterification process [51]. These important features hardly affect the final cost to biodiesel production [52].

An interesting alternative for lower biodiesel price is to produce it directly from domestic reject (i.e. waste cocking oil) and water wastes generated by food industry [53]. However, since these low-cost lipidic raw materials are normally rich in FFA, its conversion into biodiesel is not compatible with alkaline catalysts [54]. It should be mentioned that even though mineral acids are inexpensive catalysts applicable to those processes, they are also corrosive, cannot be recycled and result in generation of a large amount of acid effluents which should be neutralised at the end of the reaction, leaving greater amount of salts to be disposed off into the environment [55]. Thus, to develop recyclable alternative catalysts for FFA esterification present in low cost raw materials and food industry rejects is an option strategically certain and can make the biodiesel with more competitive price via a cleaner technology [56].

There are different Keggin-type HPAs; however, $H_3PW_{12}O_{40}$ is the most largely used, especially under heterogeneous catalysis in transesterification reactions that have been described by some authors. The activity of $H_3PW_{12}O_{40}$ catalyst impregnated on different supports (i.e. hydrous zirconia, silica, alumina and activated carbon) was assessed on transesterification of low-quality canola oil; high yields were reached at 200 °C temperature [57]. The catalytic activity of $H_3PW_{12}O_{40}$ supported on zirconia was investigated in transesterification reactions with methanol, achieving high yields of FAMEs only at high temperatures (*ca.* 200 °C) [58]. On the other hand, some few works described the use of supported $H_3PW_{12}O_{40}$ catalyst on zirconia in FFA esterification [59]. Indeed, in both reactions (esterification as well as transesterification), the catalyst leaching caused by highly polar medium can affect reaction efficiency and decreases yields obtained from their reuse, thus being the main challenge to be overcome.

As a homogeneous catalyst, the $H_3PW_{12}O_{40}$ heteropolyacid is also more extensively used than $H_3PMo_{12}O_{40}$, especially in FFA esterification reactions. For instance, recently a noticeable result was described by us: $H_3PW_{12}O_{40}$ catalyst showed a higher activity than other homogeneous acid catalysts in fatty acids esterification reactions with different alcohols at room temperature. High esters yields (*ca.* 90 %) after a 4-h reaction were obtained under mild reaction conditions (i.e. atmosphere pressure, room temperature) [60].

2 Experimental Section

2.1 Esterification Reaction Conditions

The reaction conditions used were based on typical homogeneous processes. Reactions were conducted in a 50 mL three-necked glass flask equipped and fitted with a sampling septum and a reflux condenser. In all the catalytic run, adequate alcohol was added in a large molar excess over the fatty acids to drive the equilibrium towards the ester formation. In a typical run, alcohol (15.0 mL, 155 mmol) and the fatty acid (1 mmol) were heated at 60 °C temperature and the catalyst was added ($H_3PMo_{12}O_{40}$, 0.014 mmol). Blank reactions (catalyst-free) were carried out under the same conditions.

The reaction progress was continuously monitored by taking aliquots at regular intervals and analysing them via gas chromatography. The reaction yields were calculated by matching the areas of the ethyl esters GC peaks into the corresponding calibration curves.

2.1.1 Effect of the Fatty Acids Nature on $H_3PMo_{12}O_{40}$-Catalysed Esterification with Ethanol

The effect of both carbon chain length and double bonds fatty acids in the yield of ethyl esters at 60 °C temperature was assessed. The fatty acid selects were lauric ($C_{12:0}$), myristic ($C_{14:0}$), palmitic ($C_{16:0}$) and stearic acids ($C_{18:0}$) (saturated acids) and oleic ($C_{18:1}$), linoleic ($C_{18:2}$) and linolenic acids ($C_{18:3}$) (unsaturated acids).

2.1.2 Effect of the Alcohols and Fatty Acids Nature on $H_3PMo_{12}O_{40}$-Catalysed Esterification with Oleic Acid

The $H_3PMo_{12}O_{40}$ catalytic activity on esterification of different alcohols at 60 °C temperature was also evaluated. Typically, 10 mL (155 mmol) of substrate alcohol selects (methanol, ethanol, propan-1-ol, propan-2-ol or butanol) were stirred and then reacted with oleic acid (1 mmol) at reaction temperature in the presence of dissolved $H_3PMo_{12}O_{40}$ catalyst (0.014 mmol).

2.1.3 Products Identification

The mass spectrometry analyses were carried in a GC/MS (Shimadzu GC17A gas chromatograph fitted with a DB5 capillary column (30 m length, 0.25 mm i.d., 0.25 mm film thickness) coupled with a MS-QP 5050A mass spectrometer Shimadzu). Products were also identified by GC analyses (Varian 450-GC gas chromatograph) fitted with a Carbowax capillary column (30 m length, 0.25 mm i.d., 0.25 mm film thickness and flame ionisation detector) via co-injection with the chromatography patterns. The chromatographic conditions were as follows: helium as the carrier gas at 1.0 mL·min^{-1}; the temperature profile was 80 °C for 1 min, 10 °C·min^{-1} up to 260 °C, hold time of 5 min; the GC injector and MS ion source temperatures were kept at 260 and 270 °C, respectively; the MS detector operated in the electronic impact mode at 70 eV, with a scanning range of *m/z* 50–400.

The main products obtained in the oxidation reactions of camphene and β-pinene were also characterised by ^1H and ^{13}C NMR analyses. The NMR spectra were taken in $CDCl_3$ solutions, using a Varian 300 spectrometer at 300.13 and 75.47 MHz, respectively. The chemical shifts were expressed as δ (ppm) relatively to tetramethylsilane as internal reference. Spectral simulations performed with ADC/CNMR programme were in agreement with the spectra observed.

2.2 *Oxidation Reaction Conditions*

Oxidation reactions were carried out in a stirred 50 mL three-necked glass reactor fitted with a sampling septum and reflux condenser at atmospheric pressure. The reactor temperature was adjusted to 60 °C using a water bath. The reaction progress was monitored by analysing aliquots of the reaction mixture taken at regular time intervals. Reaction products were identified as described in Sect. 2.1.3.

2.2.1 Oxidation Reactions of Monoterpenes: Catalytic Runs

In a typical run, the $H_3PMo_{12}O_{40}$ catalyst (0.075 mmol) and the adequate amount of aqueous hydrogen peroxide (3.75–15.0 mmol) were added to the CH_3CN solution (15.0 mL), stirred and heated for 5 min at 60 °C temperature. Then camphene (3.75 mmol) was added to the solution and the reaction was started. Blank reactions (catalyst-free) were carried out under the same conditions with different amounts of H_2O_2.

2.2.2 Oxidative Desulfurisation Reactions: Synthesis of the $AlPMo_{12}O_{40}$ Catalyst

Phosphomolybdate aluminium salt ($AlPMo_{12}O_{40}$) was synthesised via controlled addition (*ca.* 1 mL·min^{-1}) of CH_3CN solution containing stoichiometric amounts of aluminium nitrate (Al_3NO_3) to other solution containing $H_3PMo_{12}O_{40}$ catalyst soluble, at room temperature with constant agitation. The remaining solution was left to stand overnight at room temperature. Then, the solvent was completely evaporated at 40 °C temperature, and the resulting solid salt was collected and analysed by FT-IR spectroscopy. High yields (*ca.* 99 %) were obtained in all the catalyst syntheses.

2.2.3 Oxidative Desulfurisation Reactions: Catalytic Runs

Typically, the reaction solution (10.0 mL isooctane and 10.0 mL acetronitrile), containing DBT in the isooctane phase (*ca.* 1,000 ppm, 3.19 mmol), hydrogen peroxide (17.6 mmol) and molybdenum catalyst (5 mol%; 0.1595 mmol) in acetonitrile, was stirred and heated at reaction temperature for 2 h. The reaction progress was monitored via GC analysis of reaction aliquots taken at regular time intervals.

3 Results and Discussion

3.1 *$H_3PMo_{12}O_{40}$-Catalysed Esterification Reactions: Effect of the Fatty Acids Nature*

Fatty acid esterification is a typical reversible acid-catalysed reaction that produces ester and water as by-products (Fig. 11.1). Frequently, sulfuric acid is the catalyst used; however, it is a non-recyclable catalyst, is extremely corrosive, and provokes side reactions as alcohol dehydration. Heteropolyacids are an attractive option because they can be recycled yet used in homogeneous phase, after simple procedure distillation allowed by filtration from hexane addition (Fig. 11.2) [5].

The effects of the carbon chain length and of the presence of double bonds in fatty acids on HPW-catalysed ethanol esterification are presented in Table 11.1. The $H_3PMo_{12}O_{40}$ catalyst activity was assessed using 1.4 mol% concentration in relation to fatty acid.

Although excess of ethanol has been used, a very poor conversion of fatty acids into ethyl ester was obtained in all the blank reactions (*ca.* lower than 10 % after 8 h).

Note that such conversion and selectivity at the end reaction seem not to be affected by carbon chain length of the fatty acids. For instance, the lauric acid

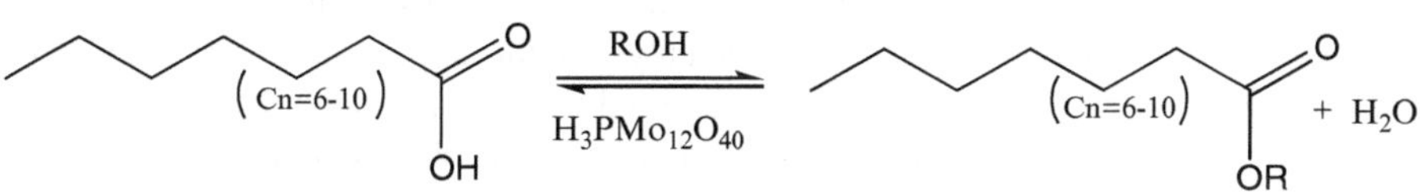

Fig. 11.1 $H_3PMo_{12}O_{40}$-catalysed fatty acid esterification reaction

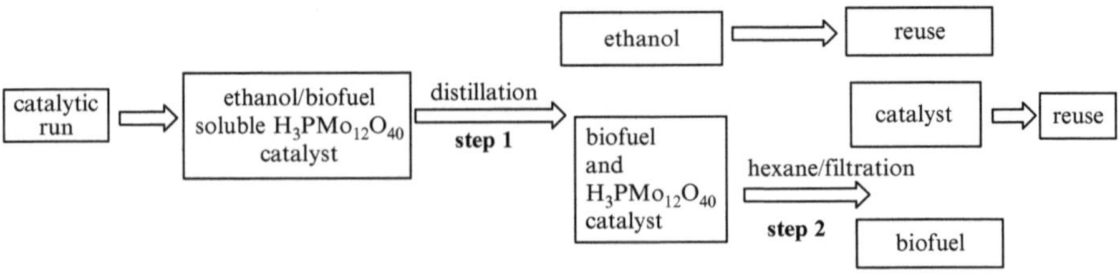

Fig. 11.2 Recovery/reuse of soluble $H_3PMo_{12}O_{40}$ catalyst in esterification reactions

Table 11.1 Conversion and selectivity of the ethyl esters obtained by the esterification of fatty acids catalysed by the $H_3PMo_{12}O_{40}$ heteropolyacid[a]

Run	Fatty acid	CN: BD[b]	Conversion (%)	Selectivity (%)
1	Lauric	12:0	96	98
2	Miristic	14:0	99	99
3	Palmitic	16:0	95	98
3	Stearic	18:0	91	98
4	Ole	18:1	99	98
5	Linoleic	18:2	92	97

[a]Reaction conditions: fatty acid (1 mmol); $H_3PMo_{12}O_{40}$ (0.014 mmol); ethanol (155 mmol); temperature (*ca.* 60 °C); 8 h. The conversion was based on both fatty acid reacted and ethyl ester formed, both determined by GC analysis via calibration curves

[b]CN:DB: carbon atoms number: double bonds

esterification (bearing a saturated carbonic chain with 12 carbon atoms) and stearic acid (bearing a saturated carbonic chain with 18 carbon atoms) high conversions were achieved.

Similarly, the presence of double bounds does not affect either conversion or selectivity of the reaction; a very close conversion was reached in ethanol esterification of saturated (stearic) or unsaturated fatty acids (oleic and linoleic acids) containing an equal carbon atoms number. A similar result was previously described by us in the investigation of the $H_3PWo_{12}O_{40}$-catalysed fatty acid esterification reactions with ethanol [5].

3.2 $H_3PMo_{12}O_{40}$-Catalysed Oleic Acid Esterification: Effect of the Alcohol Nature

The $H_3PMo_{12}O_{40}$ catalyst activity in the esterification reactions of oleic acid with different alcohols was assessed using oleic acid/catalyst molar ratio equal to 1:0.014. Kinetic curves showed in Fig. 11.3 reveal a noticeable reduction of the alcohol conversion into their respective esters when the carbon chain length of alcohol is larger.

In the first-hour reaction, a complete conversion of methanol was achieved. Conversely, the maximum conversion of ethanol was reached only after 4 h.

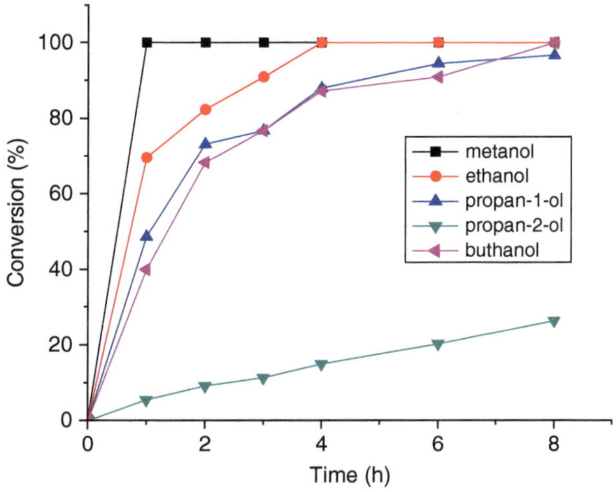

Fig. 11.3 Effects of the alcohol nature on oleic acid conversion into respective esters

Propan-1-ol and butanol have a very similar behaviour. In this case, the highest conversions were achieved after 8 h.

Contrarily, the $H_3PMo_{12}O_{40}$ catalyst faults in to promote esterification of isopropanol (propan-2-ol) with oleic acid. Only a low conversion into isopropyl oleate was reached at the end of reaction. This effect may be probably a consequence of large hindrance on secondary hydroxyl, which made the attack to oleic acid carbonyl hard, lowering the formation of esters.

3.3 $H_3PMo_{12}O_{40}$-Catalysed Camphene Oxidation Reactions by Hydrogen Peroxide

Several works have described the use of palladium or cobalt catalysts in monoterpene oxidation reactions such as camphene by dioxygen as well as hydrogen peroxide [61, 62]. Herein, the $H_3PMo_{12}O_{40}$-catalysed oxidation reactions of bicyclic monoterpenes (i.e. camphene) were carried out in CH_3CN solutions at 60 °C, using hydrogen peroxide in molar ratios (molar ratio H_2O_2/substrate = 1:1; 1:2 and 1:4). It was observed that both product selectivity and conversion rates were remarkably dependent on the $H_3PMo_{12}O_{40}$ catalyst concentration and hydrogen peroxide amount. In addition, a noticeable effect of the camphene structure in the reaction selectivity was also observed [63].

The only allylic hydrogen of camphene occupies a bridgehead position and is not easily shifted [24]. Therefore, we had no expectations in obtaining allylic oxidation products. On the other hand, the epoxy-derivatives products should be more reasonably expected [64].

Figure 11.4 displays the main oxidation products of camphene, namely, camphene-glycol (**1a**), hydroxi-camphene acid (**1b**) and aldehyde camphene,

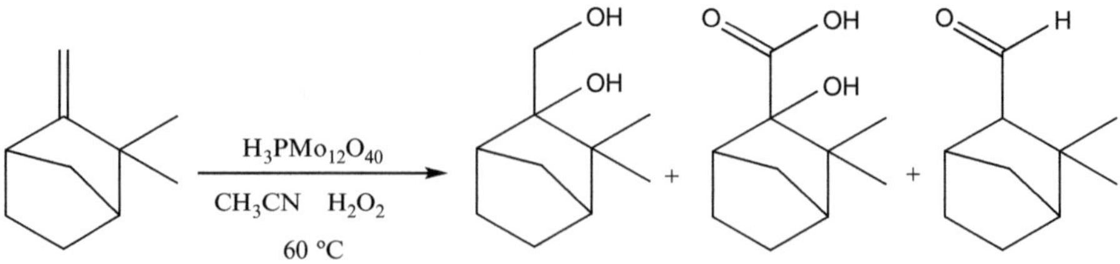

camphene **1** camphene-glycol **1a** hydroxi-camphene acid **1b** camphene aldehyde **1c**

Fig. 11.4 $H_3PMo_{12}O_{40}$-catalysed oxidation of camphene by hydrogen peroxide

Table 11.2 Conversion and product selectivity obtained in the camphene oxidation reactions by hydrogen peroxide catalysed by $H_3PMo_{12}O_{40}$ in CH_3CN solutions[a]

			Product selectivity (%)					
Ex.	H_2O_2 (mmol)	Conversion (%)	1a	1b	1c	Σoxid	Borneol	Others[b]
1	3.75	91	14	50	8	72	4	24
2	7.50	95	21	39	11	61	10	19
3	15.0	99	22	25	10	57	9	34

[a]Reaction conditions: camphene (3.75 mmol); $H_3PMo_{12}O_{40}$ (0.075 mmol); CH_3CN (15 mL); temperature (*ca.* 60 °C); 6 h. The conversion and selectivity determined by GC analysis via calibration curves
[b]Complex mixture of minority products resulting from isomerisation, skeletal rearrangement or nucleophilic addition

respectively. The formation of products **1a** and **1b** can be explained via peroxidation of exocyclic double bond of camphene, which should result in the epoxy-camphene (detected at initial period of reaction). Probably, this unstable product undergoes water addition (present in the hydrogen peroxide) which is catalysed by acidity of $H_3PMo_{12}O_{40}$. Then, the epoxy ring is broken resulting in the camphene-glycol formation (**1a**). The terminal hydroxyl of the camphene-glycol can be oxidised, resulting in the hydroxi-camphene acid (**1b**).

On the other hand, we may speculate that a possible pathway for the aldehyde camphene formation is the water molecule elimination from camphene-glycol, which may result in an unstable vinylic alcohol which undergoes tautomerisation into aldehyde camphene (**1c**).

Although undesirable acidity of $H_3PMo_{12}O_{40}$ catalyst, camphene oxidation products were obtained with high selectivity in all runs (Table 11.2). Currently, the hydroxi-camphene acid was obtained in the majority of all experiments. However, the presence of hydrogen peroxide at higher concentrations seems to favour the minority product formation that was previously identified by GC-MS analysis, as a result of isomerisation, skeletal rearrangement or nucleophilic addition reactions.

Indeed, an increase in the amount of hydrogen peroxide resulted in a decrease on camphene into oxidation products, because of the competitive reactions such as nucleophilic addition to camphene isomers as borneol (Fig. 11.5).

In Fig. 11.6, it is important to note that the conversion of camphene achieves a maximum after the 2-h reaction independently of peroxide hydrogen concentration

Fig. 11.5 Borneol structure formed from camphene skeletal rearrangement allowed by nucleophilic addition

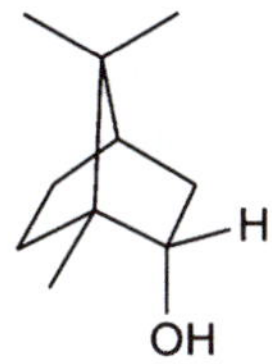

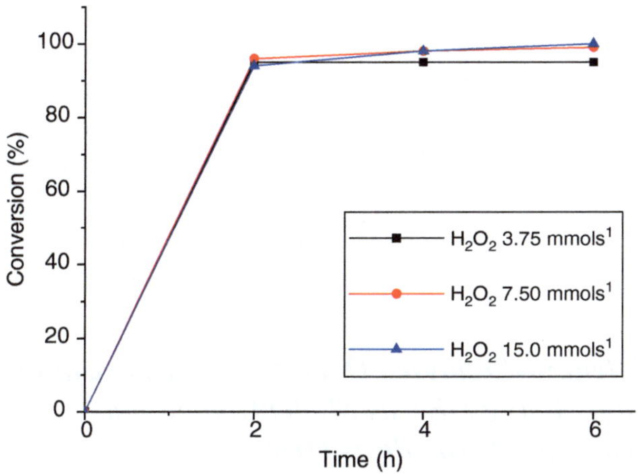

Fig. 11.6 Kinetic curves of $H_3PMo_{12}O_{40}$-catalysed camphene oxidation reaction by hydrogen peroxide

Table 11.3 Effect of $H_3PMo_{12}O_4$ catalyst concentration on conversion and product selectivity obtained in the camphene oxidation reactions by H_2O_2 in CH_3CN solutions[a]

			Product selectivity (%)					
Run	HPMo[b] (mmol)	Conversion (%)	1a	1b	1c	Σoxid	Borneol	Others[b]
1	0.075	97	26	38	9	73	7	20
2	0.060	93	24	49	7	80	7	13
3	0.045	91	23	48	3	74	6	20
4	0.030	83	21	61	1	83	6	11
5	0.015	74	12	59	0	71	5	24

[a]Reaction conditions: camphene (3.75 mmol)
[b]$H_3PMo_{12}O_{40}$ (0.075 mmol); H_2O_2 (7.5 mmol); CH_3CN (15 mL); temperature (*ca.* 60 °C); 6 h. The conversion and selectivity determined by GC analysis via calibration curves
[c]Complex mixture of minority products resulting from isomerisation, skeletal rearrangement or nucleophilic addition

employed in the reaction. The effect of catalyst concentration on selectivity and conversion of the reaction was also assessed (Table 11.3).

An increase in catalyst concentration does not affect the reaction selectivity; the hydroxi-camphene acid was a major product in all catalytic runs. However, a decrease on $H_3PMo_{12}O_{40}$ catalyst concentration resulted in an expected lowering of the camphene conversion. Nevertheless, high oxidation selectivity was reached by employing catalyst in the range concentration of 0.015–0.075 mmol (Table 11.3).

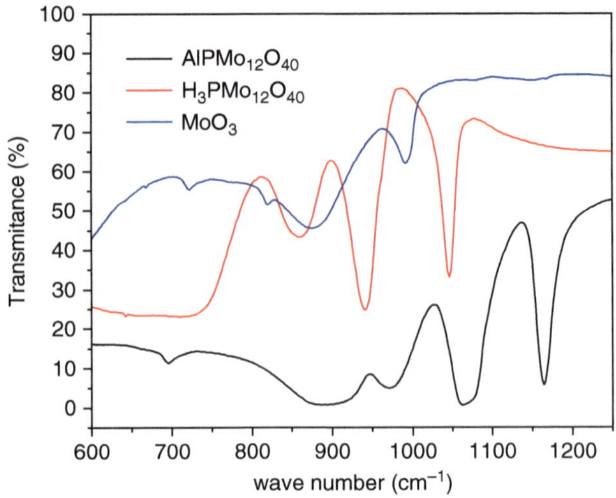

Fig. 11.7 FTIR spectra of molybdenum catalysts used in oxidative desulfurisation

3.4 Molybdenum-Catalysed Oxidative Desulfurisation Reactions Using DBT as Model Compound and Hydrogen Peroxide as Oxidant

3.4.1 FTIR Spectra Obtained in Solid Phase of the Catalysts $H_3PMo_{12}O_{40}$, $AlPMo_{12}O_{40}$ and MoO_3

The molybdenum catalysts were analysed by FTIR spectroscopy (Fig. 11.7). A comparative analysis between $AlPMo_{12}O_{40}$ FTIR spectrum, $H_3PMo_{12}O_{40}$ heteropolyacid and MoO_3 FTIR spectra allowed to confirm the presence of Keggin anion structure on salt that has been synthesised. The $PMo_{12}O_{40}{}^{3-}$ Keggin anion structure is well known and has tetrahedral PO_4 groups surrounded by four Mo_3O_{13} groups formed by edge-sharing octahedral [65]. These groups are bonded with each other by corner-sharing oxygens. In this structure, there are four types of oxygen atoms which are distinguishable via FTIR spectroscopy, being responsible for the fingerprint bands of 40 the $PMo_{12}O_{40}{}^{3-}$ Keggin ion (*ca.* 1,200–700 cm^{-1}). Figure 11.6 shows the characteristic bands for absorptions of ν (P–O) and ν (Mo–O) bonds existent on $H_3PMo_{12}O_{40}$ of all molybdenum compounds used as catalysts.

The main absorption bands observed in the FTIR spectra are summarised in Table 11.4 besides their respective attribution. The data displayed are in agreement with literature [66].

3.4.2 Molybdenum-Oxidative Desulfurisation: Catalytic Runs

The main aim of this work is to develop a promising methodology for gasoline desulfurisation using an oxidation step followed by phase transfer of oxidised sulfur without phase-transfer agent. Experiments were carried out in a three-necked glass

Table 11.4 Main absorption bands observed in the FTIR spectra of $H_3PMo_{12}O_{40}$, $AlPMo_{12}O_{40}$ and MoO_3 catalysts

Catalyst	ν(P–O)	ν(Mo–O)	ν_{as}(Mo–O$_c$–Mo)	ν_{as}(Mo–O$_e$–Mo)
$H_3PMo_{12}O_{40}$	1,056	955	880	732
$AlPMo_{12}O_{40}$	1,064	975	899	694
MoO_3	–	992	875	723

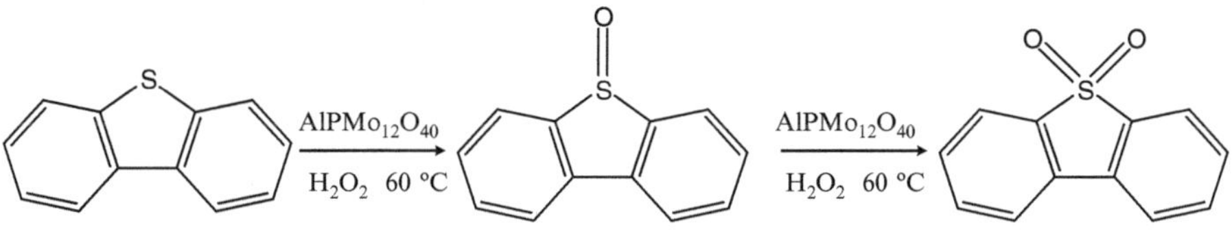

Fig. 11.8 $AlPMo_{12}O_{40}$-catalysed ODS of DBT with hydrogen peroxide

batch reactor (50 mL), fitted with a sampling septum and reflux condenser, at atmospheric pressure using isooctane as the model gasoline compound and both catalyst and oxidant in CH_3CN solution. Solid catalysts (i.e. MoO_3) or soluble catalysts (i.e. $H_3PMo_{12}O_{40}$ or $AlPMo_{12}O_{40}$) were added to biphasic mixture containing isooctane, acetonitrile and hydrogen peroxide (30 % w/w) at 60 °C temperature. The solution was stirred and heated for 2 h. Aliquots were taken off periodically and analysed via GC.

Heteropolyacids salts (i.e. $AlPW_{12}O_{40}$) were described as highly water tolerant. Lewis acid is a suitable catalyst for the condensations in which water molecules generate in the process of the reaction [67]. However, the use of homologous of this salt as catalyst in ODS reactions (i.e. $AlPMo_{12}O_{40}$) is yet novel and will be firstly described along with this work.

Here we wish to disclose a new catalytic application of aluminium dodecamolydophosphate ($AlPMo_{12}O_{40}$), as an efficient, easily available, cheap, recyclable catalyst (after solvent evaporation), for the oxidative desulfurisation reactions in CH_3CN solutions containing H_2O_2 as oxidant. With high extractability for sulfones of the acetonitrile besides their low boiling point (*ca.* 82 °C), these sulfones may be separated easily by distillation. Moreover, after this process the catalysts used may be recovered and reused.

DBT was used as model sulfur compound in isoocatane (Fig. 11.8). This novel protocol circumvented the use of phase-transfer agent, which frequently causes limitations related to mass transference rates.

We first examined the ODS reactions of model gasoline samples (isooctane) in the presence of a catalytic amount (5 mol%) of homogeneous and heterogeneous catalyst (Table 11.5). The general reaction of DBT oxidation is displayed in Fig. 11.8.

In reaction, without catalyst a very poor conversion into sulfone was reached (*ca.* <5 %). Conversely, in the presence of heteropolyacids or heteropolyacids salts,

Table 11.5 ODS processes in the presence of different catalysts using DBT as sulfur model compound in mixture biphasic isooctane/acetonitrile and hydrogen peroxide as oxidant[a]

Run	Catalyst	Conversion[b] (%)
1	–	<5 %
2	$H_3PMo_{12}O_{40}$	76
3	$AlPMo_{12}O_{40}$	100
4[c]	MoO_3	56
5	$H_3PW_{12}O_{40}$	5
6	$AlPW_{12}O_{40}$	

[a]Reaction conditions: DBT (3.19 mmol), catalyst (0.1595 mmol; 5 mol%), H_2O_2 (2. 0 mL; 30 % w/w; 17.6 mmol), 60 °C temperature, isooctane/acetonitrile (20 mL), 3 h
[b]Determined by CG analyses
[c]Solid catalyst

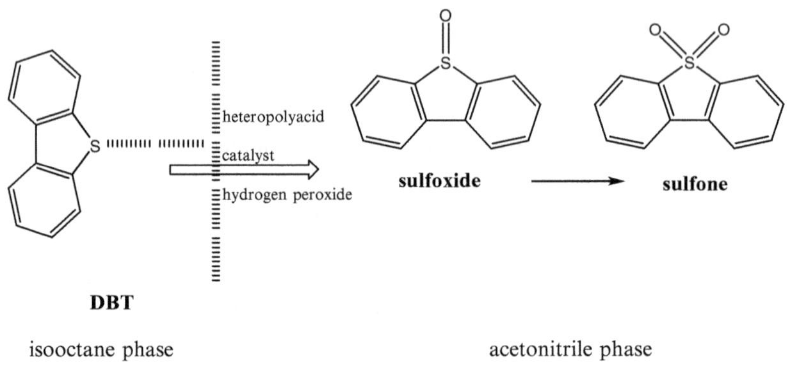

Fig. 11.9 Possible interaction for DBT in the isooctane phase with heteropoly catalyst and the oxidant in the polar phase without phase-transfer agent

DBT was oxidised into sulfoxide (detected in initial period of reaction) and into sulfone, the only product detected at reaction end.

Remarkably, the molybdenum catalysts were more active than tungsten heteropoly compounds (Table 11.5). This suggests that redox site of moldybdenum catalysts is more effective than tungsten. In special, only $AlPMo_{12}O_{40}$ catalyst achieved a total removal of DBT of isooctane phase after 2 h.

On the other hand, surprisingly, the total replacement of acid hydrogens of heteropolyacids by aluminium cations $AlPW_{12}O_{40}$ improves the activity of both catalysts significantly. The noticeable results of Table 11.5 are suggestive that the aluminium cation has a key role in this reaction.

Another notable advantage of this protocol is that it avoids the use of phase-transfer agent (Fig. 11.9).

The heteropolyacids are completely insoluble in non-polar solvents such as isooctane. Thus, we may speculate that DBT coordinates to heteropolyacids via sulfur atom when DBT molecule is present on the interfacial region; consequently, the initial reaction with hydrogen peroxide may occur in interface region between two liquids phase. Another possibility is that hydrogen peroxide and DBT are able to interact in

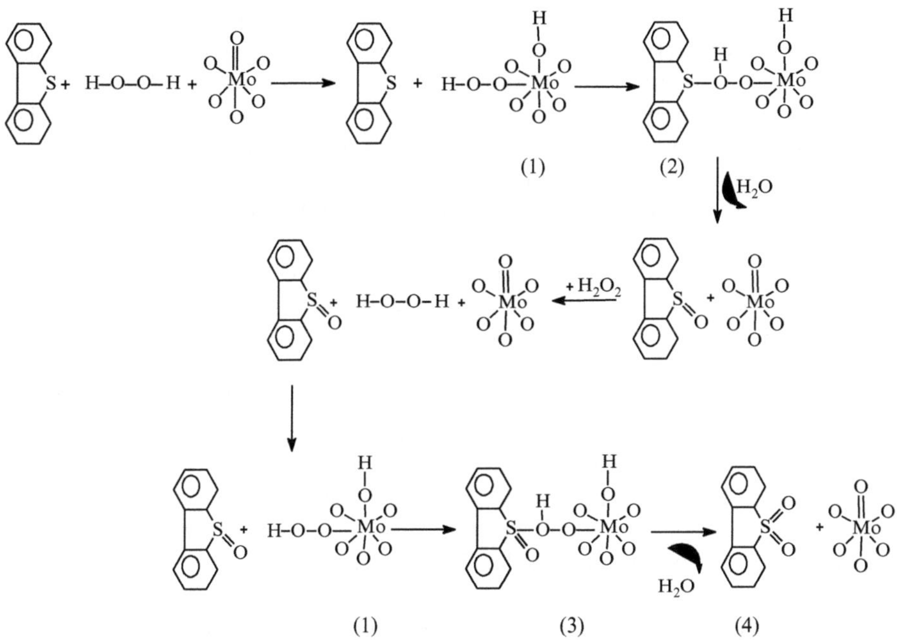

Fig. 11.10 Possible mechanism molybdenum heteropolyacid-catalysed ODS reaction without phase-transfer agent

the interfacial region of the biphasic mixture. In this region, it is possible then that molybdenum heteropolyacids coordinated to hydrogen peroxide transfer to oxygen atom to DBT (Fig. 11.10) [68].

In this free-phase-transfer agent process, mass transfer limitations commonly found in traditional processes are avoided. However, it is important to note that more deep studies about reaction mechanism yet are needed to fully understand the heteropolyacids' role in these ODS reactions.

4 Conclusion

This work summarises our recent developments for H_2O_2-based environmentally benign oxidation reactions, besides the results obtained in esterification reactions, both catalysed by $H_3PMo_{12}O_{40}$ or their salts. We have found that those catalysts are versatile bifunctional catalysts highly active in esterification reactions as well as oxidation reactions. The efficiency of molybdophosphoric acid catalyst ($H_3PMo_{12}O_{40}$) was assessed in the esterification of oleic acid with several alcohols at 60 °C temperature. High esters yield was achieved (*ca.*>90 %). Moreover, investigating the ethanol esterification reactions of the fatty acids (C_{12}–C_{18}) catalysed by $H_3PMo_{12}O_{40}$, we have found that both the increase on carbon chain size and the presence of double bonds in FA do not affect ester yields.

Molybdenum-catalysed oxidation reaction of dibenzothiophene (DBT) into dibenzosulphone (DBTO) by hydrogen peroxide was investigated in a biphasic system (i.e. isooctane/acetonitrile). A comparison between catalytic activity of molybdenum and tungsten heteropolyacids and their salts was performed. Noticeably, $AlPMo_{12}O_{40}$ was the most effective catalyst achieved 100 % removal after 2 h reaction. This highly efficient process avoided the use of the phase-transfer agent.

Finally, studying the second oxidative process, we have found that $H_3PMo_{12}O_{40}$/H_2O_2 system was also highly effective in oxidation reactions of camphene. Camphene was selectively oxidised into epoxy and allylic products with conversion rates higher than 80 %. Hydrogen peroxide, an environmentally benign reactant, was employed as an oxidant in the camphene oxidation reactions as well as ODS reactions.

Thus, it can be concluded that although yet non-finished, present methodology offers several advantages such as high yields besides high selectivity for desirable products. The development of simple procedures for recovery and reuse of homogeneous catalyst as well as synthesis process for their use in heterogeneous conditions are yet needed for that molybdenum heteropolyacids may be used in larger extension. The authors hope that with this work, a significant advancement in the field of heteropolyacids catalysts can be or has been proved.

Acknowledgements The Federal University of Viçosa and Arthur Bernardes Foundation are warmly thanked for financial support. Moreover, CAPES, CNPq and FAPEMIG deserve our special thanks.

References

1. Ferreira P, Fonseca IM, Ramos AM, Vital J, Castanheiro JE (2011) Catal Commun 12:573
2. (a) Hill CL (1998) Chem Rev 98:1–387; (b) Pope MT (2004) In: Wedd AG (ed) Comprehensive coordination chemistry II, vol 4. Elsevier, Oxford, pp 635–640
3. Timofeeva MN (2003) Appl Catal A 256:19
4. Kozhevnikov IV (2002) Catalysts for fine chemicals, catalysis by polyoxometalates, vol 2. Wiley, Chichester
5. Cardoso AL, Augusti R, da Silva MJ (2008) J Am Oil Chem Soc 85:555
6. Firouzabadi H, Iranpoor N, Jafari AA (2005) J Mol Catal A 227:97
7. Mizuno N, Kamata K, Yamaguchi K (2010) Top Catal 53:876
8. Misono M (2005) Catal Today 100:95
9. Anastas PT, Bartlett LB, Kirchhoff MM, Williamson TC (2000) Catal Today 55:11
10. Misono M (2001) Chem Commun 13:1141
11. Kim Y, Kim H, Lee J, Sim K, Han Y, Paik H (1997) Appl Catal A 155:15
12. Keigo K, Syuhei Y, Miyuki K, Kazuya Y, Noritaka M (2008) Angew Chem Int Ed 47:2407
13. Kozhevnikov IV, Matveev KI (1983) Appl Catal 5:135
14. Neumann R (2010) Inorg Chem 49:3594
15. Ishii Y, Yamawaki K, Ura T, Yamada H, Yoshida T, Ogawa M (1988) J Org Chem 53:3581
16. Misono M (2002) Top Catal 21:89
17. Murphy EF, Mallat T, Baiker A (2000) Catal Today 57:115
18. Torborga C, Beller M (2009) Adv Synth Catal 351:3027
19. (a) Gallezoti P (2007) Catal Today 121:76–91; (b) Bhatia SP, McGinty D, Letizia CS, Api AM (2008) Food Chem Toxicol 46:237

20. Gavrilov KN, Benetsky EB, Grishina TB, Tatiana B, Zheglov SV, Rastorguev EA, Petrovskii PV, Davankov VA, Macaev FZ (2007) Tetrahedron-Asymmetry 18:2557
21. Robles-Dutenhefner PA, Rocha KAS, Sousa EMB, Gusevskaya EV (2009) J Catal 265:72
22. (a) Simões MMQ, Silva AMS, Tomé AC, Cavaleiro JAS, Tagliatesta P, Crestini C (2001) J Mol Catal A 172:33; (b) Santos ICMS, Simões MMQ, Pereira MMMS, Martins RRL, Neves MGPMS, Cavaleiro JAS, Cavaleiro AMV (2003) J Mol Catal A 195:253
23. Cornell CN, Sigman MS (2007) Inorg Chem 46:1903
24. de Oliveira AA, da Silva ML, da Silva MJ (2009) Catal Lett 130:424
25. Roussel M, Mimoun H (1980) J Org Chem 45:5387
26. (a) Gomes MFT, Antunes OAC (1997) J Mol Catal A 121:145; (b) Gusevskaya EV, Gonçalves JA (1997) J Mol Catal 121:131; (c) Allal BA, El Firdoussi L, Allaoud S, Karim A, Castanet Y, Mortreux A (2003) J Mol Catal 200:177
27. (a) Babich IV, Moulijn JA (2003) Fuel 82:607; (b) Yazu K, Yamamoto Y, Furuya T, Miki K, Ukegawa K (2001) Energy Fuels 15:1535
28. Song CS, Ma XL (2004) Int J Green Energy 1:167
29. Song CS, Ma XL (2003) Appl Catal B Environ 41:207
30. (a) Lanju C, Shaohui G, Dishun Z (2007) Chin J Chem Eng 15:520; (b) De Filippis P, Scarsella M (2003) Energy Fuels 17:1452; (d) Chen LJ, Guo SH, Zhao DS (2005) J Fuel Chem Technol 33:247
31. Zhao DS, Li FT, Sun ZM, Shan HD (2009) J Fuel Chem Technol 37:194
32. Lu L, Cheng S, Gao J, Gao G, He M (2007) Energy Fuels 21:383
33. Stanger KJ, Angelici RJ (2006) Energy Fuels 20:1757
34. Campos-Martin JM, Blanco-Brieva G, Fierro JLG (2006) Angew Chem Int Ed 45:6962
35. Hirai T, Shiraishi Y, Ogawa K, Komasawa Ind I (1997) Eng Chem Res 36:530
36. (a) Prasad VVDN, Jeong KE, Chae HJ, Kim CU, Jeong SY (2008) Catal Commun 9:1966–1969; (b) Jin C, Li G, Wang X, Wang Y, Zhao L, Sun D (2008) Microporous Mesoporous Mater 111:236; (c) Gregori F, Nobili I, Bigi F, Maggi R, Predieri G, Sartori G (2008) J Mol Catal A Chem 286:124; (d) Gómez-Bernal H, Cedeño-Caero L, Gutiérrez-Alejandre A (2009) Catal Today 142:227
37. (a) Caero LC, Hernandez E, Pedraza F, Murrieta F (2005) Catal Today 107–08:564; (b) Yan X-M, Mei P, Lei J, Mi Y, Xiong L, Guo L (2009) J Mol Catal A Chem 304:52
38. (a) Kong LY, Li G, Wang XS, Wu B (2006) Energy Fuels 20:896; (b) Si X, Cheng S, Lu Y, Gao G He M-Y (2008) Catal Lett 122:321
39. (a) Te M, Fairbridge C, Ring Z (2001) Appl Catal A 219:267; (b) Fedorova EV, Zhirkov NP, Tarakanova AV, Ivanov AA, Senyavin VM, Anisimov AV, Tulyakova EV, Surin SA (2002) Pet Chem 42:253; (c) Sharipov AK, Nigmatullin VR (2005) Pet Chem 45:371
40. (a) Jin AH, Li BS, Dai ZJ (2010) Pet Sci Technol 28:700–711; (b) Yazu K, Makino M, Ukegawa K (2004) Chem Lett 33:1306
41. Hao L, Shen BX, Zhou XL (2005) Pet Sci Technol 23:991
42. (a) Campos-Martin JM, Blanco-Brieva G, Fierro JLG, Capel-Sanchez MC (2004) Green Chem 6:557; (b) Huang D, Wang YJ, Yang LM, Luo GS (2006) Ind Eng Chem Res 45:1880; (c) Huang D, Zhai Z, Lu YC, Yang LM, Luo GS (2007) Ind Eng Chem Res 46:1447; (d) Jiang X, Li H, Zhu W, He L, Shu H, Lu J (2009) Fuel 88:431
43. (a) De Filippis P, Scarsella M, Verdone N (2010) Ind Eng Chem Res 49:4594; (b) Cavani F (1998) Catal Today 41:73
44. González-García O, Cedeño-Caero L (2009) Catal Today 148:42
45. Keigo Kamata, Tomohisa Hirano, Noritaka Mizuno (2009) Chem Commun 3958–3960
46. Komintarachat C, Trakarnpruk Ind W (2006) Eng Chem Res 45:1853
47. Sachdeva TO, Pant KK (2010) Fuel Process Technol 91:1133
48. Wesseler J (2007) Energy Policy 35:1414
49. Helwani Z, Othman MR, Aziz N, Kim J, Fernando WJN (2009) Appl Catal A 363:1
50. Kawashima A, Matsubara K, Honda K (2008) Bioresour Technol 99:3439
51. Bozbas K (2008) Renew Sustain Energy Rev 12:542
52. Apostolakou AA, Kookos IK, Marazioti C, Angelopoulos KC (2009) Fuel Process Technol 90:1023

53. Haas MJ (2005) Fuel Process Technol 86:1087
54. Lou W-Y, Zong M-H, Duan Z-Q (2008) Bioresour Technol 99:8752
55. Patil PD, Deng S (2009) Fuel 88:1302
56. Dash SS, Parida KM (2007) J Mol Catal A 266:88
57. Lotero E, Liu Y, Lopez DE, Suwannakaran K, Bruce DA, Goodwin JG (2005) Ind Eng Chem Res 44
58. Kulkarni MG, Gopinath R, Meher LC, Dalai AK (2006) Green Chem 8:1056
59. Sunita G, Devassy BM, Vinu A, Sawant DP, Balasubramanian VV, Halligudi SB (2008) Catal Commun 9:696
60. Oliveira CF, Dezaneti LM, Garcia FAC, Macedo JL, Dias JA, Dias SCL, Alvim KSP (2010) Appl Catal A 372:153
61. Silva VWG, Laier LO, da Silva MJ (2010) Catal Lett 135:207
62. (a) da Silva MJ, Teixeira RR, Carari DM (2009) J Organomet Chem 694:3254; (b) da Silva MJ, Gusevskaya EV (2001) J Mol Catal 176:23; (c) da Silva MJ, Robles-Dutenhefner P, Menini L, Gusevskaya EV (2003) J Mol Catal 201:71
63. da Silva MJ, Gonçalves JA, Howarth OW, Gusevskaya EV, Pilo- Veloso D (2005) J Organomet Chem 690:2996
64. da Silva MJ, Gonçalves JA, Alves RB, Howarth OW, Gusevskaya EV (2004) J Organomet Chem 689:302
65. Brahmkhatri V, Patel A (2011) Appl Catal A 403:161
66. Komintarachat C, Trakarnpruk W (2006) Ind Eng Chem Res 45:1853
67. Firouzabadi H, Iranpoor N, Jafari AA (2005) J Mol Catal 227:97
68. Wang D, Qian EW, Amano H, Okata K, Ishihara A, Kabe T (2003) Appl Catal A 253:91

Chapter 12
Electrochemical/Electro-catalytic Applications of Polyoxometalates

B. Viswanathan

Contents

1 Introduction 246
2 Electro-catalytic Applications of POMs 248
2.1 Electro-catalytic Reduction: Hydrogen Evolution Reaction 249
2.2 Amperometric Sensors for Hydrogen Peroxide 249
2.3 Role of POMs in the Electrochemical Methanol Oxidation 249
2.4 Electrochemical Oxygen Reduction Reaction by Pt/STA–C 251
2.5 POMs as Electrochemical Super Capacitor Electrodes 251
3 Conclusion 253
References 255

Abstract Polyoxometallates, a typical metal-oxygen cluster systems have great potential for electro-catalytic applications, since the window of the redox potential of the species can be easily varied with the nature of the heteroatom and the addendum atoms. This presentation considers the application of these species for a few selected electro-catalytic reactions and demonstrates how these species can be appropriate for applications like hydrogen peroxide sensing, methanol oxidation, oxygen reduction reaction, which has great relevance for the so called energy conversion devices namely fuel cells. In addition to energy conversion devices, they can also be employed for energy storage in the form of suitable material for super capacitor electrode material.

B. Viswanathan (✉)
National Centre for Catalysis Research, Indian Institute of Technology Madras, Chennai 600 036, India
e-mail: bvnathan@iitm.ac.in

A. Patel (ed.), *Environmentally Benign Catalysts: For Clean Organic Reactions*,
DOI 10.1007/978-94-007-6710-2_12, © Panch Tattva Publishers, Pune, India 2013

1 Introduction

Polyoxometalates (POM) are a large family of discrete molecular, self-assembled nano-sized anionic metal–oxygen cluster systems composed of edge- and corner-shared MO_6 octahedra where $M=Mo^{vi}$, V^{v}, Nb^{v}, Ta^{v}s and so on [1]. They possess versatile nature in terms of structure, size, high molecular mass, rich redox chemistry, photochemistry, electron and proton transfer/storage abilities, fair thermal stability, and lability of the lattice oxygen. These systems have found applications in multidisciplinary areas such as materials [2], nanotechnology [3], medicine [4], surfaces [5], catalysis [6], colloid science [7], electronic materials [8], sensors [9] and magnetism [10]. These applications are based upon the possibility of multiple electron transfer giving rise to lacunary species.

Historically, phosphomolybdate was first reported by Berzelius in 1826 and since then, a variety of heteropoly anions have been reported. A simple classification is given in Table 12.1.

Among the available polyoxometalates, the Keggin type has received much of the attention. Salts of Keggin-type POM can be classified into two groups: group A (containing small metal ions like Na^+) and group B (containing large metal ions like Cs^+). Group A salts show characteristic features such as (1) low surface area, (2) high solubility in water and (3) capacity to absorb polar molecules in the bulk. On the other hand, group B salts possess (1) high surface area and (2) are insoluble in water, and (3) they do not absorb molecules.

The Keggin-type POM is an assembly of three kinds of structures, namely, primary, secondary and tertiary structure [11]. In Fig. 12.1, the schematic representation of the hierarchical structure of the Keggin-type POM is shown.

The important features such as multifunctionality and structural mobility make POMs as unique materials for catalytic applications. The catalytic reactions can be carried out in homogeneous as well as heterogenous systems. In the earlier years, POMs have been used both in solution and in solid state as acid and oxidation catalysts. In fact, a number of industrial processes [11] have been developed using

Table 12. 1 Classification of heteropoly anion based on the type of cluster

Type of cluster	Central group	Type of HPA	Molecular formula
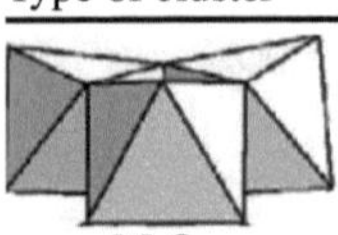	XO_4	Keggin	$X^{+n}M_{12}O_{40}^{(8-n)-}$
	XO_4	Dawson	$X_2^{+n}M_{18}O_{62}^{(16-2n)-}$
M_3O_{13}	XO_6	Waugh	$X^{n+}M_9O_{32}^{(10-n)-}$
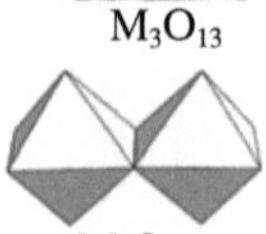 M_2O_{10}	XO_6	Anderson	$X^{+n}M_6O_{24}^{(12-n)-}$
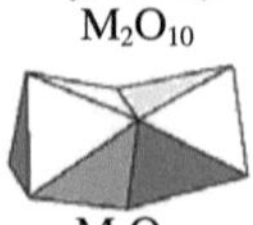 M_2O_9	XO_{12}	Silverton	$X^{+n}M_{12}O_{42}^{(12-n)-}$

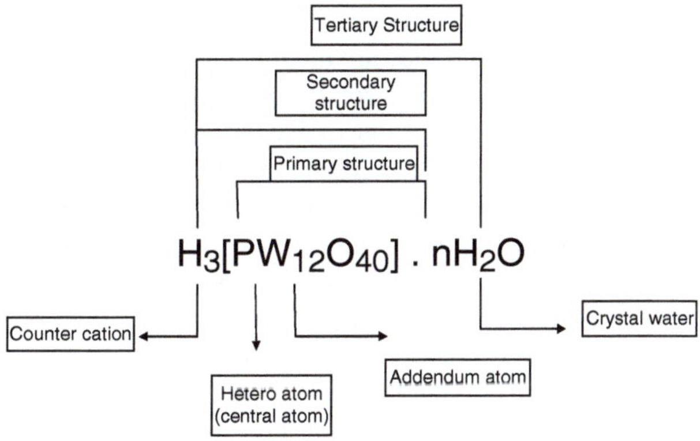

Fig. 12.1 Primary, secondary and tertiary structure of Keggin-type POM

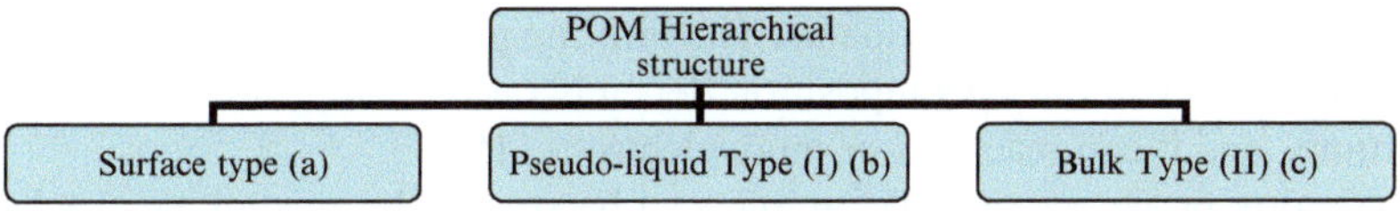

Fig. 12.2 Three types of catalysis for solid POMs: (**a**) surface type, (**b**) pseudo-liquid bulk type (I) and (**c**) bulk type (II)

POMs as catalysts. POMs have been used for a variety of reactions, for example, the formation of carboxylic acids from the corresponding carbonyl compounds as well as for the dehydrogenation of alcohols, aldehydes and carboxylic acids to form C=C and C=O bonds [12]. A number of vapour phase oxidation reactions like oxidation of methacrolein and methane have also been developed as industrial process [13].

Based on POM's hierarchical structure, it can be demonstrated that there are three completely different modes of catalysis for solid POMs. These are shown pictorially in Fig. 12.2.

POMs are promising solid acids and can replace environmentally harmful liquid acid catalysts such as sulphuric acid [14]. Liquid acids are corrosive and produce a large amount of waste; replacement of the liquid catalysts with 'water-tolerant' solid acid catalysts is desirable for developing environmentally benign processes. However, it should be remembered that a high acidity may lead to undesired side reactions and also can lead to quick deactivation due to the formation of heavy by-products. In addition to the various advantages in using solid acids, the catalytic reactions can be carried out under milder conditions. In addition, the molar catalytic activity of POMs is 10^2–10^3 times higher than that of mineral acids.

In principle, mechanisms of homogeneous catalysis by POMs and by ordinary mineral acids are of the same origin. However, since protons are in an encapsulated environment in POMs, the proton transfer reactions are amenable for control.

POMs containing electrochemically inert heteroatoms exhibit redox properties attractive for electro-catalysis. The addenda atoms are present in their highest oxidation state in d° or d¹ electronic configuration. The electrons are accepted by the addenda ions, and if they are identical, the electrons are delocalised on the structural framework by rapid electron hopping or through bridging oxygen ions. The possibility of a given metal ion to accept an electron depends on its nature. There are many examples in the literature of POM clusters with a high number of metallic electrons, and hence, even termed as reservoir of electrons [1], POMs can thus undergo electrochemical reduction with retention of the geometry.

Electrochemical investigations of Keggin-type POMs in aqueous or nonaqueous solutions have revealed sequence of reversible one- or two-electron reduction steps [12]. On reduction, the POMs are converted into heteropoly-blues – mixed valence systems which are intense blue coloured as a result of intervalence electronic charge transitions. The feature of heteropoly-blue formation is a rapid and reversible reduction process. Electronic spectra of reduced POMs show intensified d–d bands in the visible region and intervalence charge transfer (IVCT) bands in the near IR region.

POMs are many electron oxidants. The limiting number of electrons accepted by POM molecule without decomposition depends on its composition and reduction conditions, and it can reach six or more electrons [15].

The redox processes of POM anions can be studied by several electrochemical methods. By a continuous change of their composition, their redox potentials can be chosen to span a wide range. In solution, the reduction potentials of POMs containing Mo and V are high as these ions are easily reduced. Oxidative ability decreases generally in the order V → Mo → W-containing POMs. The nature of the heteroatom affects the overall charge of the POM.

2 Electro-catalytic Applications of POMs

The features of POMs such as high stability of most of their redox states, tunable redox potentials without affecting their structure and possibility of multiple electron transfer make POMs attractive as redox catalysts (mediators) for indirect electrochemical processes. In indirect electrochemical reactions, a mediator is activated by a heterogeneous redox step at the electrode surface in order to react homogeneously with the substrate in the bulk solution regenerating the inactivated mediator. Heterogeneous electron transfer between the electrode and the substrate is sometimes slow because of poor interaction. In these cases the electrode reaction occurs only at high over-potentials. Electro-catalysts can minimise the activation energy and hence allow such an electrode reaction to occur at high current density close to the equilibrium potentials or even considerably below it. Furthermore, suitably designed electro-catalysts can improve not only the reactivity but also the product selectivity. Passivation or film formation of/on the electrode surface can also easily be avoided. Thus electro-catalytic methods are important for the development of both preparative electrolysis and electrochemical sensors. A large number of organic and inorganic compounds and metal complexes including biomolecules have been successfully used as electro-catalysts. However, the number of highly selective and long time stable redox catalysts is still limited. In this sense, POMs and their transition metal substituted derivatives have the potential to fill this gap. POMs tend to undergo spontaneous adsorption from aqueous solutions on various electrode substrates like Hg, Au, Ag, glassy carbon and highly oriented pyrolytic graphite (HOPG). Toth and Anson [16] evaluated the electrochemical reduction of nitric oxide and nitrite to ammonia with $(Fe^{III}H_2O)XW_{11}O_{39}^{n-}$ where X = Si or Ge (n = 5)P, As (n = 4) and found the catalysts to have long-term durability. Rong and Pope [17] reported the electro-catalytic oxidation of sulfoxide to sulfone by $Ru^{V}(O)PW_{11}O_{39}^{4-}$ with greater than 90 % current efficiency. Rong and Anson [18] found that benzyl or ethyl alcohol was slowly oxidised to aldehyde by $(CrO)PW_{11}O_{39}^{4-}$ generated electrochemically. Steekhan and Kandzia [19] reported the electrochemical cleavage of styrene derivatives to benzaldehydes catalysed by $RuSiW_{11}O_{39}^{5-}$ in a multiphase system.

2.1 Electro-catalytic Reduction: Hydrogen Evolution Reaction

POMs have been used as reductive electro-catalysts for (1) homogeneously dissolved in electrolyte solution and (2) attached to the electrode surface. Keita and Nadjo [20] have reported that $SiW_{12}O_{40}^{4-}$ can catalyse the hydrogen evolution reaction and oxygen reduction in acidic aqueous and organic solutions. The reactive species for the hydrogen evolution reaction are two or four electron reduced species and the four electron reduced species being more active. For oxygen reduction, the first one-electron species is active.

2.2 Amperometric Sensors for Hydrogen Peroxide

POMs have also been used as electro-catalysts for hydrogen peroxide reduction and determination. H_2O_2 sensors are of biological importance. $H_3PW_{12}O_{40}$ is immobilised on the Pt electrode by sol–gel method which is formed using tetraethoxysilane, polyethylene glycol, benzyl ethyl ether and $H_3PW_{12}O_{40}$. The electrochemical behaviour of the modified electrode is characterised using cyclic voltammetry and found that they show three two-electron redox couples and the electrode is stable for several runs.

2.3 Role of POMs in the Electrochemical Methanol Oxidation

In the last few years, the research on POMs towards electro-catalytic applications has been a topic of widespread technological importance. Nakajima and Honma [21, 22] have employed POM-coordinated Pt catalyst for methanol oxidation.

Table 12.2 Comparison of electro-catalytic activities of various electrodes towards methanol oxidation [28]

Catalyst	EAS ($m^2 \cdot g^{-1}$)	Onset potential (V)	I_f/I_b	Mass activity ($mA \cdot g^{-1}$ Pt)	Specific activity ($mA \cdot cm^{-2}$ Pt)
Pt/C	23.3	0.41	0.80	191	0.81
Pt/STA–C	33.0	0.31	1.11	37	1.12
Pt–Ru/C	17.2	0.30	0.91	204	1.18
Pt–Ru/STA–C	25.6	0.24	1.05	53	1.96
Pt–Ru/C (J.M.)	22.5	0.25	0.95	271	1.20

Table 12.3 Specific capacitances of various electrodes

Electrode material	Specific capacitance (Fg^{-1})
0 % RuO_2 (only Vulcan $XC_{72}R$)	23
20 % RuO_2/C	200
10 % RuO_2/STA–C	325
20 % RuO_2/STA–C	453
40 % RuO_2/STA–C	557

Chojak et al. [23] have utilised layer-by-layer method to link Ru-stabilised Pt nanoparticles with POMs to form electro-catalytic network films. It was found to be efficient towards methanol oxidation due to the presence of tungstate units from $H_3PW_{12}O_{40}$ which can provide additional –OH groups capable of facilitating oxidation of intermediate (CO_{ads}) on Pt. Pan et al. [24] have utilised the chemisorptive ability of POMs to modify the carbon nanotubes (CNTs) with POMs to form POM–CNT composites. Pt and Pt–Ru nanoparticles were electrochemically deposited on POM–CNT composites and employed as electro-catalysts for methanol oxidation. High specific activity was observed for the catalysts containing POMs when compared to the POM-free electro-catalysts. Keita et al. [25] have utilised Mo^V–Mo^{VI} mixed polyoxometalates for the preparation of POM-modified Pt and Pd nanoparticles and shown them as efficient electro-catalysts for methanol and ethanol oxidation. In addition, Pt–POM composites were found to be resistant to poisoning during methanol oxidation. Seo et al. [26] have reported 50 % higher mass activity and improved stability for POM modified Pt/CNT catalyst than the POM-free catalyst. Ferrel et al. [27] have modified Pt/C with POMs and found a higher performance towards methanol oxidation. This was attributed to the reduced resistance towards charge transfer in Pt/C when modified with highly conducting POM. Due to their electrochemical redox properties and oxidising abilities, the POMs can be potential oxidising agents. They have proven to be efficient catalysts towards various kinds of oxidation reactions. The data given in Table 12.2 compares the activity of Pt/C with other electrodes modified by silicotungstic acid (STA) [28].

The points that emerge from the data presented in Table 12.3 are:

1. The onset potential for methanol oxidation on Pt/STA is 100 mV less positive than Pt/C electrode and 60 mV less than Pt–Ru/C electrode showing the better electro-catalytic activity of STA-containing electrodes.

2. The mass activity follows the order Pt–Ru/STA–C > Pt/STA–C > Pt–Ru/C (J.M.) > Pt–Ru/C > Pt/C.
3. The values of I_f/I_b show that the STA-containing systems show effective removal of the poisoning species on these catalysts.

2.4 Electrochemical Oxygen Reduction Reaction by Pt/STA–C

Kulesza et al. [29] utilised chemisorptive ability of POMs on metal surfaces to form anionic monalayers of POMs on Pt surfaces and immobilised within ultrathin polyaniline film using layer-by-layer method. The formed network films containing POMs were employed as promising electro-catalysts for oxygen reduction reaction. Giordano et al. [30] employed POMs as surface promoters for oxygen reduction. Lu et al. [31] have mixed $H_3PW_{12}O_{40}$ solution with Pt/C, and the resulting slurry was coated on glassy carbon electrode and employed as electrode for oxygen reduction reaction in acid solution. The activity was fond to be enhances by about 38 % when compared to Pt/C electrode. Karnicka et al. [32] have fabricated multilayer network films of conducting polymer, POMs and Pt metal particles on electrodes to combine the attractive physicochemical properties and reactivity of these materials. Wlodarczyk et al. [33] have functionalised Pt nanoparticles coated on glassy carbon with ultrathin films of POM by dip coating method. The resulting electrode was utilised for electro-reduction of dioxygen in acid solution. The activity was found to be enhanced when compared to bare Pt nanoparticles on glassy carbon. The concept of functionalising Pt nanoparticles with POM was extended to functionalise Nafion-stabilised carbon-supported Pt nanoparticles. The ORR activity of these electrodes was found to be better than conventional electrodes. Kurys et al. [34] have achieved enhanced ORR activity on PAni–$H_3PW_{12}O_{40}$–V_2O_5 composite electrode. In Fig. 12.3, the comparison of the activity of Pt/STA–C and Pt/C electrodes under oxygen saturated 0.5 M H_2SO_4 is shown [28]. It is seen that the half-wave potential for oxygen reduction on Pt/STA–C is relatively more positive compared to that in Pt/C. The specific activity on Pt/STA–C (7.6 mA cm^{-2}) is greater than that on Pt/C (4.25 mA cm^{-2}).

2.5 POMs as Electrochemical Super Capacitor Electrodes

The solubility of POMs in typical solvents has caused them to be ignored as active compounds for solid-state applications. Integration of POMs into a conducting polymer matrix to form a hybrid material is an effective way to harness the electrochemical activity. By anchoring them into conducting polymer, the reversible redox chemistry of the POM clusters could be combined with that of the conducting polymer. POMs anchored to conducting polymer networks were examined as electrode materials for lithium rechargeable batteries and super capacitors [35]. An inexpensive electrochemical capacitor system with an asymmetric configuration was prepared using POMs and a proton exchange membrane. The prepared

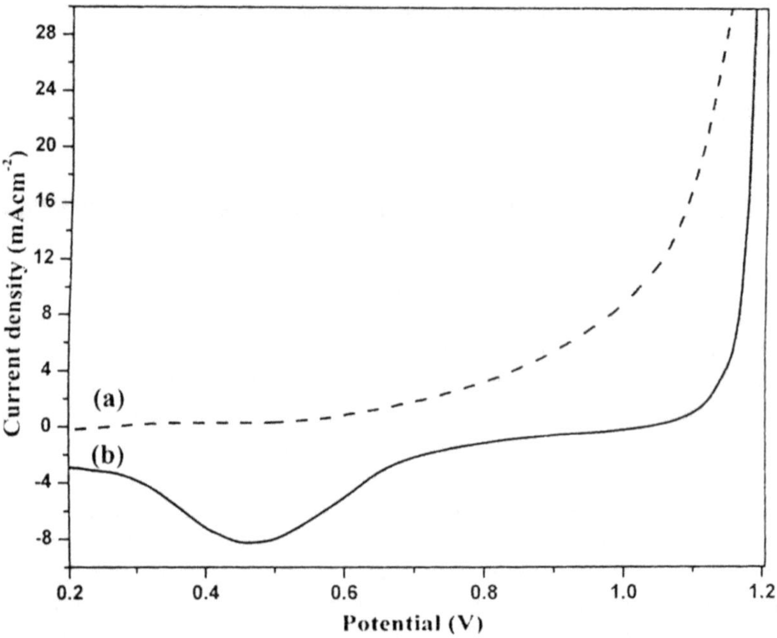

Fig. 12.3 Overlay of linear sweep voltammograms of (*a*) Pt/C and (*b*) Pt/STA–C in O_2 saturated in 0.5 M sulphuric acid at a scan rate of 5 mV s^{-1} [32]

electrode showed specific capacitance of 112 F g^{-1} [36]. Polypyrrole doped with 10-molydo-2-vanadophosphoric acid, chemically fabricated via vapour transport of monomer, showed a specific capacitance of 33.4 F g^{-1} [37]. Chemically synthesised polyaniline-doped phosphomolybdic acid materials in electrochemical capacitor cells with acidic electrolyte membranes showed remarkable improvement in their cyclability [38]. Keggin tpe were effectively doped (3,4-ethylenedioxythiophene) (PEDOT) to form organic/inorganic hybrid films by electro-deposition. Other systems generated by layer-by-layer growth technique of POM on PEDOT and hooked on to carbon or indium-tin oxide conductive glass electrodes also showed capacitance of 0.6 mF cm^{-2} [39]. The hybrid material polyaniline-doped phosphomolybdic acid prepared with the aid of H_2O_2 showed specific capacitance value of 168 F g^{-1}, and PEDOT-doped phosphomolybdic acid prepared by using a similar strategy showed specific capacitance of 130 F g^{-1} [40]. However, the polymer-containing POM electrodes are not suitable for long-term application since the polymers undergo degradation. Hence, the idea of anchoring POM on a carbon network would be a desirable strategy to enhance the stability for cycling when used as electrodes for super capacitor applications.

Typical charge–discharge behaviour obtained for RuO_2 present in a variety of composites is shown in Fig. 12.4 in the potential range of 0.0–0.7 V in 1.0 M H_2SO_4 at a current density of mA cm^{-2}. The absolute slope value calculated from the discharge curves of the composites decreased with increasing RuO_2 weight percentage

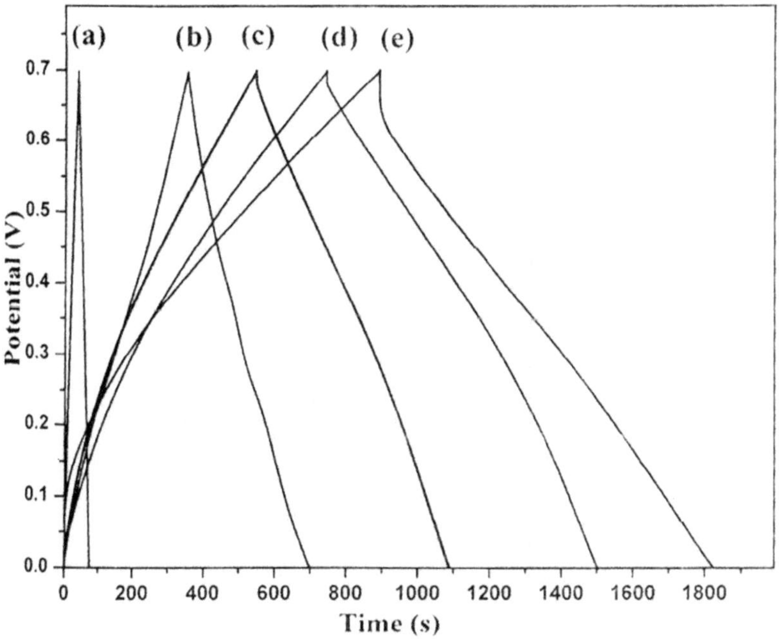

Fig. 12.4 Galvanostatic charge–discharge studies measured in 1.0 M H_2SO_4 solution between 0.0 and 0.7 V at 3 mA cm^{-2} for (*a*) Vulcan XC 72R, (*b*) 20 % RuO_2/C, (*c*) 10 % RuO_2/STA–C, (*d*) 20 % RuO_2/STA–C and (*e*) 40 % RuO_2/STA–C [28]

in RuO_2/STA–C. The specific capacitance of each composite was calculated using the expression

Specific capacitance C (F g^{-1}) = '*I*' (d*t*/*m*d*V*) where *I* is the current used for charge discharge, d*t* is the time elapsed for the discharge cycle, *m* is the mass of the active electrode and d*V* is the voltage interval of the discharge. The calculated specific capacitance values of various electrodes are given in Table 12.3 .

The data given in table show that the POM-containing RuO_2 systems are better than simple RuO_2/C systems, and these systems showed long-term stability or life cycle.

3 Conclusion

It is known that biological electron transfer processes are facile due to the possibility of modulating and tuning of the intermediate environment. This results in appropriate redox potential conditions to effect the electron transfer in a facile downhill pathway. Any synthetic system which can mimic the biological systems should be capable of being tuned in terms of redox potentials with minimum alteration of the nature and environment of the species. This appears to be possible in POMs.

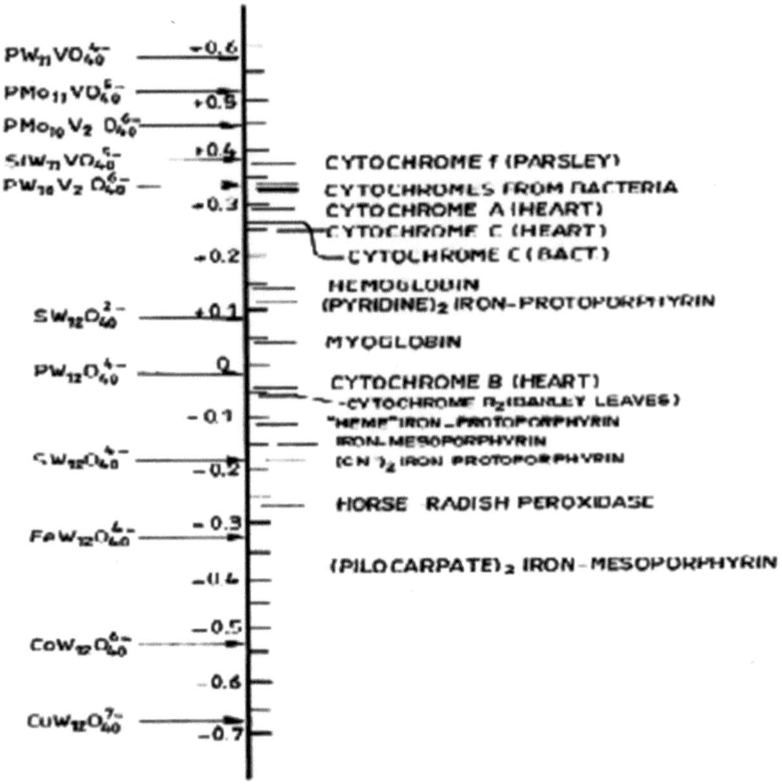

Fig. 12.5 Redox potentials (in Volts) of Fe^{2+}/Fe^{3+} in different biological systems (*right side* of the scale) and heteropoly anions (*left side* of the scale) with varying composition (Values were collected from literature [41])

In Fig. 12.5, this has been amplified in terms of the redox potential similarities. These excellent possibilities were considered in an earlier publication [41] by us. It appears that the electrochemical applications of POM are on a take-off stage and many exciting applications will be revealed in the near future.

It appears that complex biological electrochemical reactions can be mimicked in the laboratory by heteropoly anions by appropriate modulation of the electron energy levels of the cluster species brought forth by substitution or by alteration of the environment. This may provide a pathway to study a variety of chemical transformations of biological relevance. It is also possible that these systems lead to some devices which can mimic a few of the natural process. The level of this comparison and the extent to which these systems can run parallel to natural processes are issues that await imagination and execution.

Acknowledgements The author wishes to record his grateful thanks to the Department of Science and Technology, Government of India, for supporting the National Centre for Catalysis Research.

References

1. Pope MT (1983) Heteropoly and isopoly oxometalates. Springer, Berlin
2. Khan M (2000) J Solid State Chem 152:105
3. Long D, Cronin L (2006) Chem Eur J 12:3698
4. Na H, Peng J, Han ZG, Yu X, Dong BX (2005) J Solid State Chem 178:3735
5. Errington RJ, Petkar SS, Horrocks BR, Houlton A, Lie LH, Patole SN (2005) Angew Chem Int Ed 44:1254
6. Vasylyev MV, Neumann R (2004) J Am Chem Soc 126:884
7. Liu T, Diemann E, Li HL, Dress AWM, Muller A (2003) Nature 426:59
8. Chaidogiannos G, Velessiotis D, Argitis P, Koutsolelos P, Diakoumakos CD, Tsamakis D, Glezos N (2004) Microelectron Eng 73–74:746
9. Liu S, Volkmer D, Kurth DG (2004) Anal Chem 76:4579
10. Luban M, Borsa F, Bud'Ko S, Canfield PC, Jun D, Jung JK, Kogerler P, Mentrup D, Nuller A, Modler R, Procissi D, Suh BJ, Torikachvili M (2002) Phys Rev B Condens Matter 66:1
11. Okuhara T, Mizuno N, Misono M (1996) Adv Catal 41:113
12. Pope M, Muller A (1991) Angew Chem Int Ed Engl 30:34
13. (a) Misono M (1987) Catal Rev Sci Eng 29:269; (b) Mater Chem Phys 17:103
14. Kozhevnikov IV (1998) Chem Rev 98:171
15. Pope M, Popaconstantinou E (1967) Inorg Chem 6:1147
16. Toth JE, Anson FC (1989) J Am Chem Soc 111:2444
17. Rong C, Pope MT (1992) J Am Chem Soc 114:2932
18. Rong CY, Anson FC (1996) Inorg Chim Acta 242:11
19. Steckhan E, Kandzia C (1992) Synlett 139
20. Keita B, Nadjo L (1987) J Electroanal Chem 217, 287; 227, 77
21. Sadakane M, Steckhan E (1998) Chem Rev 98:219
22. Nakajima H, Honma I (2004) Electrochem Solid State Lett 7:A135
23. Chojak M, Mascetti M, Wlodarczyk R, Marassi R, Karnicka K, Miecznikowski K, Kulesza PJ (2004) J Solid State Electrochem 8:854
24. Pan D, Chen J, Tao W, Nie L, Yao S (2006) Langmuir 22:5872
25. Keita B, Zhang G, Dolbecq A, Mialane P, Secheresse F, Miserque F, Nadjo L (2007) J Phys Chem C111:8145
26. Seo M, Choi SM, Kim HJ, Cho BK, Kim WB (2008) J Power Sources 179:81
27. FerrellIII J, Kuo MC, Turner JA, Herring AM (2008) Electrochim Acta 53:4927
28. Kishore S (2008) PhD thesis, Indian Institute of Technology, Madras
29. Kulesza P, Chojak M, Karnicka K, Miecznikowski K, Palys B, Lewer A (2004) Chem Mater 16:4128
30. Giordano N, Hocevar S, Staiti P, Arico AS (1996) Electrochim Acta 41:397
31. Lu Y, Lu T, Liu C, Xing W (2005) J New Mater Electrochem Syst 8:251
32. Karnicka K, Chojaka M, Miecznikowskia K, Skunika M, Baranowskaa B, Kolarya A, Piranskaa A, Palysa B, Adamczykb L, Kulesza PJ (2005) Bioelectrochemistry 66:79
33. Wlodarczyk R, Chojak M, Miecznikowski K, Kolary A, Kulesza PJ, Marassi R (2006) J Power Sources 159:802
34. Kurys Y, Netyaga NS, Koshechko VG, Pokhodenko VD (2007) Theor Exp Chem 43:334
35. Gallegos A, Cantu M, Pastor NC, Romero PG (2005) Adv Funct Mater 15:1125
36. Yamada A, Goodenough JB (1998) J Electrochem Soc 145:737
37. White A, Slade RCT (2003) Synth Met 139:123
38. Romeo PG, Chojak M, Cuentasgallegos K, Asenso JA, Kulesza PJ, Casan-pastor N, Lira-Cantu M (2003) Electrochem Commun 5:149
39. Skunik M, Baranowska B, Fattakhova D, Miecznikowski K, Chojak M, Kuhn A, Kulesza PJ (2006) J Solid State Electrochem 10:168
40. Vaillant J, Lira-Cantu M, Cuentas-Gallegos K, Casan-Pastor N, Gomez Romero P (2006) Prog Solid State Chem 34:147
41. Rajeswari J, Viswanathan B, Varadarajan TK (2005) Bull Catal Soc India 4:109

Index

A

Albonetti, S., 26, 32
Ammonium salt of 12-molybdophosphoric acid (AMPA)
  ammoxidation, phosphate precursor
    BET surface area and acidic strength values, 35
    calcination temperature, 33–34
    ^{31}P NMR spectra, 32–33
  Keggin unit of, 17
  vs. MPA
    activity and selectivity variations, 31–32
    FTIR spectra, 26–27
    ^{31}P NMR spectra, 27–29
    product distribution of, 31
    TPD profiles, 29–30
    XRD patterns, 26, 27
  Nb_2O_5, SiO_2, ZrO_2, and Al_2O_3 supports, 35–36
  preparation of, 19
  in situ synthesis of, 17–18, 20–21
    AMPA/$FePO_4$ catalysts, 42
    AMPA/$NbOPO_4$ catalysts, 40–42
    AMPA/$VOPO_4$ catalysts, 42–45
  TiO_2-supported AMPA catalysts
    BET surface area and acidic strength values, 36
    half-bandwidth and ammoxidation activity, 40
    ^{31}P NMR spectra, 36, 38
    TPDA spectra, 39
    XRD and FTIR patterns, 36, 37
  V, Sb, and Bi modified AMPA-based systems
    calcination temperature, 46
    MP conversion, 45–46
    thermal stability, 47
Anderson, M.W., 117
Anson, F.C., 249

B

Benzyl alcohol oxidation
  benzaldehyde formation, 210
  PMo_{11}/ZrO_2
    amount of catalyst, 218, 219
    control experiments, 220
    effect of loading, 217, 218
    effect of temperature, 217, 218
    heterogeneity test, 221
    mole ratio, 217
    reaction time effect, 219–220
    recycled catalyst, 221–222
Berzelius, J.J., 2, 16
Biodiesel
  vs. conventional diesel, 190
  definition, 190
  feedstocks, 191
  vs. petroleum-based diesel, 190
  production of, 190
  transesterification and esterification
    enzymatic catalysts, 192
    heterogeneous catalysts, 192
    homogeneous catalysts, 191
    12-tungstosilicic acid (*see* 12-Tungstosilicic acid (TSA))
  triglycerides transesterification, *Jatropha* oil
    catalyst amount, 203, 204
    reaction temperature, 204, 205
    reaction time, 203, 204
    weight ratio, 203
Bondareva, V.M., 26, 32
Bordoloi, A., 105

DOI 10.1007/978-94-007-6710-2,

Bradley, 3
Brahmkhatri, V., 189
Bruckman, K., 48
Brunauer, S., 79

C
Caiado, .M., 153
Cal-Ad method, 170
Camphene oxidation, $H_3PMo_{12}O_{40}$ catalyst
  allylic hydrogen, 235
  hydrogen peroxide
    Borneol structure, 236, 237
    catalyst concentration, 237
    conversion and product selectivity, 236
    kinetic curves, 236–237
    oxidation products, 235–236
Cardoso, L.A.M., 66
Castanheiro, J.E., 153
Catalysis, 1–2
Cavani, F., 46
Chojak, M., 250
Cs salt of Pd substituted lacunary phosphotungstate-supported MCM-41
  characterization
    ^{31}P MAS NMR spectrum, 85–86
    Raman spectroscopy, 86–87
    XPS spectrum studies, 87
    X-ray powder diffraction studies, 84–85
  inorganic supports, 83–84
  p-aminophenol hydrogenation, 84
    catalytic activity, 87–88
    mechanism, 88
  synthesis of, 84

D
Damyanova, S., 26
da Silva, M.J., 225
da Silva Xavier, R., 225
de Macedo, J.L., 165
de Souza, R.L., 141
Devassy, B.M., 105
Dias, J.A., 165
Dias, S.C.L., 165
Differential thermal analysis (DTA)
  monolacunary phosphomolybdate, 213, 214
  supported $H_3PW_{12}O_{40}$, 144–146
dos Santos, L.F., 225

E
Essayem, N., 141

F
Fe-modified lacunary phosphotungstate-supported MCM-41
  bromination of phenol, 82–83
  characterization
    FTIR studies, 81–82
    surface area and pore size distribution, 79–80
    X-ray powder diffraction studies, 81
  Cs salt, 79
  Fe, Ni-substituted Keggin-type heteropoly anion, 78
  sodium salt, 79
  trans-stilbene oxidation, 78, 83
Ferrel, J., 250
Fourier transform infrared spectroscopy (FTIR) spectra
  Fe-modified lacunary phosphotungstate-supported MCM-41, 81–82
  silica-occluded tungstophosphoric acid, 155–156
  supported H_3PW, 169
  titania-supported TPAV1 catalyst, 96, 97
  12-tungstosilicic acid (TSA), 194
  zirconia-supported heteropoly acids, 111–112
Fournier, M., 26

G
Garte, J.H., 25
Giordano, N., 251
Gomez Sainero, L.M., 49
Gonzales, W.A., 141

H
Halligudi, S.B., 105
Hari Babu, B., 11
Hensen, K., 158
Heteropolyacids (HPAs)
  acidic properties of, 62–63
  acidity, 59
  advantages, 2
  alkoxylation of terpenes (*see* Silica-occluded tungstophosphoric acid (PW/S))
  ammonium salts of, 16, 17

as ammoxidation catalysts (*see* Ammonium salt of 12-molybdophosphoric acid (AMPA))
ammoxidation of MP, 22–23
carbon-entrapped $H_3PW_{12}O_{40}$ catalyst, 167
catalysts preparation
AMPA, 19–21
supported vanadium incorporated AMPA, 20
vanadium incorporated MPA, 19–20
characterization of, 21
chemicals and supports, 19
classification, 13–14
vs. conventional liquid acid catalysts, 226
design of, 5
dioxygen, 227
disadvantages, 4–5, 154
as heterogeneous catalysis
advantages, 64–65
supported HPA, 65–66
heteropolyanions
Cs-salt-type, 5
discovery of, 2–3
Keggin-type, 3–4
polyhedral representation of, 4
reoxidation of, 4
types of, 3
as homogeneous catalysis
acid strength and oxidation potential, 64
mineral acids, 63
neutralization, 64
thermal and hydrolytic stability, 64
HPA-supported catalyst dispersion, 37–39
hydrogen peroxide, 227
Keggin-type HPAs
Brønsted acidity of, 226
redox properties, 226
modification of
intrinsic acidity and oxidizing property, 15
transition metal addition, 15
vanadium incorporated systems, 16
palladium(II)-copper(II)-heteropolyacid combinations, 227
properties, 59
redox properties of, 63
structure, 13–14
Anderson structure, 61
Dawson structure, 61
Keggin structure, 60
reorganization of, 2
Silverton structure, 61
Waugh structure, 62
support and supporting method, 5–7
supported
advantages of, 7–8
catalytic activity of, 65
salts of, 16–17
silica, 65–66
synthesis, 59–60
thermal stability of, 106–107
TiO_2-supported MPA, VMPA, and VOMPA, 24–25
as useful catalysts, 14–15
VMPA and VOMPA catalysts, 23–24
Holderich, W.F., 58
Honma, I., 249
$H_3PMo_{12}O_{40}$ heteropolyacid
fatty acids esterification
biodiesel production, 230–231
catalyst recovery, 233, 234
with ethanol, 231
ethyl ester conversion and selectivity, 233–234
with oleic acid, 231, 234–235
products identification, 232
monoterpenes oxidation
camphene (*see* Camphene oxidation, $H_3PMo_{12}O_{40}$ catalyst)
catalytic runs, 232
dioxygen oxidant, 228
hydrogen peroxide, 227–228
non-catalytic oxidation reactions, 227
Wacker catalyst, 228
oxidative desulfurisation (ODS) process
$AlPMo_{12}O_{40}$ catalyst synthesis, 232–233
catalytic runs, 233
dibenzothiophene, 228–229
heterogeneous recyclable catalyst, 229
hydrodesulfurisation (HDS), 228
hydroperoxide, 229
molybdenum-catalysed (*see* Molybdenum-catalysed oxidative desulfurisation reactions)
molybdenum catalysts, 229
polyoxometalates-catalysed ODS, 229
transition metal catalysts, 229

I

Illingworth, 3
Ito, T., 17

K
Kamala, P., 66
Kandzia, C., 249
Kanno, M., 120
Karnicka, K., 251
Kasztelan, S., 65
Kaur, J., 66
Keita, B., 249, 250
Kholdeeva, O.A., 107
Kim, H., 119
Kozhevnikov, I.V., 16, 66
Kulesza, P.J., 251
Kumar, A., 78
Kurys, Y., 251
Kuznetsova, N.I., 210

L
Lacunary polyoxometalates (LPOMs), 8, 210–211
Lapham, D., 17
Lingaiah, N., 11, 91
Loridant, S., 113
Lu, Y., 251

M
Marchal-Roch, C., 15, 26
Marignac, C., 2
Maurya, M.R., 78
McGarvey, G.B., 16
McMonagle, J.B., 26
Mesoporous silica
  MCM-41
    FT-IR spectra, 195, 196
    synthesis, 193
    TEM images of, 198
    textural properties of, 197
    total acidity, 197, 198
    XRD patterns, 195–197
  SBA-15
    FT-IR spectra, 195, 196
    synthesis, 193–194
    TEM images of, 198–199
    textural properties of, 197
    total acidity, 197, 198
    XRD patterns, 195–197
Methanol oxidation, 249–251
1-Methyl-4-[alpha-alkoxy-isopropyl]-l-cyclohexenes, 154, 155
Miolati, A., 2
Misono, M., 16
Mizuno, N., 78
Moffat, J.B., 16, 17, 26, 30, 65
Mohan Reddy, K., 11
Molybdenum-catalysed oxidative desulfurisation reactions
  dibenzothiophene, H_2O_2
    aluminium dodecamolydophosphate, 239
    gasoline desulfurisation, 238–239
    isooctane phase, 239, 240
    mechanism, 240–241
    solid/soluble catalysts, 239
  FTIR spectroscopy, 238, 239
Molybdovanadophosphoric acid immobilization
  on amine-modified mesoporous silica
    adamantane oxyfunctionalization, 125–129
    anthracene oxidation, 124–125
    mass NMR spectroscopy, 120–124
  inorganic–organic hybrid materials
    mesoporous carbon functionalization, 129–130
    2-methylnaphthalene oxidation, 129–131
    N-MCF-C support, 129
    organo SBA-15 (OSBA-15), 129
  ionic liquid-modified mesoporous silica
    alcohol oxidation, 133, 135–136
    ^{31}P mass NMR spectroscopy, 133, 134
    SILP strategy, 132
Monolacunary phosphomolybdate
  DRS of, 215
  FT-IR spectra, 213, 214
  hydrous zirconia
    synthesis, 211
    total surface area for, 215, 216
  PMo_{11}/ZrO_2
    benzyl alcohol oxidation (*see* Benzyl alcohol oxidation)
    catalytic reaction, 212–213
    DRS of, 215
    FT-IR spectra, 213, 214
    leaching of, 213
    synthesis of, 212
    TG-DTA for, 213, 214
    total surface area for, 215, 216
    XRD of, 215, 216
  synthesis of, 211–212
  TG-DTA for, 213, 214
  total surface area for, 215, 216
  XRD of, 215, 216

N
Nadjo, L., 249
Nakajima, H., 249
Narasimha Rao, K., 11
Nitrogen-containing mesostructured cellular foam carbon (N-MCF-C), 129

O
Oleic acid esterification
 TSA_3/MCM-41
  catalyst amount, 200
  mole ratio effect, 199, 200
  reaction time, 200–202
 TSA_3/SBA-15, 202
Organo SBA-15 (OSBA-15), 129

P
Pan, D., 250
Pandurangan, A., 66
Parida, K., 57
Patel, A.U., 1, 189, 209
Patel, K., 78
Pathan, S., 209
Pauling, L.C., 2, 3
Pito, D.S., 153
^{31}P MAS NMR spectroscopy
 supported 12-tungstophosphoric acid (H_3PW), 169
  on silica–alumina, acetic acid esterification, 175
  on silica, benzene transalkylation, 172–173
  on zirconia, oleic acid esterification, 179
 zirconia-supported heteropoly acids, 113–115
Poly (3,4-ethylenedioxythiophene) (PEDOT), 252
Polyoxometalates (POMs)
 biological electron transfer process, 253–254
 catalysis types, 246
 characteristics, 245
 classes of, 59, 166
 electro-catalytic applications
  amperometric sensors, 249
  electro-catalysts, 248
  electro-catalytic reduction, 249
  electrochemical methanol oxidation, 249–251
  indirect electrochemical reactions, 248
  oxygen reduction, Pt/STA–C, 251
 as electrochemical super capacitor electrodes
  galvanostatic charge–discharge behaviour, 253
  poly (3,4-ethylenedioxythiophene), 251–252
  specific capacitance, 253
 Keggin-type, 166, 246
 liquid acids, 247
 multifunctionality and structural mobility, 246
 redox properties, 247
 structure of, 246
Pope, M.T., 249

R
Raman spectra
 titania-supported TPAV1 catalyst, 96–98
 zirconia-supported heteropoly acids, 112–113
Rana, S., 57
Renewable Directive Energy (RED), 142
Ressler, T., 16
Rocchiccioli-Deltcheff, C., 26
Rong, C., 249
Rong, C.Y., 249
Rosenheim, A., 2

S
Sahoo, S., 105
Sai Prasad, P.S., 11, 91
Seo, M., 250
Shanbhag, G.V., 105
Shi, X., 132
Silica-occluded tungstophosphoric acid (PW/S)
 α–pinene alkoxylation
  catalytic activity of alcohols, 158
  catalytic stability, 159, 160
  temperature effect, 158, 159
  α-terpinyl alkyl ether selectivity, 158, 159
 β–pinene alkoxylation
  catalytic activity of, 160
  catalytic stability, 160, 161
  α-terpinyl alkyl ether selectivity, 160, 161
 FT-IR spectrum of, 155–156
 gas chromatography, 157
 limonene alkoxylation, 161–162
 preparation of, 155
 stability tests, 156
 XRD patterns of, 156
Solid acid catalyst
 vs. Bronsted and Lewis acid catalysts, 58
 characteristics, 58
 industrial processes, 58

Srilakshmi, C.H., 11
Steekhan, E., 249
Supported $H_3PW_{12}O_{40}$
  catalytic tests, 144
  characterisations
    differential thermal analysis, 144–146
    X-ray diffraction patterns, 146
  composition of, 143
  glycerol etherification with *tert*-butyl alcohol
    acid sites, 149
    Amberlyst 35, 148
    ether formation, 149–150
    HPA leaching, 151
    temperature effect, 148, 149
    water tolerance, 150–151
  NH_3 adsorption microcalorimetry
    acidic properties, 143–144
    calorimetric curves, 147
    support interactions, acid strength, 147
  preparation, 143
Supported 12-tungstophosphoric acid (H_3PW)
  characterization, 169–170
  free fatty acids esterification, 168
  Keggin ion structure, 166
  on mesoporous MCM-41, citronellal cyclization, 167–168
    catalytic results, 183, 184
    double bond nucleophilic attack, 184
    experimental setup, 171–172
    intramolecular cyclization, 184
    kinetic studies for, 184
    textural properties, 183
    XRD of, 182, 183
  preparation of, 168–169
  on SBA-15, 168
  on silica–alumina, acetic acid esterification
    acetic acid:ethanol molar ratio, 176
    acid site distribution, 177
    BET-specific surface area, 174, 175
    catalyst mass effect, 177, 178
    ethyl acetate yield, 174, 175
    experimental setup, 171
    impregnation, 175
    leaching, 177
    loading and catalytic activities, 175–176
    ^{31}P MAS-NMR spectrum, 175
    short-chain esters, 175
  on silica, benzene transalkylation
    C9+ aromatics, 172
    experimental setup, 170–171
    physical–chemical properties, acidity, and selectivity, 173
    ^{31}P MAS-NMR, 172–173
    thermal stability, 173
    xylenes formation, 173–174
  on zirconia, oleic acid esterification, 178
    acidity, 181
    catalytic activity, 180
    catalytic tests, 181
    experimental setup, 171
    FTIR spectra, 179
    leaching, 180–181
    ^{31}P MAS-NMR, 179
    surface area and pore size, 179, 180
    temperature, 181–182

T

Tanabe, K., 58
Titania-supported $TPAV_1$ catalyst ($TPAV_1/TiO_2$)
  characterisation of
    FT-IR patterns, 96, 97
    H2–TPR technique, 98
    Raman spectra, 96–98
    XPS measurement, 98–99
    X-ray diffraction patterns, 96
  preparation of, 93
  reaction procedure, 94
  recycling results of, 102–103
  styrene oxidation
    catalysts screening for, 94
    cleavage of, 95
    mole ratio, 100
    reaction temperature, 100
    solvent effect, 100
    styrene derivatives and aliphatic olefins, 100–101
    vanadium location, 95
Toth, J.E., 249
TPD profile
  ammonium salt of 12-molybdophosphoric acid (AMPA), 29–30
  TiO_2-supported AMPA catalysts, 39
  zirconia-supported heteropoly acids, 110–111
Transition metal-substituted polyoxometalates (TMS POMs), 8
Tsigdinos, G.A., 26
Tungsten-based polyoxometalate-modified mesoporous silica
  cesium salt, anisole acylation
    catalytic activity, 68–70
    liquid phase of, 70
    scanning electron micrograph, 70
    substrate effects of, 70–72
    synthesis, 66–67

characterization
FTIR spectra, 68, 69
SEM studies, 68
surface area and pore size distribution, 67–68
Cu salt, Heck vinylation reaction
aryl halide, 76, 77
catalyst preparation, 72
catalytic evaluation, 74
$Cu_{1.5}PW_{12}O_{40}$/MCM-41 catalyst, 76
mechanism for, 75, 76
temperature-programmed reduction profiles, 73–74
transmission electron microscopy, 74, 75
X-ray diffraction, 72–73
synthesis, 66
12-Tungstosilicic acid (TSA)
FT-IR spectra, 195, 196
n-butylamine acidity determination, 194
oleic acid esterification, 195
recycling of, 202
triglyceride feedstocks transesterification, 195
TSA_3/MCM-41
FT-IR spectra, 195, 196
oleic acid esterification, 199–202
synthesis, 194
TEM images of, 198
textural properties of, 197
total acidity, 197, 198
vs. TSA_3/SBA-15, 205–206
XRD patterns, 195–197
TSA_3/SBA-15
FT-IR spectra, 195, 196
oleic acid esterification, 202
synthesis, 194
TEM images of, 198, 199
textural properties of, 197
total acidity, 197, 198
XRD patterns, 195–197

V
Vanadium incorporated AMPA systems
FTIR spectra, 48
powder X-ray diffraction patterns, 47–48
thermal stability and surface area, 47
TPR measurements, 48–49
TPR pattern, 49
Vanadium-substituted tungstophosphoric acid
characterisation of
FT-IR spectra, 93
Raman spectra, 93
temperature-programmed reduction, 93–94
XPS measurements, 94
preparation of, 92–93
titania-supported (*see* Titania-supported $TPAV_1$ catalyst ($TPAV_1/TiO_2$))
Van Veen, J.A.R., 27
Venkateswara Rao, K.T., 91
Viswanathan, B., 245

W
Weng, Z., 210
Wlodarczyk, R., 251

X
X-ray diffraction (XRD) patterns
monolacunary phosphomolybdate, 215, 216
silica-occluded tungstophosphoric acid, 156
supported 12-tungstophosphoric acid (H_3PW), 182, 183
TiO_2-supported AMPA catalysts, 36, 37
titania-supported $TPAV_1$ catalyst, 96
12-tungstosilicic acid (TSA), 194–197
zirconia-supported heteropoly acids, 108–109

Y
Yang, H., 74

Z
Zirconia-supported heteropoly acids, 107
catalytic activity measurements, 108
catalytic activity studies
allylation of anisole, 115–117
benzoic anhydride, benzoylation of veratrole, 117
linear alkyl benzene synthesis, 115
phenol alkylation, 118
characterization, 108
physicochemical characterization
FTIR pyridine adsorption spectra, 111–112
N_2 sorption studies, 109–110
^{31}P MAS NMR spectroscopy, 113–115
Raman spectroscopy, 112–113
TPD profile, 110–111
X-ray diffraction, 108–109
preparation, 108
vs. tungstated zirconia, 118–119

MIX
Papier aus verantwortungsvollen Quellen
Paper from responsible sources
FSC® C105338

If you have any concerns about our products,
you can contact us on
ProductSafety@springernature.com

In case Publisher is established outside the EU,
the EU authorized representative is:
Springer Nature Customer Service Center GmbH
Europaplatz 3, 69115 Heidelberg, Germany

Printed by Libri Plureos GmbH
in Hamburg, Germany